Werner Meske (Ed.)

From System Transformation to European Integration

Werner Meske (Ed.)

From System Transformation to European Integration

Science and technology in Central and Eastern Europe
at the beginning of the 21st century

LIT

Bibliographic information published by Die Deutsche Bibliothek
Die Deutsche Bibliothek lists this publication in the Deutsche
Nationalbibliografie; detailed bibliographic data are available in the
Internet at http://dnb.ddb.de.

ISBN 3-8258-7290-4

© LIT VERLAG Münster   2004
Grevener Str./Fresnostr. 2   48159 Münster
Tel. 0251-23 50 91   Fax 0251-23 19 72
e-Mail: lit@lit-verlag.de   http://www.lit-verlag.de

Distributed in North America by:

Transaction Publishers
New Brunswick (U.S.A.) and London (U.K.)

Transaction Publishers   Tel.: (732) 445 - 2280
Rutgers University   Fax: (732) 445 - 3138
35 Berrue Circle   for orders (U. S. only):
Piscataway, NJ 08854   toll free (888) 999 - 6778

CONTENTS

List of Figures v

List of Tables vii

Notes on the editor and contributors xi

List of Abbreviations xiii

Preface 1

I SCIENCE AND TECHNOLOGY IN CENTRAL AND EASTERN EUROPEAN COUNTRIES IN THE SECOND HALF OF THE 20th CENTURY

1. Science and technology in CEECs in the socialist era 7
   Werner MESKE

2. Science and technology in CEECs at the end of the socialist era . . 27
   Werner MESKE

3. Analyzing the transformation of S&T in CEECs – the theoretical and methodological approach 47
   Werner MESKE

II THE TRANSFORMATION OF S&T SYSTEMS IN INDIVIDUAL CEECS IN THE 1990S

4. Russia: towards a national innovation system – institutional changes and funding mechanisms 61
   Nadezhda GAPONENKO

5. Ukraine: institutional changes in S&T in a period of economic decline 91
   Lidiya KAVUNENKO

6. Belarus: transformation of the S&T system 113
   Gennady A. NESVETAILOV/Anton A. SLONIMSKI

7. Estonia: transformation of the R&D system 135
   Helle MARTINSON

8. Latvia: transformation of the S&T system 151
    Janis KRISTAPSONS

9. Lithuania: the science system from 1989–2001 165
    Ina DAGYTE

10. Poland: restructuring S&T without radical transformation 185
    Jan KOZLOWSKI

11. Czech Republic: transformation of R&D – from research policy to
    a national S&T policy 197
    Karel MÜLLER

12. Slovakia: S&T transformation without a strategy 215
    Stefan ZAJAC

13. Hungary: from transformation to European integration 235
    Judith MOSONI-FRIED

14. Romania: transformation of the S&T system 259
    Steliana SANDU

15. Bulgaria: the long road to a new innovation system 283
    Kostadinka SIMEONOVA

16. Federal Republic of Yugoslavia: restructuring the S&T system –
    indicators of transformation 307
    Duro KUTLACA

17. Slovenia: transformation of the S&T system 337
    Peter STANOVNIK

III COMMON FEATURES, PARTICULARITIES AND RESULTS: A COMPARATIVE ANALYSIS OF S&T TRANSFORMATION BY COUNTRY

18. The reorganization of S&T systems in CEECs during the 1990s . . . 357
    Werner MESKE

19. The reduction in scientific resources during the 1990s 381
    Werner MESKE

20. Publication activity in CEECs during the 1990s 407
    Werner MESKE

IV SCIENCE AND TECHNOLOGY IN CEECS AT THE BEGINNING OF THE 21st CENTURY

21. A provisional appraisal: the transformation of S&T during the 1990s and the challenges of the 21st century 419
    Werner MESKE

22. What future for S&T in the CEECs in the 21st century? 443
    Slavo RADOSEVIC

LIST OF FIGURES

1.1 Scientists and engineers engaged in R&D in the USSR and USA
1.2 Groups of countries in the socialist world system
2.1 Share of the individual republics of the USSR in selected science resources
2.2 CEECs: real industrial output levels
3.1 S&T system in society
3.2 Model of the institutional transformation of the S&T system in CEECs
4.1 Russia: national innovation system – an overview
4.2 Russia: R&D organizations and R&D personnel
4.3 Russia: R&D cooperation by industrial enterprises
4.4 Russia: the benefits of ITCs for spin-offs
4.5 Russia: R&D cooperation by spin-offs
4.6 Russia: labor mobility between firms
4.7 Russia: R&D expenditure
5.1 Ukraine: the organization of the S&T system in 1990 and 2000
10.1 Poland: annual growth rate in GDP
13.1 Hungary: R&D personnel and GERD
13.2 Hungary: National Scientific Research Fund (OTKA) spending at current and real prices
13.3 Hungary: Central Technology Development Fund (KMÜFA) spending at current and real prices
13.4 Hungary: direct expenditure by the Ministry of Education on R&D in HE institutions
13.5 Hungary: R&D expenditure (GERD) as percentage of GDP
13.6 Hungary: R&D expenditure by financial source
13.7 Hungary: R&D expenditure by sector of performance
16.1 FRY: share of income from R&D in the total income of R&D organizations
16.2 FRY: R&D personnel
16.3 FRY: allocation of R&D personnel
16.4 FRY: structure of the R&D workforce by function
18.1 Dissolution and fragmentation of the former socialist S&T system
18.2 CEECs: real GDP levels since 1989
18.3 East Germany: development of newly founded R&D-intensive enterprises
18.4 East Germany: structure of turnover in R&D-oriented innovative SMEs

19.1 CEECs: general trends in R&D personnel
19.2 East Germany: methodological and real changes in R&D personnel
19.3 European CIS countries – R&D personnel
19.4 Baltic states – R&D personnel
19.5 Six CEECs – R&D personnel
19.6 Successor states of the SFR Yugoslavia – R&D personnel
19.7 Romania: R&D personnel by sector
19.8 Poland: R&D personnel by sector
19.9 Hungary: scientists and engineers by R&D sector
19.10 Lithuania: age structure of scientists
20.1 CEECs: Publication activity and co-authored papers with EU15 countries
20.2 CEECs: share of papers co-authored with scientists from EU15 countries
20.3 Six Central Eastern European countries: papers co-authored with German scientists and scientists from these six CEECs
21.1 Phases in the transformation of the S&T system
21.2 Comparison of R&D structures in West and East Germany
22.1 Employment in high-tech manufacturing vs. BERD/GDP

LIST OF TABLES

2.1 The starting conditions for transformation in the individual CEECs
4.1 Russia: R&D expenditure in state research centers by source of funds
4.2 Russia: trends in R&D spending by source of funds
4.3 Russia: national R&D expenditure
4.4 Russia: innovation expenditure
4.5 Russia: the main directions of Russian innovation policy
5.1 Ukraine: trends in economic activity
5.2 Ukraine: SFBR grants by organization
5.3 Ukraine: SFBR funding and results
5.4 Ukraine: State Innovative Foundation projects
5.5 Ukraine: the AoS and international cooperation
5.6 Ukraine: S&T potential by science sector
5.7 Ukraine: distribution of R&D organizations by type
5.8 Ukraine: distribution of R&D organizations by science sector
5.9 Ukraine: S&T budgeting by science sector
5.10 Ukraine: sources of funds for R&D
5.11 Ukraine: S&T personnel by function
5.12 Ukraine: scientists and engineers by R&D sector
5.13 Ukraine: doctors and candidates of science by R&D sector
5.14 Ukraine: finished experimental developments
6.1 Belarus: changes in S&T resources
6.2 Belarus: manpower in S&T
6.3 Belarus: distribution of researchers by scientific field
6.4 Belarus: number of postgraduate students by scientific field
6.5 Belarus: technological structure of S&T activity
6.6 Belarus: patent activity indicators
6.7 Belarus: small scientific innovation enterprises
7.1 Estonia: R&D financing by source of funds
7.2 Estonia: number of researchers and engineers
7.3 Estonia: number of universities and R&D institutes
7.4 Estonia: R&D expenditure by type of institution
7.5 Estonia: R&D expenditure by kind of R&D activity
8.1 Latvia: basic indicators
8.2 Latvia: main events in the transformation of the R&D system

9.1 Lithuania: gross domestic product
9.2 Lithuania: number and structure of R&D institutions
9.3 Lithuania: R&D personnel and researchers
9.4 Lithuania: distribution of researchers by type of organization
9.5 Lithuania: scientists, scientific degrees and academic titles
9.6 Lithuania: scientists by field of science
9.7 Lithuania: age structure of scientists
9.8 Lithuania: PhD students by field of science
9.9 Lithuania: expenditure on R&D
9.10 Lithuania: sources of finance by sector
10.1 Poland: number of S&T units
10.2 Poland: total employment in S&T units
10.3 Poland: total number of researchers
10.4 Poland: current expenditure on R&D activities
10.5 Poland: sectoral structure of current expenditure on R&D activities
11.1 Czech Republic: selected R&D indicators
11.2 Czech Republic: R&D personnel by sector
11.3 Czech Republic: R&D workforce by selected manufacturing industries
11.4 Czech Republic: general economic and social indicators
11.5 Czech Republic: R&D personnel by scientific discipline and type of activity
11.6 Czech Republic: gross expenditure on R&D
11.7 Czech Republic: government funded programs supporting industrial R&D and innovation
11.8 Czech Republic: gross expenditure on R&D by source of funds and sector of performance
11.9 Czech Republic: distribution of R&D sources by sector of performance
12.1 Slovakia: basic macro-economic indicators
12.2 Slovakia: R&D personnel
12.3 Slovakia: expenditure on R&D
12.4 Slovakia: structure of activities in R&D organizations
12.5 Slovakia: gross expenditure on R&D in the SAS
12.6 Slovakia: R&D personnel in the SAS
12.7 Slovakia: R&D personnel in the higher education sector
12.8 Slovakia: gross expenditure on R&D in the higher education sector
12.9 Slovakia: R&D personnel in the business enterprise sector
12.10 Slovakia: gross expenditure on R&D in the business enterprise sector
12.11 Slovakia: the structure of R&D personnel and expenditure by sector
13.1 Hungary: number of R&D units
13.2 Hungary: scientists and engineers in R&D institutions
14.1 Romania: R&D units and personnel

14.2 Romania: R&D units and personnel by form of ownership
14.3 Romania: R&D expenditure
14.4 Romania: structure of R&D expenditure
14.5 Romania: structure of current R&D expenditure in enterprises by CANE activities
15.1 Bulgaria: GDP and GERD
15.2 Bulgaria: R&D organizations by sector
15.3 Bulgaria: R&D staff and scientists by sector
15.4 Bulgaria: scientists by sector
15.5 Bulgaria: BAS and HE staff
15.6 Bulgaria: researchers by sector
15.7 Bulgaria: researchers with scientific degrees by sector
15.8 Bulgaria: R&D personnel total and in the enterprise sector
15.9 Bulgaria: researchers total and in the enterprise sector
16.1 FRY: selected macroeconomic indicators, 1987–1991
16.2 FRY: selected macroeconomic indicators, 1992–2000
16.3 FRY: structure of the S&T system by republic, organization and personnel
16.4 FRY: structure of the S&T system by scientific field
16.5 FRY: R&D financing
16.6 FRY: selected R&D indicators
16.7 FRY: R&D in industrial firms
16.8 FRY: sources of ideas/information for innovation activities
16.9 FRY: sample of firms in the Innovation Survey II
16.10 FRY: expenditure on innovation activities
16.11 FRY: innovation activities and acquisition of technology
17.1 Slovenia: R&D personnel
17.2 Slovenia: human capital in R&D sectors
17.3 Slovenia: product and process innovations in manufacturing industries
18.1 Institutional changes in S&T sectors in individual CEECs in the first half of the 1990s
18.2 Institutional changes in S&T policies in individual CEECs in the first half of the 1990s
18.3 Changes in the economy in individual CEECs in the first half of the 1990s
18.4 Classification of individual CEECs by degree of institutional transformation in the national S&T system in the first half of the 1990s
18.5 CEECs: economic situation and resources devoted to R&D in the second half of the 1990s
19.1 CEECs: trends in R&D intensity: GERD/GDP
20.1 CEECs: publication activity
21.1 Candidate countries: level and trends in R&D expenditure relative to EU average

NOTES ON THE EDITOR AND CONTRIBUTORS

THE EDITOR

Werner Meske has been Head of the Research Group 'Transformation of Science Systems' at the Social Science Research Center Berlin (WZB) since 1992; before he was Professor for the field of Theory and Organization of Science at the Academy of Sciences of the GDR, Berlin

THE CONTRIBUTORS

Nadezhda Gaponenko is Head of Department, Russian Research Institute of Economics, Policy and Law in Science and Technology (affiliated to the Ministry of Industry, Science and Technology of the Russian Federation), Moscow

Lidiya Kavunenko is Professor and Deputy Director of the Dobrov Center for S&T Potential and Science History Studies, Ukrainian National Academy of Sciences, Kiev

Gennady Nesvetailov[†] was Professor and Head of the Sociology of Science Department, Belarussian National Academy of Sciences, Minsk

Anton A. Slonimski is Senior Fellow, Research Institute of Economics, Ministry of Economy of the Republic of Belarus, Minsk.

Helle Martinson is Member of the Board of the Estonian Science Foundation, Tallinn

Janis Kristapsons is Head of the Center for Science and Technology Studies, Latvian Academy of Sciences, Riga

Ina Dagyte is Professor of Vytautas Magnus University, Institute of Political Science and Diplomacy, Kaunas, and Head of Science Management Program, Baltic Management Foundation, Vilnius

Jan Kozlowski is Research Fellow, Center for Science Policy and Higher Education Studies, Warsaw University, Warsaw and Adviser, State Committee on Scientific Research (KBN), Warsaw

Karel Müller is Associated Professor and Vice-Dean, Faculty for Humanistic Studies, Charles University, Prague.

Stefan Zajac is Director of the Institute for Forecasting, Slovak Academy of Sciences, Bratislava

Judith Mosoni-Fried is Deputy Director of the Institute for Research Organization, Hungarian Academy of Sciences, Budapest

Steliana Sandu is Professor, Head of S&T Policy Department, Institute of National Economy, Romanian Academy of Sciences, Bucharest

Kostadinka Simeonova is Professor and Director of the Center for Science Studies & History of Science, Bulgarian Academy of Sciences, Sofia

Duro Kutlaca is Head of the Science and Technology Policy Research Center at the Mihajlo Pupin Institute, Belgrade

Peter Stanovnik is Director of the Institute for Economic Research, Ljubljana

Slavo Radosevic is Reader in Industrial and Corporate Changes at University College London, the School of Slavonic and East European Studies, London

LIST OF ABBREVIATIONS

AoS	Academy of Sciences
acquis communautaire	the entire body of laws, policies and practices that have evolved up to the present day in the EU, full compliance with which is one requisite for accession
BERD	business enterprise expenditure on R&D
BES	business enterprise sector
CEECs	Central and Eastern European Countries
CERN	European Organization for Nuclear Research
CIS	Commonwealth of Independent States
CMEA (COMECON)	Council for Mutual Economic Assistance
CoCom	coordinating committee (intergovernmental body controlling the export of high technology to communist countries during the Cold War)
EiT	economies in transition
EBRD	European Bank for Reconstruction and Development
EC	European Commission
EU	European Union
FDI	foreign direct investment
FIG	financial and industrial group
FTE	full-time equivalence
GERD	gross domestic expenditure on R&D
GDP	gross domestic product
GDR	German Democratic Republic
HE	higher education
IMF	International Monetary Fund
NATO	North Atlantic Treaty Organization
NGO	non-governmental organization
NIS	newly independent states
OECD	Organization for Economic Cooperation and Development
OSCE	Organization for Security and Cooperation in Europe
PNP	private non-profit (sector/organization)
RF	Russian Federation

R&D	research and development
S&T	science and technology
SME	small and medium-sized enterprise
STS	science and technology system
UN	United Nations
UNDP	United Nations Development Program
UNIDO	United Nations Industrial Development Organization
USD	US dollar
WTO	World Trade Organization

PREFACE

The situation of science and technology in the former socialist Central and East European countries (CEECs) in the last decade of the 20th century was determined largely by processes of transformation, since the collapse of the socialist systems in these countries had also led to the more or less complete breakdown of their science and technology systems.

This period of transformation in the CEECs was and remains characterized by a number of changes taking place concurrently. Negative implications in the rapid development of their science and technology systems in the 20th century had to be corrected at the same time as adjustments to new social and economic structures at national and international level had to be made. With respect to the past, this period of transformation represents the break-up of the old system and its restructuring; with respect to the 21st century, however, it can also be seen as the turbulent starting phase in the transition to modern science and innovation systems, management of which, against the background of globalization, requires active intervention and shaping.

Thus the real – and far more difficult – task in the CEECs is not the correction of errors and misdevelopments but the search for and shaping of forms of science and innovation and of development processes that meet the demands of the 21st century. Those of a pessimistic disposition will point out that it will be particularly difficult for these countries to build up new, more efficient science and innovation systems; however, it might be argued, more optimistically, that it is precisely because of the break-up and removal of the old structures that these countries now have many new options at their disposal and that, as a result, it may even be easier for them than for the established 'Western' countries to rise to the new challenges.

However, the initially widespread optimism that liberalization, privatization and the introduction of the market economy would in themselves be sufficient to bring about economic renewal and to give science and technology a befitting role once again has now vanished in most of these countries. Experiences in East Germany have shown that the actual 'transformation' is followed by a lengthy period during which new processes unfold inside and outside the scientific sphere itself, leading to a new integration of scientists, scientific institutions and communities, areas of research and spheres of application, and so on – often also internationally.

In view of the changes taking place in Europe and across the entire world, which will be alluded to here only by citing the catchwords 'EU enlargement' and 'globalization', on the one hand, and 'the research system in transition', on the other, this raises questions about the preconditions for, objectives of and approaches to the reordering of science and technology systems in the CEECs as a strategic task for the new century.

Without an analysis of the actual situation in the countries in question as they enter the 21st century, and in particular one that takes account of their often highly specific development paths and traditions in the previous century, and especially their experiences in the last decade of that century, it will scarcely be possible to take stock of the challenges and opportunities and demonstrate the real prospects for science and technology in the CEECs.

The present book is intended as a contribution to the debate on these issues, which concern both the past and the future of science and technology in these countries. It is based on the authors' many years of research in this area, much of it conducted jointly. In particular, the collaborative work carried out in the course of the EU-TSER project on 'The institutional transformation of S&T systems in the European economies in transition' between 1996 and 1998 demonstrated the advantages to be gained from exchanges of experience and led, among other things, to the development of a 'three-phase model' of the transformation of S&T systems in the CEECs. In view of the rapid pace of change and of the different paths the changes have taken in the various countries, the authors updated their analyses once again in 2001/2002 and took greater account of the future development opportunities and paths that were beginning to emerge. The decided and planned enlargement of the European Union plays a particular role in this respect, since for many of the countries under investigation here it offers real prospects for development but also constitutes a major challenge.

Consequently, the book begins with an outline of the general characteristics of science and technology in the CEECs in the second half of the 20th century (Part I) before proceeding to an analysis of the changes in the various national science and technology systems in the 1990s in their respective national contexts (Part II). By means of a comparative assessment of these changes and the factors influencing them, the common features and differences in the transitional phase from the 20th to the 21st century are revealed (Part III) and the prospects for the development of science and technology in Central and Eastern Europe inferred (Part IV), together with the policy implications of those prospects for the CEECs.

With its wide-ranging contents, the book is written for a relatively broadly-based readership from the scientific, political and economic spheres, particularly in the CEECs and in the EU member states, with an interest in the theoretical and practical aspects of the formation of new science and technology systems in both the national and European context.

Thanks are due to the Social Science Research Center Berlin (WZB) for supporting research and international collaboration on the topic of S&T transformation and for financing this book.

The editor is particularly grateful to the authors from the CEECs who have made their contributions to the present book under sometimes very difficult conditions, and to Slavo Radosevic who, besides writing his chapter, undertook specialized editorial work. The following individuals have given sterling service in preparing the manuscript for publication: Andrew Wilson, who was responsible for the English language editing of all chapters, which varied considerably with respect to the standard of English used and placed considerable demands on his language and subject-related skills and Margret Arzt and Christine Nait for organizational work and the copy editing of the manuscript with its very varied texts, figures and tables.

Werner Meske

Part I

Science and technology in Central and Eastern European countries in the second half of the 20th century

1. SCIENCE AND TECHNOLOGY IN CEECs IN THE SOCIALIST ERA

Werner MESKE

As we advance into the 21st century, an awareness of the various aspects of the development of science during the 20th century is necessary if we are to be able to identify essential trends in the new one. The last century was characterized more than the previous one by rapid advances in science and technology (S&T) and their impact on economy and society. These advance were based in turn on fundamental changes within science and technology itself.

1. INTRODUCTION: GROWTH, INTERNATIONAL PROLIFERATION AND DIVERSIFICATION OF SCIENCE IN THE 20th CENTURY

Around the year 1800, there were fewer than 10 000 scientists in the entire world, most of them concentrated in Central and Western Europe. By 1900, the number had increased tenfold, although the total figure was still only around 100 000. The early years of the 20th century saw the start of a rapid expansion in the number of scientific and technological organizations. In the first half of the century, this expansion was concentrated largely in the leading countries of Western Europe, the USA and Japan. As early as the 1920s and 1930s, however, the Soviet Union also began to built a significant science base. The growth of science and, in particular, technology accelerated in the second half of the century – not least because of the division of the world into two political-military blocs and the ensuing arms race, which was based particularly on nuclear and rocket weapons research (Jungk, 1982). Additionally, S&T activities began to spread to more and more countries and continents, including the former colonial areas in Asia and Africa. Because of the increased intensity as well as the geographical extension of science and technology, the number of scientists and engineers in the world gradually rose until it had reached about 10 million in the year 2000, a hundredfold increase over 1900. (All data are own estimations, based on Bernal, 1967, p. 453 and other authors: cf. Meske, 1986, p. 73 – 83.) In the 1990s, there were more than four million scientists and engineers working in R&D alone (own compilation based on UNESCO, 1998).

The growing demand for *researchers* is one reason for the increase in the number of scientists and engineers. This increased demand was due, in turn, to the emergence of more and more fields and sub-fields or topics in basic and applied research and experimental development (this is reflected in the data on the number of scientific journals and publications over a longer period, cf. Price, 1974). Another reason was new challenges in *teaching*. In 1900, only 0.5 per cent of the relevant age cohorts graduated from university, even in the most developed countries; by the 1950s, the share of university graduates had already reached 5 per cent of the relevant age cohorts. In the most developed countries this share has increased since the 1980s to more than 30 per cent (Engel, 1981) and in some countries the entry rates for tertiary education reached more than 50 per cent in the 1990s (OECD, 2000, p. 150). This was accompanied by an increase of the number of higher education institutions, their teaching staff and other scientific, technical and administrative personnel (NSB, 2000, Chapter 4). The OECD reflected this situation by giving greater emphasis to indicators of tertiary education, "which is now replacing secondary education as the focal point of access to rewarding careers" (OECD, 2000, p. 9). Thus we have to be aware that not all students finish tertiary education, for a variety of reasons (OECD, 2000, pp. 161–176).

Fundamental shifts in the disciplinary structure and in curriculum content in universities and other HE institutions were related to the changing professional careers of most *graduates*. Until the end of the 19th century, about 90 per cent of them remained in scientific institutions and activities. However, as a result of institutional changes in science and society, in more and more scientific disciplines only 10 per cent now stay in scientific institutions while about 90 per cent of these graduates 'leave' the science system once they have graduated. Most of them work as highly qualified experts in various areas, which also creates opportunities for non-graduates. The range of such areas is increasing because of the penetration of science and technology into more and more sectors and activities. This is related to a large extent to the use of 'scientific' equipment and working methods as well as to the frequently science-based goods and services that are produced.

This is a characteristic feature of the shift to science or knowledge-based societies in highly developed countries. Modern society needs new knowledge, inventions, innovations and experts creating them, not only in the production of goods and services but also in their use and application. Thus in more and more areas of work and daily life, there is a need for experts or other highly-qualified people with the ability to understand and use the knowledge embodied in modern equipment, structures and services. The rapidity with which research results are obtained and their often immediate application in practice have led to a process of continuous change in knowledge-based innovations in all areas and to the creation and growth of the so-called 'mode 2' of research production wherein knowledge is produced in the context of application (cf. Gibbons et al., 1994). At the same time, more and more people are having to deal with these changes in their work and life. As a result, they are having to engage in life-long learning and keeping pace with advances in science and technology in order to avoid or overcome skills mismatches (Ducatel and Burgelman, 1999; Gavigan et al., 1999).

Long-term processes of differentiation and specialization of both activities and organizations within the S&T system itself are both a precondition for and a result of scientific and technological progress in the modern world. In the 19th century, the dominant type of scientific institution was the university based on Humboldt's ideal of the unity of teaching and research. In the 20th century, however, there were many institutional innovations in this field and a widespread diversification of scientific and technological organizations.

At the beginning of the 20th century, these developments were mainly concentrated in the leading capitalist industrial nations of Western Europe and the USA, as well as in Japan (cf. Miyabayashi, 1997), which still have the most developed S&T systems as well as the experience and capabilities to further develop and adapt them to new conditions (cf. Cozzens et al., 1990; Nelson, 1993; Reger and Schmoch, 1996; Edquist, 1997). In all these countries, the most important elements in the S&T system are those set out below.

The *higher education* sector, which is today only partly comprised of universities with interlinked teaching and research functions, and also includes numerous diverse institutions, many of which are oriented largely or even exclusively toward teaching.

A diversified system of (publicly funded) *research institutes and organizations*, specializing in a wide variety of research.

An expanding *industrial R&D sector*. This sector has its roots in the technological changes which were already occurring in individual factories in the 19th century as a result of experience and experimentation (Bernal, 1967). At the beginning of the 20th century, it consisted largely of in-house R&D units in large enterprises in the chemical and electrical engineering industries. It spread later to encompass most industries and many enterprises, including small and medium-sized enterprises (SMEs). R&D was, and still is, conducted by specialized independent private (or partly publicly financed) institutes, too.

Consequently, Lundgreen et al. identify three emerging (ideal-typical and in practice partially overlapping) sectors of national S&T systems, namely academic research (self-regulated/theory-driven), state-sponsored research (legally regulated/service-driven) and industrial research (market-regulated/product-driven) (Lundgreen et al., 1986, pp. 17–26).

In connection with these developments, an institutional structure of public bodies and regulatory and financing mechanisms has developed in the political and economic spheres, as well as within the realm of science itself (Braun, 1997; Krohn and Küppers, 1989). All these institutions and their activities form different national systems of R&D (OECD, 1972) and, in a wider sense, national innovation systems (NIS) (Freeman, 1987; Nelson, 1993). In the second half of the last century, such national systems became a characteristic part of the institutional landscape in an increasing number of countries.

Although the tripartite division of national S&T systems outlined above is a basic pattern that can be observed in most countries at least, there are, for various reasons, greater or smaller divergences in the form it takes in reality. Krishna et al. (1998) point to the historical conditions under which S&T systems emerged and progressed in developing countries, with their particular colonial, national and private modes of scientific development. Similarly, the development of science and technology in the USSR (and to some extent in other socialist countries) can be regarded as an attempt to catch up with and even overtake the advanced countries. In this 'socialist mode' of scientific development, the advancement of science and technology as a matter of top priority at a pace that outstripped the rate of economic development was regarded as an investment in the future and an attempt was made consciously and single-mindedly to shape the entire scientific and technological landscape and its development.

The emergence and growth of 'national science systems' – accompanied and supported by institutional diversification and a high degree of specialization as well as increasing networking and interactions between the main actors in such national systems (cf. Meske, 1989) – were the main features of scientific development in developed and also in developing countries during the 20th century. The number of such countries increased in the second half of the last century but remains limited. Now it seems highly probable that there will be a shift in this basic pattern of development in the world science system in the 21st century. Indeed, such a shift was already evident at the end of the 20th century and we now need to put the changes in the S&T system (STS) in Central and East European countries (CEECs) into this wider context.

2. THE DEVELOPMENT OF SCIENCE AND TECHNOLOGY IN THE SOCIALIST COUNTRIES AND THE EMERGENCE OF AN 'ALTERNATIVE' WORLD SYSTEM

S&T in former socialist countries was (and, to a great extent, still is) characterized by certain particularities that set it apart from its counterparts in developed capitalist societies. In some cases, these particularities are specific to the socialist system and typical of it and sometimes they are nationally specific, that is they are explicable only in the historical context in which a particular system was shaped. For this reason alone there are essential differences between the STS in Russia, the other successor states of the former Soviet Union, the other CEECs and the non-European socialist countries, in particular China, Vietnam and Cuba. There is not only "a common structural heritage in the research system of Central and Eastern Europe rooted in the shared past", as mentioned by Balazs, Faulkner and Schimank (1995, p. 615), but there are also important differences in the structure and functioning of the STS in each country, rooted as each one is in the country's particular history before and during the socialist era!

This has a lot to do with the fact that the 'socialist model' of STS was first developed in Russia. Only at a later date was this 'Soviet model' applied to other countries both within and outside the USSR (it had some features that are typical of the socialist system itself and some that are specific to Russia, for example the weakness of universities in research, the lack of 'in-house' R&D capacities in enterprises and the vast and closed military R&D sector). Depending on the general and especially the scientific and technological state of development attained by the country in question, the (Russian-based) 'Soviet model' was more or less heavily adapted and modified. Over time, however, the model underwent a number of further changes. Again, these varied from country to country. As a result, although the inherited basic type was the same in all cases, the transformation of S&T was undertaken under widely varying starting conditions in the individual CEECs (Meske, 1990).

For a better understanding of this situation, it might be useful briefly to review the particular features of the institutional framework of S&T in the former socialist countries.

In the first half of the 20th century, but somewhat later than in Western Europe and the USA, a new center of scientific activity emerged in Russia, the core of the Soviet Union (Bernal, 1954; Kröber and Lange, 1975). Its aim was to catch up with the leading industrialized nations and to establish a fairly unique institutional approach to S&T, one in line with its political system. Graham found that as early as the 1920s the Soviet Union was providing significantly greater support for the development of science and technology than any other country, a policy that led to the achievement of world record results (Graham, 1967, p. 209). Bernal noted in his investigation of the development of the organization of science across the world that a qualitatively new relationship between science and the state first began to emerge in the USSR and subsequently became established, with various modifications and in various forms, in other countries (Bernal, 1954, p. 504). This special 'Soviet model' of an STS evolved as a hierarchically structured and politically governed system. It included strong research institutes within the Academy of Sciences (AoS) (Graham, 1975), newly-founded universities and other higher education organizations (with an emphasis on large size and the strictly organized training of qualified personnel and, right from the outset, a weakness as regards research), the establishment of a network of R&D institutes in most industries in order to compensate for the lack of 'in-house' R&D capacities in enterprises and, last but not least, a vast and closed military R&D sector. Figure 1.1 shows the pace and extent of this development relative to the USA, using the number of scientific personnel as an indicator. This model was extended from Russia to other countries both within and, after World War II, outside the USSR. The strategic aim was to build up a fairly independent 'alternative (socialist) world S&T system', which was later expected to take on the role of global leader. In the political environment of the Cold War, its focus was on S&T relevant to military and armament purposes, combined with such prestigious fields as rocket and space research. Despite considerable success in the latter areas, the main objective could not be achieved, as shown by the collapse of this system around 1990 and the processes of system transformation, national restructuring and international re-integration that have been taking place in the formerly socialist countries since then. On the other hand, the realization of this 'model' was in some respects far advanced and had fundamentally shaped not only the political and economic systems of all socialist nations but also their S&T systems and international links. Thus in socialist countries, a supranational STS coexisted alongside various national subsystems, each with it own specific attributes that deviated to a greater or lesser extent from the original 'Soviet model' established in Russia.

The *supranational character of S&T* under socialism resulted in particular from the following conditions and factors (cf. Figure 1.2). (For a detailed account of the supranational character of S&T under socialism, see Lavigne, 1995.)

SUPRANATIONAL STATE FORMATION

Russia created a confederation in the shape of the USSR, which had a centralist leadership and permitted the separate Soviet republics only very limited

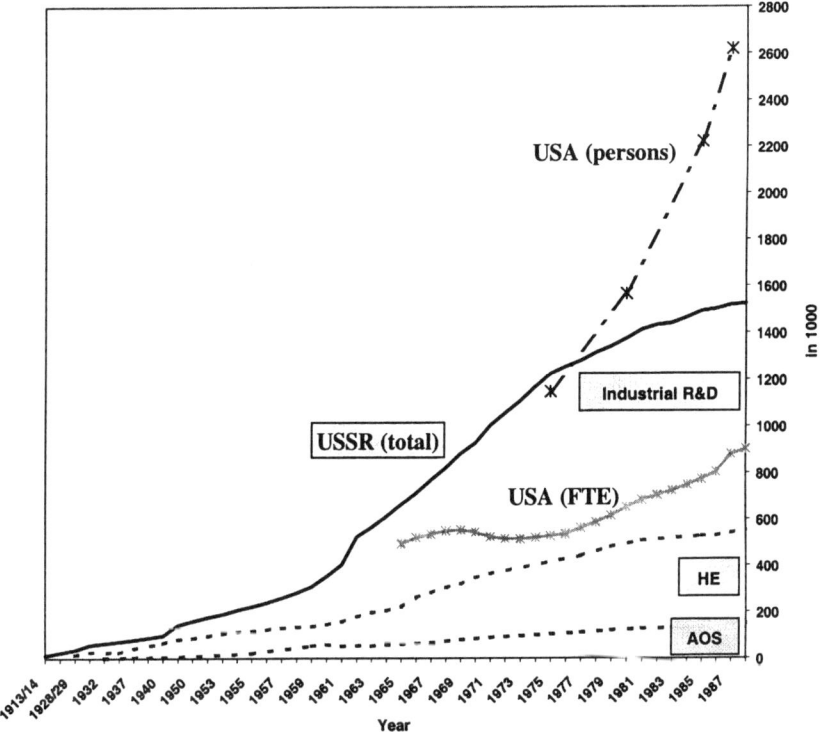

HE = Higher Education; AoS = Academy of Sciences
Fig. 1.1 Scientists and engineers engaged in R&D in the USSR (persons; by sectors) and USA (persons and FTE)
Source: Compilation by Meske; based on data from (for USSR) CSO (1971, p. 260); Goscomstat (1987 and 1991) and (for USA) FTE: NSB (1989, p. 260; 2000, p. A-204); Persons: Šokareva (1991, p. 41-42).

autonomy. Not only politics, the economy and the military but also S&T were all centrally organized within this confederation. Aside from functional integration through plans and programs, the all-union institutes and their branches in the various republics constituted an organizational network of state research facilities. In addition, there were joint scientific communities in the form of the AoS of the USSR and all-union societies for specific scientific fields (constituted primarily as professional associations) (cf. Nadiraschwili, 1994). Further links resulted from the supranational training of scientists in leading USSR universities (through the allocation of 'student quotas' to the individual republics and nationalities) and from the supranational workforce in leading research and teaching facilities as well as in science policy bodies (cf. Kannengießer and Meske, 1982). Such arrangements also existed on a smaller

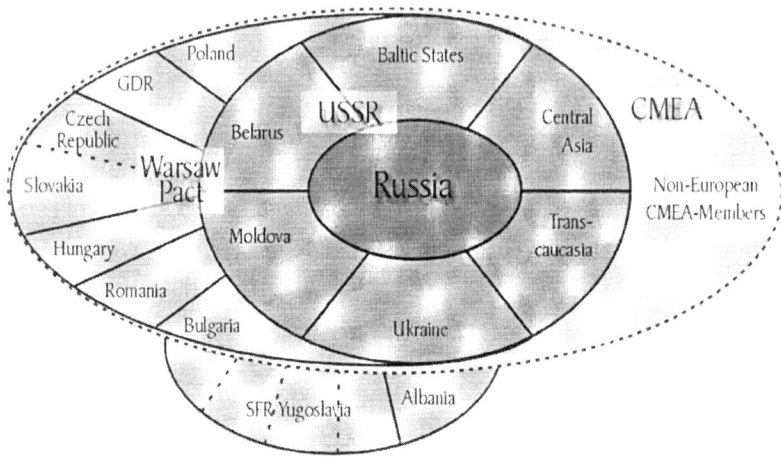

Fig. 1.2 Groups of countries in the socialist world system

scale in the Socialist Federal Republic of Yugoslavia (SFRY) and in socialist Czechoslovakia (CSSR).

POLITICAL, MILITARY AND ECONOMIC ALLIANCES

The USSR was linked to the other socialist countries – and they in some cases to one another – by various types of bilateral state agreements, through which significant influence was exerted on the international coordination and harmonization of science policy and science structure plans. Multilateral agreements also played an important role, in particular the Warsaw Pact, signed in 1955, which served to bind the CEECs to the USSR, and the coordinating body 'Council for Mutual Economic Assistance' (CMEA/COMECON – formed in 1949), which fostered economic and scientific cooperation between these countries as well as with other non-European member states. The SFR Yugoslavia was also fairly loosely affiliated to CMEA by way of a special agreement.

BILATERAL AND MULTILATERAL AGREEMENTS ON SCIENTIFIC AND TECHNOLOGICAL COOPERATION

Apart from various bilateral agreements between states as well as between scientific organizations (AoS, universities) and institutes, multilateral agreements also had a considerable influence on research programs and S&T facilities. The greatest influence was exerted by the 'Complex Program for the Further Extension and Perfection of the Cooperation and Development of the Socialist Economic Integration of the Member States of the CMEA' (RGW, 1971).

There were other multilateral agreements, such as the 'Interkosmos' agreement on space research, and international facilities for advanced education in various disciplines were also established in some countries.

This internal integration process in the socialist countries was further intensified by active (striving for autarchy) and passive (embargoes/CoCom list) dissociation from the OECD nations. This dissociation manifested itself in the socialization processes that scientists underwent, in the collaborative relationships that were established among scientists and in research institutions and the ensuing publications, which were predominantly in Russian. It is also evident in the publications recorded in the Science Citation Index (SCI). The USSR, Czechoslovakia, GDR, Hungary, Bulgaria and Poland, for example, formed a distinct cluster in the early 1980s with regard to the level of 'international cooperative links in the sciences'; of all these countries, only Poland had approximately the same level of links with another country, namely the Federal Republic of Germany (FRG) (cf. Braun and Schubert, 1990). The formation of an international STS for the socialist countries was facilitated by the formally similar organizational structures and modes of operation in national S&T systems, while the differences between socialist and Western countries in these regards often made cooperation with them more difficult.

3. COMMON INSTITUTIONAL CHARACTERISTICS OF S&T IN THE SOCIALIST CEECs

This is not the place for an analysis of the socialist system of S&T. However, our understanding of the changes ensuing from the transformation of the system will be much enhanced by an outline of the most significant features that shaped the Soviet model institutionally and played a part in shaping the STS in all CEECs.

Science played a specific role in the ideology and politics of socialist societies. The 'scientific-technical revolution' was seen as "an important condition for the development of socialist society" and as the "main area of the historical competition between capitalism and socialism" (RGW, 1971, p. 3). In addition, there was a political expectation that science would also be capable of solving problems in all areas of society. Science was to be harnessed as a means of confronting military and economic competition on the international level. The impact of this politicizing of science was ambivalent. On the one hand, the state support that flowed as a result enabled a wide range of research to be carried out on a relatively large scale in international terms (cf. Figure 1.1). At the same time, the high status given to science and research intensified politicians' desire to control and manage this sphere as effectively as possible. Not least for this reason, science was integrated into national planning, with the planning of

S&T being based in particular on the 'linear model of innovation', which had also been dominant in Western nations for a long time (cf. Mayntz, 1997). This model was based on the notion of a linear chain consisting of basic research – applied research – development – production – sales, which was to be realized through a division of labor between science and production. With this concept as its foundation, the STS was built up in the 1920s and 1930s in the Soviet Union (Kröber and Lange, 1975) and later in all other CEECs.

a) As far as the institutional structures of S&T were concerned, *hierarchical systems* predominated. For their part, the STS and its component elements or sectors were subordinate to the political hierarchies. The STS itself was structured hierarchically and in a way that was compatible with the structures in other sectors, especially in the economy (Meske, 1994). This integration in and subordination to political and economic structures led, among other things, to a fundamental distinction being made in the USSR between military and civil S&T. The military also took precedence in S&T tasks in the other socialist countries, but because their military-industrial complexes were smaller and acted primarily as suppliers for the Soviet military this did not lead to such extensive organizational differentiation as in the USSR. In the civil sector, on the other hand, there was a 'sectoral' differentiation in all CEECs between universities/higher education, the Academy of Sciences and industrial R&D.

All S&T facilities were linked to the political-administrative direction and planning system via the ministries responsible for these respective sectors. In the Academy sector, the top management body, the Presidium, assumed the function of a de facto ministry, thereby internalizing the reciprocal flow of information as well as conflicts between politics and science. Thus the AoS management bodies functioned as mediators between politics and science and are to be regarded as 'intermediary organizations' in the institutional spectrum of differentiated societies; this has been examined in detail in the case of the GDR's AoS by Gläser and Meske (1996, pp. 176 – 181). In the largely state-owned industrial sector, this function was taken on by the many 'industry ministries', which were responsible for both the productive enterprises and the centralized R&D facilities of an entire industry. These ministries were primarily political bodies but were also a component of the industrial management system and thus also had an intermediary character. These specific management and organizational structures led to the centralization of the entire STS, all the way to the government; the associated coordination was the responsibility of a special ministry (or committee) for 'science and technology' (which usually had very limited R&D facilities of its own, if any at all). Although the principle of 'democratic centralism' allowed for feedback from bottom to top, S&T institutions ultimately enjoyed no real autonomy at all and actors at subordinate levels had very limited leeway in decision-making. In the course of what were usually very long development processes, the system-related character of S&T had become pronounced and been turned into the determining feature of its organization and mode of operation.

These dominant vertical structures led to centralization within the STS and to specialization based on the management bodies already mentioned above. However, the weakness of horizontal relationships led to the development of a relatively complex R&D system in each of these management segments, which in turn gave rise to the duplication of capacities. In the USSR, the extensive separation of the military and civil sectors, the large territorial distances and the partial jurisdictions of the Union republics meant that this duplication was particularly pronounced. It also reached significant proportions in the other countries as well, particularly in the industrial sector. Since resources were allocated through these management structures, the vertically organized STS structures dominated the process and project-related structures and there was little connection between the various scientific and industrial units. This aspect was intensified in the 1970s and 1980s by a reduction in scientists' mobility; a 'job for life' in the first position or facility following graduation became the rule and a change of job and facility even within the STS increasingly became the exception (cf. Kannengießer and Meske, 1982; Meyer et al., 1981).

b) Processes within the STS were shaped by the 'linear model of research and innovation'. The underlying concept was that science could be utilized as a productive force along the whole chain from basic research to production. This was essentially organized and coordinated by the individual industry ministries. Primacy in this chain was assigned to production and its needs, which for their part were always influenced by political decisions. Thus the original linear innovation model was subsequently modified to include 'feedback' in the formulation of research tasks by the economic and political spheres. However, this feedback involved only the immediate users of R&D results, namely the 'producer' enterprises. In particular, final users' reactions to the newly developed products did not make it past the producers back to R&D. (The military sector was to a certain extent an exception to this, which is probably one reason why it was relatively successful.) There were also strong administrative barriers between individual industries that hindered technology diffusion and transfer (even via products), both between industries and, especially, from the largely 'closed' military sector to the civil sector. As a result of these industry-oriented performance structures, not only was there unnecessary duplication of R&D capacities, particularly in the economic sector, but their practical impact was restricted because of a failure to innovate at enterprise level.

c) In accordance with the 'linear model', with its sequence of steps, and abetted by the centralized hierarchical management system, individual *tasks* within this chain were assigned to particular *scientific institutions*. This division of labor resulted in their specialization in a particular function. Since both science and industry were supposed to avoid duplication of tasks wherever possible, this gave rise not only to specialization but also to the development of monopolies on both sides. In the Central European nations, relations between enterprises and R&D institutions were often relatively close and usually direct. In the USSR, on the other hand, they were mostly indirect and mediated by the ministry responsible. A general disadvantage, however, was that there was scarcely any *networking* between enterprises and a variety of different R&D institutions (Gläser and Meske, 1996). On the one hand, this hampered competition, transfer and innovation but, on the other, promoted mutual understanding among a relatively small number of partners from science and industry involved in a series of work relationships that became firmer and more varied as they progressed. This often led to a cooperative style of working and a far closer relationship between the two sides than was normal in the formal contractual buyer-seller relationships that prevailed in OECD countries. It had greater similarities with the relationship between centralized in-house R&D capacities and production units in large companies (often multinationals) in OECD countries. In some cases, this even led to the successful integration of basic and applied research. This was e.g. one of the significant legacies inherited from the AoS of the former GDR by the newly founded 'blue-list' institutes in East Germany after unification (Meske et al., 1997).

d) Funding for all S&T institutes and processes was usually provided directly or indirectly by state sources on a per-institution basis (basic or institutional funding). With few exceptions, even research financed by enterprises on a contract basis was funded by the state, which either placed funds for this purpose at the disposal of the enterprises or recognized it as an expense. Concepts such as competition, market orientation, 'value for money' and especially the 'project' (with fixed duration and costs) as the centerpiece of the organization of work in S&T systems were neither well developed nor widespread. Although in the 1970s and the 1980s, in particular, attempts were made in almost all socialist countries to link R&D more closely to enterprise-level innovation processes, they were not very successful because the main problem, namely the low interest in and capacity for innovation in the enterprises, had not been resolved (cf. AdW-WIZ, 1976 and 1984; Hanson and Pavitt, 1987).

e) The structure, function and size of the three main R&D sectors as well as of individual organizational components of the STS displayed certain characteristic patterns in all CEECs.

Universities and other *higher education* (HE) institutions were made to prioritize the education of students and the output of the graduates required in the economy. At the same time the emphasis on university studies with a more practical orientation resulted in universities and analogous institutions being linked to industry and other sectors. However, in several Central European countries, such as Poland, Hungary and Czechoslovakia, the long tradition of combined research and teaching in important universities was continued rather than broken with. Nevertheless, in general, the Academy sector was given preference when it came to the promotion and consolidation of research. Responsibility for research in its full breadth (according to research field) and depth (from basic to applied research) was thus shifted into the non-university (governmental non-profit) sector. The central institutions charged with this task were the *Academies of Sciences* with their various institutes, which had been created in line with the Soviet model. They were supplemented in part by specialized research academies (for medicine, education, agriculture, and so on) or by departmental research institutes attached to individual specialized ministries. Thus this 'Academy' or public sector constituted the main element of the research infrastructure in qualitative and often in quantitative terms and frequently combined basic and applied research in the same organization.

In the *industrial sector* too, a specific R&D structure was set up as an element of and a prerequisite for product and process innovations in industry. It was here that there were the greatest deviations, both functionally and organizationally, with respect to the OECD countries. R&D activity was concentrated in industry-level research institutes; R&D capacities in individual enterprises were usually fairly weak in comparison and found primarily in those CEECs that had already had such a tradition prior to World War II. Here too, from a Western vantage point, the dual nature of industrial R&D institutes is noteworthy. It resulted from the fact that these institutes, although they had industrial R&D tasks to perform in close contact with production, generally had the character of state departmental research institutes as far as organization, management structure and funding were concerned. They were therefore not integral parts of an enterprise, but committed to act on the instructions of the ministries above them, which commissioned their work. This meant they were mainly financed from central funds (from the state or from taxes paid by production enterprises). This often led to an overemphasis on research and the neglect both of developments ready to go into production and of continuous improvements to products and production processes. This not only resulted in a generally low level of innovation activity at enterprise-level but also, and even more seriously, meant that the preconditions necessary for innovation (in terms of capacity, experience and motivation) barely existed within enterprises. Innovation activity at enterprise-level was, therefore, not only the last but also usually the weakest link in the basic-research-to-production chain.

f) Most scientific facilities, however, even in the HE and Academy sectors, did have a stock of practically-oriented experience and capabilities. All facilities had to develop their research results into products and processes ready for application because it was only then that they would be taken on by the enterprises, which often had no development capacities of their own. The scientific support provided for this process of product development meant that many scientists gained experience of production processes and, especially, innovation processes and problems. In connection with this, some institutes built up their own facilities for the practical utilization of their research results, in which they produced test models and prototypes, in some cases even small batches in experimental workshops or pilot plants. These 'result-oriented' facilities were often linked to those developed for the purpose of providing special materials and equipment for the institutes' own research needs, which served to compensate for the lack of imports and domestic industrial supplies (cf. Meske, 1986, p. 405–411; Gläser and Meske, 1996, p. 188–193). As a result, most S&T facilities were relatively complex organizations (similar to the production combines). In contrast to Western R&D institutes, they produced a significant amount of the materials and services required for their actual R&D activities themselves; in the West, most such materials and services were bought in. Such structural differences explain, for example, the coexistence in the socialist countries of relatively high levels of R&D personnel with low levels of funding and indicate the difficulties faced in comparing statistics.

g) As a result of these specific political goals and conditions and the structural particularities of S&T in the socialist countries, a relatively high share of GDP was allocated to science and research (Radosevic, 1996). Consequently, industrialization was accelerated and the scientific standard brought into line with that of the leading nations. Especially in the military sector and related fields, such as atomic and space research, internationally top-class performance was achieved because they were given high priority and the best possible conditions for R&D work. On the other hand, from the 1970s onwards, the economic situation was such that it became increasingly difficult to sustain the conditions for the continuous expansion of STS with its growing material and financial requirements, especially given the embargoes then in place. This situation necessarily led to the stagnation and differentiation of scientific activity. Thus by the mid-1970s, the socialist countries had already reached their peak levels, both in terms of the absolute number and the world ranking of their publications recorded in the SCI in all scientific fields. After this time, they began to lag behind global developments (Meske, 1990, pp. 9–15) and by the end of the 1980s (that is at the beginning of the transformation) S&T had reached a state that might be described as 'sick science in a sick society' (Nesvetailov, 1990). A thorough evaluation by the West German *Wissenschaftsrat* (Scientific Council) confirmed that this was indeed the case with the science base in the GDR. It "... was bad in areas in which Western embargo restrictions on the one hand and lack of hard currency on the other hand were effective. In chemistry, for example. But where political circles had certain desires – as in space research – science was very good. It was comparable to ours in the original federal states in the areas in which it was dependent neither upon extrinsic conditions nor special political interests" (Simon, 1991, p. 5; own translation). Moreover, all the socialist countries experienced difficulties with technological development and the practical application of scientific output on a broad basis. Consequently, the proportion of technological innovations continually decreased and the STS became more and more 'science-heavy' and science itself overloaded with staff relative to its financial and material resources. A contributing factor was the curtailment of productive investments as a result of the downward trend in economic growth and the increasing burden of military expenditure, on the one hand, and social programs, on the other. In the 1970s and 1980s, this led to innovative weaknesses not only in industry (Berliner, 1988) but also in science. In contrast, for many industries in the FRG of the 1980s, for example, it was not their own efforts at technological innovation that were decisive but rather those of other industries that were used *indirectly* through technology transfer via preliminary work, investments, material, etc. Thus a number of industries innovated mainly through the acquisition of technology and did so to a much greater extent than their own R&D expenditure would lead one to assume (DIW, 1988). In contrast, analyses showed that in the Soviet Union, for example, the lag behind international standards lay less in the introduction of the latest equipment and methods than in their dissemination from the institutes conducting basic research to the industry-level research institutes and thence into R&D units within enterprises and the production process itself (Nguyen Shi Lok and Kara-Murza, 1979). In the 1970s and 1980s, the gap between the USSR and the Western nations grew significantly because in the USSR modern methods remained stuck in the research institutes, while in the USA, for example, they were widely disseminated, all the way into routine quality controls in industrial enterprises.

These contradictions intensified in the 1980s due to the stagnation of the economy, politics and scientific endeavor in the socialist countries. In the USA and other OECD countries, on the other hand, there was not only a new increase in the number of scientists and engineers in R&D after the crisis in the early 1970s (cf. Figure 1.1 – USA –), but also a fresh wave of innovative activities in new fields. At the same time, economic competition in the world market, in which Asian and other countries were now increasingly participating, became much stronger. This lent new impetus to international competition and led to completely new configurations (cf. Hanson and Pavitt, 1987). As a result, S&T systems throughout the world were not only strongly impelled to grow quantitatively but also infused by the Asian 'tiger states' with a qualitatively new kind of dynamism in the areas of technology transfer and adaptation, reverse engineering, independent improvement and creation of completely new innovations, based above all on progress in the technological application of (available) scientific output in industry (with state support) and less on their own basic and applied research.

This, in conjunction with ecologically determined limits to growth, faced all countries and their various social, economic, science and innovation systems with new problems and demands in the last decades of the 20th century. It was characterized globally as 'science in a steady state' (Ziman, 1987) and as the 'research system in transition' (cf. Cozzens et al., 1990).

As global economic competition intensified and the arms race between NATO and the Warsaw Pact continued apace, the type of society that had come into being in the socialist countries, especially the Soviet Union, proved less and less competitive, not only in scientific and technological matters but in political and economic terms as well. This was chiefly because its innovative capacities were insufficient or too superficial (Freeman et al., 1991). Accordingly, the see-saw effect of mutual overburdening on the part of the political sector, the military, the productive sector and science, resulted in the collapse of the socialist world system. It was accompanied by the disintegration of its main pillar – the Soviet Union – and by fundamental transformations in each individual CEEC as well as by new approaches to socio-political and S&T matters in the remaining socialist countries China and Vietnam (cf. Andreff, 1993; Yang, 1998; Gu, 1999; Meske and Dang Duy Thinh, 2000).

REFERENCES

AdW-WIZ (1976) *Die Entwicklung von Wissenschaft und Technik in den Dokumenten und Beschlüssen der Parteitage der kommunistischen und Arbeiterparteien der sozialistischen Staatengemeinschaft (1974–1976)*. Berlin: Akademie der Wis-

senschaften der DDR, Wissenschaftliches Informationszentrum Berlin (AdW-WIZ).
AdW-WIZ (1984) *Erfahrungen bei der Organisation, Planung, Finanzierung und ökonomischen Stimulierung der Schaffung und Überleitung neuer Technik in den Mitgliedsländern des RGW*. Berlin: Akademie der Wissenschaften der DDR, Wissenschaftliches Informationszentrum Berlin (AdW-WIZ).
Andreff, W. (1993) "The Double Transition from Underdevelopment and From Socialism in Vietnam." *Journal of Contemporary Asia* 23, no. 4: 515–531.
Balazs, K., W. Faulkner, and U. Schimank, (1995) "Transformation of the Research Systems of Post-Communist Central and Eastern Europe: An Introduction." *Social Studies of Science* 25: 613–632.
Berliner, J. S. (1988) *Soviet Industry from Stalin to Gorbachev. Essays on management and Innovation*. Ithaca, NY: Cornell University Press.
Bernal, J.D. (1954) *Science in History*. London: Watts.
Bernal, J.D. (1967) *Die Wissenschaft in der Geschichte*. Berlin: Deutscher Verlag der Wissenschaften.
Braun, D. (1997) *Die politische Steuerung der Wissenschaft. Ein Beitrag zum 'kooperativen Staat'*. Series by the 'Max-Planck-Institut für Gesellschaftsforschung', Cologne, Vol. 28. Frankfurt a.M.: Campus Verlag.
Braun, A., and T. Schubert (1990) "International collaboration in the sciences 1981–1985." *Scientometrics* 19, no. 1–2: 3–10.
Cozzens, S. E., P. Healey, A. Rip, and J. Ziman (Eds.) (1990) *The Research System in Transition*. Dordrecht: Kluwer Academic Publishers.
CSO (Central Statistical Office of the Council of Ministries of the USSR) (Ed.) (1971) *Narodnoe obrazovanie, nauka i kul'tura v SSSR* (Education, science and culture in the USSR). Moscow: Statistica.
DIW (1988) "Industrielle Forschung und Entwicklung kommt vor allem dem Export zugute." *Wochenbericht* 47/1988. Berlin: Deutsches Institut für Wirtschaftsforschung (DIW), 631–635.
Ducatel, K., and J.-C. Burgelman, (1999) *Employment Map. The Futures Project*. EC JRC, EUR 19033 EN. Seville: Institute for Prospective Technological Studies (IPTS).
Edquist, C. (Ed.) (1997) *Systems of Innovation. Technologies, Institutions and Organizations*. London: Pinter Publishers.
Engel, P. (1981) *Japanische Organisationsprinzipien – Verbesserung der Produktivität durch Qualitätszirkel*. Zürich: Verlag moderne Industrie.
Freeman, C. (1987) *Technology Policy and Economic Performance: Lessons from Japan*. London: Pinter Publishers.
Freeman, C., M. Sharp, and W. Walker (Eds.) (1991) *Technology and the Future of Europe. Global Competition and the Environment in the 1990s*. London: Pinter Publishers.
Gavigan, J. P., M. Ottitsch, and S. Mahroum (1999) *Knowledge and Learning – Towards a Learning Europe. The Futures Project*. EC JRC, EUR 19034 EN. Seville: Institute for Prospective Technological Studies (IPTS).
Gibbons, M., C. Limoges, H. Nowotny, S. Schwartzman, P. Scott, and M. Trow (1994)

The New Production of Knowledge: The Dynamics of Science and Research in Contemporary Societies. London: Sage Publishers.

Gläser, J., and W. Meske (1996) *Anwendungsorientierung von Grundlagenforschung? Erfahrungen der Akademie der Wissenschaften der DDR*. Series by the 'Max-Planck-Institut für Gesellschaftsforschung', Cologne, Vol. 25. Frankfurt a.M.: Campus Verlag.

Goscomstat (Ed.) (1987) *Narodnoe chozjajstvo SSSR za 70 let: jubilejnyj statisticeskij ežegodnik* (National economy of the USSR during 70 years: jubilee statistical yearbook). Moscow: Finance and Statistics.

Goscomstat (Ed.) (1991) *Nauka SSSR v cifrach: 1990* (Science in the USSR in figures: 1990). Moscow: Finance and Statistics.

Graham, L. R. (1967) *The Soviet Academy of Sciences and the Communist Party, 1927–1932*. Princeton, NJ: Princeton University Press.

Graham, L. R. (1975) "The Formation of Soviet Research Institutes: A Combination of Revolutionary Innovation and International Borrowing." *Social Studies of Science* 5: 303–329.

Gu, S. (1999) *China's Industrial Technology. Market reform and organizational change*. London and New York: The United Nations University Press, Institute for New Technologies.

Hanson, P., and K. Pavitt (1987) *The Comparative Economics of Research Development and Innovation in East and West: A Survey*. Chur: Harwood Academic Publishers.

Jungk, R. (1982) *Heller als tausend Sonnen. Das Schicksal der Atomforscher*. Reinbek bei Hamburg: Rowohlt Verlag.

Kannengießer, L., and W. Meske (Eds.) (1982) *Das Kaderpotential der Wissenschaft im Sozialismus*. Berlin: Akademie-Verlag.

Krishna, V., R. Waast, and J. Gaillard (1998) "Globalization and scientific communities in developing countries." *World Science Report 1998*. Paris: United Nations Educational, Scientific and Cultural Organization (UNESCO), 273–287.

Kröber, G., and B. Lange (Eds.) (1975) *Sowjetmacht und Wissenschaft. Dokumente zur Rolle Lenins bei der Entwicklung der Akademie der Wissenschaften*. Berlin: Akademie-Verlag.

Krohn, W., and G. Küppers (1989) *Die Selbstorganisation der Wissenschaft*. Frankfurt a.M.: Suhrkamp Verlag.

Lavigne, M. (1995) *The Economics of Transition. From Socialist Economy to Market Economy*. New York: St. Martin's Press.

Lundgreen, P., B. Horn, W. Krohn, G. Küppers, and R. Passlack (1986) *Staatliche Forschung in Deutschland 1870–1980*. Frankfurt a.M.: Campus Verlag.

Mayntz, R. (1997) "Forschung als Dienstleistung? Zur gesellschaftlichen Einbindung der Wissenschaft." *Berichte und Abhandlungen Vol. 3*. Ed. Berlin-Brandenburgische Akademie der Wissenschaften. Berlin: Akademie Verlag, 135–154.

Meske, W. (1986) *Technik für die Wissenschaft. Entwicklungsprozesse und -probleme der materiell-technischen Basis der Wissenschaft. Studien und Forschungsberichte No. 20*. Berlin: Akademie der Wissenschaften der DDR, Institut für Theorie, Geschichte und Organisation der Wissenschaft (AdW-ITW).

Meske, W. (1989) *Nationale Wissenschaftspotentiale. Studien und Forschungsberichte No. 27*. Berlin: AdW-ITW.
Meske, W. (Ed.) (1990) *Wissenschaft der RGW-Länder. Länderberichte zur Situation am Ende der 80er Jahre aus der DDR, Polen, der Tschechoslowakei, Ungarn, Bulgarien, der Sowjetunion, der Mongolischen VR, Vietnam und Kuba. Studien und Forschungsberichte No. 30*. Berlin: AdW-ITW.
Meske, W. (1994) *Veränderungen in den Verbindungen zwischen Wissenschaft und Produktion in Ostdeutschland*. WZB-Paper P 94–402. Berlin: Wissenschaftszentrum Berlin für Sozialforschung (WZB).
Meske, W., and Dang Duy Thinh (Eds.) (2000) *Vietnam's Research & Development System in the 1990's – Structural and Functional Change*. Research Report P 00–401. Berlin: WZB.
Meske, W., J. Gläser, G. Groß, M. Höppner, and C. Melis (1997) "Die Integration von ostdeutschen Blaue-Liste-Instituten in die deutsche Wissenschaftslandschaft – Forschungsbericht." Unpublished paper. Berlin: WZB.
Meyer, H., G. Groß, A. Krause, and C. Waltenberg (1981) *Struktur und Dynamik des Kaderpotentials in der Wissenschaft. Studien und Forschungsberichte No. 14*. Berlin: AdW-ITW.
Miyabayashi, M. (1997) "Present Status and Future Trends of the National R&D System of Japan." Unpublished Paper for the National Institute for Science and Technology Policy (NISTEP) International Workshop 'Strategic Models for the Advancement of National R&D Systems'. Tokyo: NISTEP.
Nadiraschwili, A. (1994) *Die Transformation der Wissenschaft in den Ländern der ehemaligen UdSSR*. WZB-Paper P 94–401. Berlin: WZB.
Nguyen Shi Lok, and S.G. Kara-Murza (1979) "Technologija naucnych issledovanij. Izucenie rasprostranenija spektralnych metodov s pomoscju analiza publikacij." (Technology of Research. Investigation in the proliferation of spectral methods by publication analysis). *Naucno-techniceskaja informacija* (Science and technology information) 1, no. 12: 5–9.
Nelson, R. (Ed.) (1993) *National Innovation Systems – A Comparative Analysis*. New York: Oxford University Press.
Nesvetailov, G.A. (1990): "Bol'naja Nauka v bol'nom obscestve" (Sick science in a sick society.) *Sociologiceskie issledovanija* no. 11: 43–55.
NSB (National Science Board) (1989) *Science & Engineering Indicators – 1989*. Washington DC: US Government Printing Office.
NSB (2000) *Science & Engineering Indicators – 2000*. Arlington, VA: National Science Foundation (NSF).
OECD (1972) *The research system. Vol 1: France, Germany, United Kingdom*. Paris: OECD.
OECD (2000) *Education at a Glance*. Paris: OECD.
Price, D.J. de Solla (1974) *Little Science, Big Science. Von der Studierstube zur Großforschung*. Frankfurt a.M.: Suhrkamp Verlag.
Radosevic, S. (1996) "Divergence or Convergence in Research and Development and Innovation between 'East' and 'West'." Unpublished Paper presented at the 5th Freiberg Symposium on 'Economics, Innovation and Transformation', August 29–31.

Reger, G., and U. Schmoch (Eds.) (1996) *Organisation of Science and Technology at the Watershed.* Vol. 3 of the 'Academic and Industrial Perspective, Technology, Innovation, and Policy' Series by the 'Fraunhofer-Institute for Systems and Innovation Research' (ISI). Heidelberg: Physica-Verlag.

RGW (1971) "Komplexprogramm für die weitere Vertiefung und Vervollkommnung der Zusammenarbeit und Entwicklung der sozialistischen ökonomischen Integration der Mitgliedsländer des RGW." (RGW – Rat für gegenseitige Wirtschaftshilfe) *Neues Deutschland* August 7: 3–11.

Simon, D. (1991) "Kritik aus dem Osten ist berechtigt." *Berliner Zeitung* August 10/11: 5.

Šokareva, T.A. (1991) "SSSR i SŠA: Kadrovoe obespecenie issledovatel'skoi dejatel'nosti" (USSR and USA: Personnel in research activities). *Vestnik AN SSSR* no. 9: 40–43.

UNESCO (1998) *World Science Report 1998.* Paris: United Nations Educational, Scientific and Cultural Organization (UNESCO),

Yang, Q. (1998) "The Structural Reform of the R&D System in China." *Transforming Science and Technology Systems – The Endless Transition? NATO Science Series 4: Science and Technology Policy – Vol. 23.* Ed. W. Meske, et al. Amsterdam: IOS Press, 153–159.

Ziman, J. (1987) *Science in a 'steady state': The research system in transition.* London: Science Policy Support Group.

2. SCIENCE AND TECHNOLOGY IN CEECS AT THE END OF THE SOCIALIST ERA

Werner MESKE

Reference was made in Chapter 1 to different groups of (now independent) countries that were integrated in different ways into the socialist world system (cf. Figure 1.2). The collapse of the Soviet Union and of the 'socialist world system' it dominated inevitably produced different starting conditions for the reforming and development of the countries and states emerging from this system.

On the basis of their history up to 1990 and the resultant differences in the starting conditions for the construction of a new national science and technology system (STS), the now independent countries can be divided into at least three broad categories. The first group includes the European countries that were constituent, non-independent parts of the USSR until its dissolution in December 1991. Of these, Russia, the Ukraine, Belarus and Moldova are still member states of the Commonwealth of Independent States (CIS) and in this respect are still relatively closely linked to each other. On the other hand, the Baltic states of Estonia, Latvia and Lithuania, which were also previously part of the USSR, explicitly distanced themselves from Russia and right from the early days of the newfound independence began to fashion their own development paths, looking primarily to their Scandinavian neighbors for assistance. The second group is made up of Central Eastern European countries that were independent states prior to 1990 but closely linked to the USSR as members of the Warsaw Pact and the CMEA. The third and final group consists of the successor states to the former SFR Yugoslavia. They form a group on their own, since they were not so closely tied to the Soviet system either in their external relations or in their institutional structures.

1. THE DIFFERENCES BETWEEN FORMERLY SOCIALIST COUNTRIES IN CENTRAL AND EASTERN EUROPE

1.1. THE EUROPEAN CIS COUNTRIES: RUSSIA, UKRAINE, BELARUS AND MOLDOVA

Russia and its Slavic neighbors, the Ukraine and Belarus, formed the core of the former USSR. Around 1990, the STS in the USSR represented a unique complex that included not only research institutions but also pilot plants and service organizations. There were 1.32 million researchers in the 'science and science services' sector in 1991 and more than 3 000 research institutes. A main characteristic of the STS was its extreme militarization; up to 70 per cent of R&D funds in the country were spent (directly or indirectly) on defense research[1]. After the collapse of the USSR, each republic inherited fragmented parts of the military industrial complex (MIC). However, most were able to produce only separate, although sometimes complicated, components of military equipment as suppliers to the Russian MIC. Russia was – and remains – the only country to consume and produce these components and assemble them into modern weapons systems. Institutionally, the STS was characterized by the 'Soviet model' and extreme similarity in the forms and methods of the social organization of scientific research in all Soviet republics. However, there were also great differences between these republics in terms of the amount of science resources and their structure (cf. Nadiraschwili, 1994), on the one hand, and in terms of the functions and level and quality of research, on the other. This was the result of differences in population sizes and level of industrialization in the various republics, as well as of their differing positions in the STS of the USSR. A low resource stock (primarily in material and currency) and difficult access to scientific communication channels gave rise to social problems and a relatively narrow-minded and unsophisticated approach to science in all the republics except Russia and Ukraine. As a result, the gap between the level in the center (some parts of Russia) and in the periphery grew ever wider. By the time the USSR collapsed, the ratio between employment in scientific activity and total employment in the economy was highest

---

[1] According to official government estimates, about 70 per cent of the technology Russia processed was of a dual-use character. This estimate was given by president Boris Yeltsin in 1995. It is not, however, absolutely clear how the figure was estimated (Bzhilianskaya, 1996, p. 7). The Soviet Military-Industrial Complex (MIC) included a wide range of different production units supplying for example 100 per cent of television sets and sewing machines in Russia (Rouvez, 1995, p. 1–2). Nevertheless, the direct R&D expenditures within the defense budget amounted, in 1989, to about 15 billions Roubel or 36 per cent of GERD (Nauka Rossii, 1992, p. 168–169, and Nadiraschwili, 1994, p. 26).

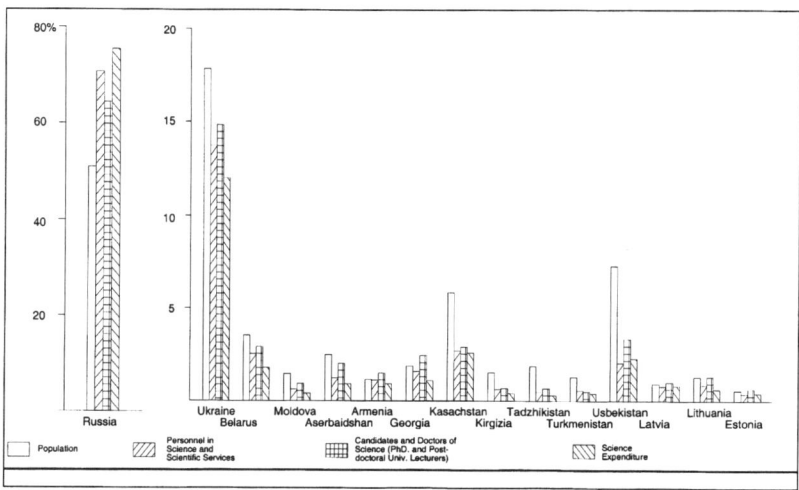

Fig. 2.1 Share of the individual republics of the USSR in selected science resources (in %, USSR = 100)
Source: Nadiraschwili, 1994, p. 21.

in Russia, more or less balanced (near to the average of the USSR) in Ukraine and Belarus and lowest in Moldova (cf. Figure 2.1).

Russia was the dominant user of science resources, both in absolute terms (with a share of 60–80 per cent for all relevant indicators) and relative to population size. The scale of this great power enabled it to be active in almost all areas of science and technology, although a specific characteristic was the dominance of military and industrial research.

The *Ukraine* had a share of between 13 and 15 per cent of the entire research potential of the USSR. Important particularities were an increased concentration of experimental bases in USSR research organizations, a developed regional structure of the science potential, the dominance of the industrial research sector (which accounted for about 50 per cent of R&D personnel in science), a high percentage of applied research even in Academy science and the active search for new forms of research organization. In terms of size of population (about 60 million) and R&D potential, Ukraine was more prepared for independent development than the other (mostly very small) Soviet republics, which were characterized by greater specialization and their role as 'suppliers' for Russia.

Belarus had a population of 10 million and, with 4.6 per cent of all research organizations and 3 per cent of all researchers was the third-placed USSR republic in terms of science and technology. The Belorussian AoS (with 15 per cent of all researchers in the republic) was the center of science, forming leading scientific schools and conducting research in optics and spectroscopy, theoretical physics and mathematics. Industrial R&D was characterized by a high share of design offices and organizations engaged in development for metal-intensive industries. The last two decades of the USSR's existence were characterized by the development of applied defense R&D financed mainly by contracts from outside Belarus.

Moldova (until 1940 a part of Romania, with a population of 4 million) had a small science potential (0.5 – 0.7 per cent of research personnel in the USSR); almost one third of researchers in Moldova were concentrated in biological and agricultural research, in line with the republic's economic specialization. The sectoral structure of science was dominated by the higher education (HE) sector, accounting for 55 per cent of all scientists. Moldova was low in the skill structure and in many research areas personnel levels did not reach the 'critical mass' necessary for efficient research [2].

The specific conditions of S&T in the various Soviet republics show that, following the dissolution of the USSR, they all had to combine institutional *reform* with the extensive *restructuring* of their own (national) S&T systems; the only exception was Russia, which absorbed most of the USSR's S&T legacy. Changes in the 'Soviet model' of S&T had already occurred in the USSR in that 'scientific and production associations' and 'inter-branch scientific and technological complexes' had been formed to better integrate the various parts of the STS. With the introduction of 'perestroika', the pace of change accelerated and major reforms were introduced (cf. Gaponenko, 1998). These reforms included conversion from military to civil science, the relaxation of state, party and departmental control over enterprises' activities, increased democratic freedom and openness ('glasnost'), the partial elimination of economic and scientific autarchy and the establishment and development of an alternative, non-state sector of the economy. However, the changes made to the economic system and the STS during the Gorbachev era caused instability. The greatest impetus came from the reckless destruction of the planned economy at that time, as institutions were dismantled without anything being put in their place. The demise of centralized management totally disrupted the interaction between science and industry, traditionally channeled through the ministries. Moreover, the accumulation of financial resources in ministries and departments was also eliminated (cf. Mindeli and Pipiia, 1998). As the old system of coordination disappeared, no provision was made for the creation

---

[2] Since we were unable to continue the collaboration with experts from Moldova that began in the mid-1990s (cf. Kramarenko, 1998), this country could not be incorporated into the country analyses in Part II.

of market mechanisms through which innovation, the funding of industrial research, credit transactions and so on could be mediated. Legal chaos further aggravated the situation. Indeed, the Gaidar team carried out its radical market reforms without an appropriate legal base. The disintegration of the USSR in 1991 led to the breakdown of economic, technological and informational ties between former republics, despite the founding of the CIS in December 1991. The all-union ministries, which had regulated and financed STS, disappeared, but the formation of new central executive powers took time in all successor states. Thus the sovereign development of S&T in most of the individual CIS countries began in an *administrative* and *legislative vacuum.*

1.2. THE BALTIC STATES: ESTONIA, LATVIA AND LITHUANIA

Economic, social, cultural and educational development in the Baltic states has always been greatly influenced by their geopolitical location on the border between Western and Eastern Europe. The characteristic features of their development changed several times during the 20th century as a result of changes in their political status as they moved from being a border region of a large empire to become small, independent states and vice versa.

Unlike other post-Soviet countries, *the Baltic states were not an original part of the former USSR*. Up until 1918, they were part of the Russian Empire and from 1918 to 1940 they were independent democratic or semi-democratic states. The Soviet Union was founded in 1917–1922, while the Baltic states enjoyed independence. During World War II, from 1940 to 1945, the Baltic states were denied even the formal independence and were forcefully re-incorporated into the Soviet Union after a period of dominance by Nazi Germany. This incorporation lasted until the collapse of the USSR in 1991. Research in the Baltic countries was closely linked to the research community of the USSR and the successful establishment of a science and technology research base in the Baltic countries was driven by the interests of the USSR's military-industrial complex.

Dramatic political changes and the following transition from a command to a market economy brought about major changes in the R&D system as well as in scientific research in Estonia, Latvia and Lithuania (Dagyte et al., 2000). The process of transformation was marked by a number of specific features that set the Baltic countries apart from the other countries in the region. Thus in the traditional parts of the USSR (now the Commonwealth of Independent States or CIS) and in the independent Central East European countries, the traditional bureaucratic R&D management structures were partly preserved, while this was not the case in the Baltic states. Estonia, Latvia and Lithuania had to start from scratch in rebuilding their scientific management structures in the aftermath of the successful struggle for independence in 1990–1991. The countries of Central Europe had to some extent been able to implement science policies of their own during the socialist era and were thus less dependent on the USSR in this respect. The Baltic states could not even dream of a science policy of their own until 1990, since science in those countries had developed as a part of the mighty USSR science system. To some extent, science in the Baltic countries remained peripheral to the USSR science establishment during the 45 years of communist rule (1946–1990). Baltic science did not acquire the character of 'small-country science' until 1990–1991, and then only as part and parcel of the struggle for independence.

Despite the appearance of similarity to outside observers, the cultural heritage of the three Baltic states actually differs quite radically, with noticeable effects on their science and innovation systems. There is significant diversity between their ethnic roots, historical heritage, language, religion and national identity. These differences have led the Baltic states to take distinctly different approaches toward the reform of their R&D systems and set them apart from the CIS countries, on the one hand, and the other CEECs, on the other.

1.3. THE SIX CENTRAL EAST EUROPEAN COUNTRIES: POLAND, CZECH REPUBLIC, SLOVAKIA, HUNGARY, ROMANIA, BULGARIA

This group of countries has in common their integration into the socialist system after World War II and the strong influence exerted by membership of the Warsaw Pact and the CMEA. On the other hand, the separate histories of these countries both before and during the socialist period mean that they can by no means be regarded as a homogenous bloc. We can only allude here briefly to some of the events that marked their histories during the period of Soviet domination, such as the 1956 upheaval in Hungary, the 'Prague Spring' of 1968, the Solidarity movement in Poland or, more broadly, the Ceausescu regime in Romania. Thus the duality of a common structural heritage rooted in socialism and individual development paths due to differences in their history before and during the socialist period applies absolutely to this group of countries. Because of this diversity and in order to avoid duplication, we will not examine the individual countries in any greater detail here but refer the reader to the relevant country chapters 10–15 in Part II.

1.4. COUNTRIES OF THE FORMER SFR YUGOSLAVIA

The former Socialist Federal Republic of Yugoslavia (SFRY) consisted of six republics, namely the socialist republics of Serbia, Slovenia, Montenegro, Croatia, Bosnia and Herzegovina, and Macedonia. The constitution of the SFRY gave all the republics broad rights and obligations to draw up their own constitutions and organize political, economic and cultural life as they saw fit (cf. Kutlaca, 1996). The republics were responsible for S&T and organized their STS in very similar ways. The STS in the various republics were organized separately and self-sufficiently (in terms of funding, monitoring and evaluating processes) and were made up of universities, independent institutes, R&D units in industry and the S&T infrastructure. The federal government made several, mostly unsuccessful attempts to bring together R&D institutions and researchers from different republics in joint research projects.

The individual republics of the SFRY shared certain similarities and also had their own particularities in S&T. All the science and technology systems suffered from over-manning, while R&D was strong in independent institutes but weak in the universities and in industry (with Slovenia being an exception in some cases). Each republic tended to be active in all fields of S&T but the contribution of their R&D efforts to the country's development (as measured by the number of patents granted, for example) was inadequate.

Serbia had the highest concentration of R&D activity, accounting for 36 per cent of all researchers in the SFRY. The share of R&D workers in the population as a whole was similar to levels in Spain and Austria. R&D was concentrated in independent institutes and industrial R&D was weak.

In *Slovenia*, the share of researchers in the population as a whole was comparable with most developed OECD countries. Research in technical sciences was strong and Slovenia led the other republics of the SFRY in industrial R&D.

Kutlaca assumed that, at the beginning of the transformation, Yugoslavia's economy should have been be able to keep pace in science and technology with the developed countries (Kutlaca, 1996, p. 15). Innovation activities in industry were financed and realized by the firms themselves and focused on new products or processes likely to improve firms' market position. Firms used ideas from clients, customers, professional conferences, fairs, exhibitions, the academic sector and so on as sources of inspiration for innovations. The dissolution of the SFRY has had devastating effects on most of the republics. These effects were exacerbated by the war in Croatia and Bosnia and Herzegovina, and the international isolation of Serbia and Montenegro. "The constitutions of all Yugoslavian successor states declare their commitment to a market economy. However, in the face of the differing basic conditions and basic political stance of each, they pursued divergent strategies in the restructuring of their former 'yugosocialist' system into a market oriented system. The progress made by each in dealing with the relevant key problems, the macroeconomic stabilization and the microeconomic restructuring thus differs. The collapse of Yugoslavia and the difficulties which ensued have brought about phases of socio-economic depression. Among other things, this is evident in a dramatic reduction in the net domestic product, caused mainly by the crash in industrial production, in the rapid increase in unemployment, in the considerable reduction in real earnings and, as a consequence, in a significantly lowered standard of living" (Büschenfeld, 1997, p. 488; own translation).

The dissolution of the SFRY also impacted in various ways on the STS.

All the republics are badly affected by a brain drain. Estimates based on immigration statistics for OECD countries show that between 10 and 30 per cent of the total research population has left the former SFR Yugoslavia. As a result, all the republics are suffering from shortages of skilled researchers. A strong 'shift' from research institutes to the universities, a trend common to all the former Yugoslav republics, may eventually lead to the establishment of a 'critical mass' of R&D activity in the universities.

The development of national S&T systems in the SFRY successor states has been underway since 1991, except in *Bosnia and Herzegovina* (due to the war and its consequences). For the *Federal Republic of Yugoslavia* (that is the republics of *Serbia* and *Montenegro*) and *Slovenia* more detailed descriptions follow in Part II.

2. DIVERGENT CONDITIONS FOR S&T IN THE INDIVIDUAL CEECS AT THE END OF THE SOCIALIST ERA

Regardless of their common 'socialist heritage', the differing histories of the individual CEECs give grounds for supposing that each of them found itself in a different situation at the end of the socialist era. Consequently, the starting conditions for the transformation of the STS also differed from country to country, and this definitely exerted considerable influence over the processes of change. The main influencing factors were
– each country's position and role in the socialist world S&T system,
– the degree of institutional deviation from the 'Soviet model' of STS, and
– the basic societal conditions in each country when the socialist system collapsed.

THE POSITION OF THE INDIVIDUAL COUNTRIES UNDER SOCIALISM

The socialist STS was by no means a globally homogenous entity, as it often appeared to be from the outside. Rather, it had both common characteristics *and* internal differences, with the individual countries being integrated into the system as a whole to varying degrees. As the core of the former USSR, Russia certainly represented not only the political, military and economic, but also the scientific-technical center of the global socialist STS. This was the result not only of the dominant weight of Russia's own capabilities but also of its central management and coordination function with regard to all other parts.

All the other republics in the USSR, as well as the formerly socialist European countries, played a peripheral role compared with that of Russia. This arose primarily out of their political and military dependency but also out of the (qualitative and quantitative) significance of economic and scientific relations within the socialist bloc. With the exception of Ukraine, Poland and Romania, the other countries generally had a population of considerably less than 20 million and in some cases less than 5 million. For this reason alone, they carried far less weight than Russia, with its population of 148 million. This was, however, not the only reason behind the clear gradations in the relationships between Russia or the Soviet Union and the other countries, which can be represented schematically as a number of concentric circles, with Russia at the center and the other countries located in one of the other circles at a greater or lesser distance from the center (cf. Figure 1.2).

The circle closest to Russia contained the Soviet republics belonging to the confederation of the USSR; these states were completely integrated into the USSR's Russian-dominated state administration, legislation, etc. In contrast, the other socialist European countries tended to be located in the outer circles. Despite all their dependencies, their state sovereignty afforded them more or less scope to act autonomously. Relations between the individual CEECs and the USSR were strictly controlled by their membership of the Warsaw Pact and the CMEA. In contrast, the non-member states, SFR Yugoslavia and Albania, tended to have a looser relationship to the USSR. This gave rise to a basic differentiation between the countries depending on whether they were located in the center (Russia), circle 1 (Soviet republics), circle 2 (member states of the Warsaw Pact and CMEA) or circle 3 (other socialist European and non-European countries). There were also differentiations within each group. Even at the center, in Russia, there were considerable differences between regions in terms of development level. By way of example, we can mention the variation between the largest regions of Russia, namely the European part, Siberia and the Far East. Of the (European) Soviet republics in circle 1, Ukraine and Belarus were once very closely linked historically, ethnically, culturally and economically to Russia and had built up their STS in line with the 'Soviet model' right from the start. In contrast, the Baltic republics had only been incorporated into the Soviet Union in 1940 following 20 years of independence during which they had belonged to the West and North European economic and cultural sphere. The situation was similar in Moldova, which had been part of Romania until 1940; as a small agrarian region, however, Moldova had no scientific or technological tradition like those of the Baltic states.

The other Central and East European countries had only entered the socialist system after 1945 and had brought very different economic and scientific-technical traditions with them, which usually continued to have an effect throughout the entire socialist period. They led to variations in implementation of the 'Soviet model' and not only influenced scientific and technical relations with the Soviet Union and the other socialist countries but also contributed to the continuation of intensive contacts beyond this bloc, in particular between Poland and Hungary and the Western nations. The SFR Yugoslavia and Albania in particular were able largely to escape the Soviet influence and develop their 'own' variations of the socialist model in S&T as well. This applied, for different reasons, to most of the non-European socialist countries as well.

INSTITUTIONAL COMPLEXES IN S&T

By the end of the socialist era, under the influence of their respective 'pre-socialist' histories and the (frequently leading) roles they took on in certain areas of S&T, the individual countries had institutional complexes that deviated to a greater or lesser extent from the 'Soviet model' of STS. Moreover, these institutional complexes often harbored within themselves the seeds of transformation towards democracy and the market economy.

The traditional modes of behavior and organizational forms within the sphere of S&T that the individual countries brought with them into the socialist era played a particular role here. As verified by the country analyses (cf. Part II of this volume) and also by our own experiences and other studies in East Germany (cf. Gläser and Meske, 1996; Kocka and Mayntz, 1998), this legacy had a considerable long-term effect, often over generations, due to the special training and developmental processes scientists underwent.

This explains the clear differences between those countries that developed S&T under socialist conditions (Russia, Ukraine, Belarus and Bulgaria) and those, particularly the Central East European and Baltic states, that had already developed their own national STS along Western lines. This legacy was also an important precondition for retaining relationships with the non-socialist scientific world. These relationships in turn had considerable repercussions on the national STS. This is verified by the considerable differences in the number of co-authorships with scientists in Western countries that are revealed by bibliometric studies (cf. Braun and Schubert, 1990; Czerwon, 1998) as well as by the high proportion of R&D-intensive industrial branches and the deviations from the 'Soviet model' in the funding and organization of science in those socialist countries that were advanced in S&T. One of the major divergences from the Soviet model was the continuing strength of universities within the science system, including in research, for example in the Baltic countries and in Poland. Another lay in the retention and subsequent reinforcement of enterprise R&D capacities, for example in Czechoslovakia, where branch R&D institutes were assigned to combines, and in Poland, as well as in the establishment of direct contractual relationships between enterprises and branch R&D institutes, universities and Academy institutes.

In connection with the expansion of such contractual research, the largely institutional funding of R&D was supplemented by forms of project and performance-related financing of individual facilities, research groups and scientists. Hungary went furthest down this road with the introduction of the OTKA fund for competitive project selection and funding even in basic research (cf. Chapter 13). The SFR Yugoslavia never had such close ties with the Soviet Union that it fully adopted the 'Soviet model'; rather, the individual republics in the SFRY built up their own systems and thus, for example, had neither Academies of Sciences nor branch R&D institutes along Soviet lines (cf. Chapters 16 and 17).

These national particularities in the institutional complexes of S&T were significant for the transformation in several respects. In some countries, even in the socialist era, new ideas and critical attitudes to their own performance had arisen or been given greater urgency by contacts with Western countries. Attempts to come to grips with the developmental problems and processes of modern innovation systems had generated awareness of and interest in new solutions, particularly among scientists but also among research administrators. Especially in times of radical change, this led to considerable 'bottom-up' activity in science (in the Baltic countries, for example, even prior to their independence). After all, the significance of science in these countries and its connection, in Ukraine, Romania and Poland, for example, with earlier periods of national independence had helped to create a more positive attitude to (national) science among the intelligentsia, and later among politicians as well, than was the case in other countries, such as Bulgaria, in which science had been experienced only under socialism.

Account also has to be taken of the basic societal conditions at the time when the system changed. In most countries, the system change can hardly be said to have occurred smoothly or uneventfully. Individual countries faced considerable conflicts in achieving national independence through the dissolution of the former confederations of the USSR, the SFR Yugoslavia and Czechoslovakia; in some cases, this was even associated with military confrontations and wars. This was a major setback for the consolidation processes of many of the new states on a political level and in particular for the restructuring of their societal systems. In addition, their economies were destroyed, which intensified the general problems of economic change and decline even further. These effects could be observed, to a certain degree, in the Baltic states before the withdrawal of Soviet troops, but they were most apparent in the former SFRY, where even now in some parts neither political nor economic stability has been achieved and the STS is still in disarray. Clearly, all the former socialist countries suffered the cumulative effects of the collapse of the socialist economic system and the ensuing loss of foreign trade within the Soviet bloc, which had accounted for such a high share of their economic activities. From 1989 onwards (or later, as the case may be), GDP generally fell in each country by about 10 per cent and in some cases by as much as 20 to 30 per cent in comparison with the previous year (cf. Stern, 1997, Table 2; Figure 18.2 in this volume). In most countries, the decline in real industrial output was much greater (cf. Figure 2.2) and was one important reason for the decline in indus-

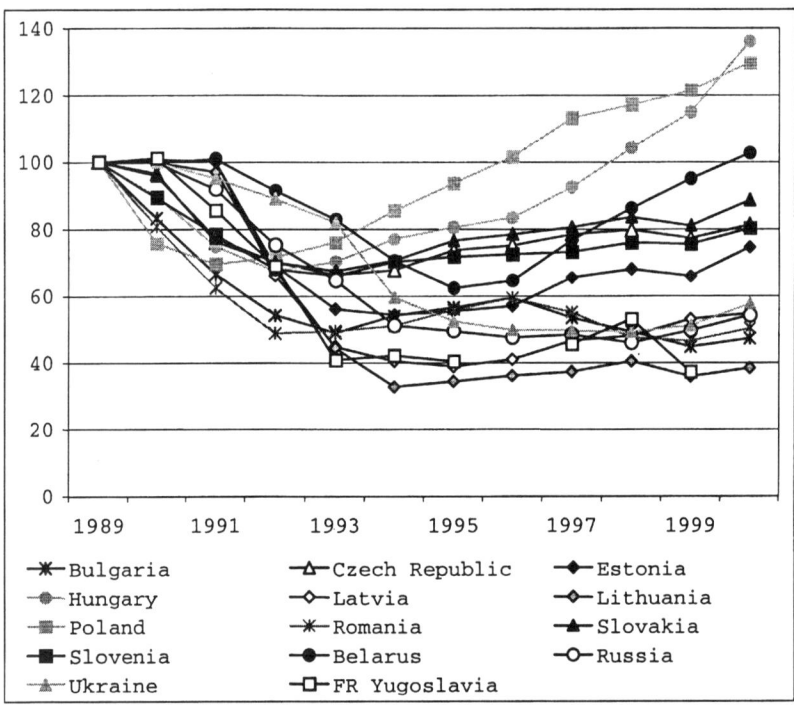

Fig. 2.2 CEECs: real industrial output levels (1989 = 100)
Source: Weber et al., 1999; EC, 1998-2001; EBRD, 1998 and 2001 (for Belarus, Russia and Ukraine); and Tab. 16.1 and 16.2 in this volume (for FR Yugoslavia).

trial R&D. These economic difficulties clearly had a direct impact on S&T. However, S&T was even further marginalized in terms of societal importance by the fact that economic and political priorities were clearly being determined by other, more pressing concerns. As a result, those countries in particular that first gained independence and had to build up their own STS found themselves in a vacuum, and in all countries the creation of new administrations, laws and other S&T regulations was delayed (on the changes in the funding of S&T, cf. Fedorko, 1995).

Together with the basic societal conditions, the aims underlying the reorganization of the STS and the models on which the reorganization was based also played a fundamental role. In this respect, the individual countries can be arranged along a spectrum. At one of the poles lie those countries that opted to seek EU and NATO membership as soon as possible and thus adopted the rules and forms of S&T applicable in the EU. At the other pole lies Russia, which opted to retain or to restructure, largely independently, the 'super power' STS it had inherited from the USSR and to safeguard its role as an international center for S&T. Both poles differ not only in terms of their aims but above all in the lack of clarity in the latter as to the concrete character of a future STS in Russia and its relationship to other countries. For this reason, there were, from the very beginning of the transformation, no clear ideas as to the aims of transformation and the means by which they might be achieved.

In the first case, scientists and politicians single-mindedly attempted to adjust the institutional complex of S&T in line with the patterns and experiences of the EU and its member states (in particular Germany and the Scandinavian countries). This was particularly true of Slovenia, Poland, Hungary and the Czech Republic, but also of the Baltic states. As a result, all the actors involved had a clear set of strategic goals but no guidelines by which to operate. In contrast, Russia was dogged by uncertainties and instabilities with regard to its future STS, with chaotic consequences for its functioning and the changes that were already underway. Practically all the other CEECs were positioned somewhere between these two poles in terms of the guidelines they adopted in transforming their STS. In particular, the CIS countries were closely linked to Russia and the former 'Soviet model' and began restructuring hesitantly, not least because the key actors in politics and science lacked clear objectives. Its rapid incorporation into the Federal Republic of Germany (and hence into the EU and NATO) in October 1990 made the GDR a special case. For this reason, the GDR/East Germany is not included in the country analyses. Nevertheless, account will be taken of East Germany's transformation and integration into the FRG in assessing the paths taken by CEECs on the road to transformation, the results achieved to date and the prospects for the future.

When all these historical and current factors are taken into account, it is clear that the initial conditions for transformation in the individual CEECs varied from country to country in the last decade of the 20th century (cf. Table 2.1). The actual conditions in each country were strongly influenced by its relative position within the global socialist STS.

Given these very different and often inconsistent starting conditions in the individual countries at the outset of transformation, it is essential in our view to consider the transformation processes in each country in order to discern not only the basic tendencies but also the factors and actors influencing these processes.

In view of the wholly unexpected collapse of the socialist bloc and the rapid changes that followed, the theoretical and methodical bases for a well-founded analysis of the transformation of the S&T systems in the CEECs had to be developed in the last decade of the 20th century at more or less the same time as these changes were being observed and described.

Table 2.1 The starting conditions for transformation in the individual CEECs*

Countries	Position in Socialism			Population (million)
	Affiliation to	(socialism since)	Category	
I USSR				
– Russia	USSR	(beginning)	C	148
– Ukraine	"	"	P1	51
– Belarus	"	"	P1	10
– Moldova	"	(1940)	P1	4
– Baltic Rep.	"	(1941)	P1	
. Latvia	"	"	"	3
. Estonia	"	"	"	2
. Lithuania	"	"	"	4
II CEC				
– Poland	Warsaw Pact/ CMEA	(after '45)	P2	38
– CSSR	"	"	P2	
. Czech Rep.				10
. Slovakia				5
– Hungary	"	"	P2	11
– Romania	" / isolation	"	P2	23
– Bulgaria	"	"	P2	9
III Yugoslavia (SFRY)	loose ties to USSR, CMEA	(after 1945)	P3	
– Slovenia	"	"	P3	2
– Croatia	"	"	P3	5
– Bosn.-Herz.	"	"	P3	5
– FRY	"	"		
. Serbia			P3/C	10
. Montenegro			P3	1
– Macedonia	"	"	P3	2
IV Albania	loose ties to USSR, autarchy	(after 1945)	P3	3

* This is an attempt to provide a comparative overview; the resilience of the individual indicators and their (in part) roughly estimated attributes require further testing and more precision by historical studies.

Table 2.1 continuing

S&T Institutions (deviations from the Soviet model)			
Pre-socialist S&T tradition*[1]	Organization	Financing (beside budget sources)	International relations (to West)
		military research	centralized
+	AIR*[3]	"	centralized
+	organizational experiments / AIR	" /contract R&D	partial
−	AIR	" / " /Russ. enterprises	weak
−	.	"	"
		contract R&D/ Russ. enterprises	weak/ traditional
++	Univ. (strong)	"	"
++	"	"	"
++	"	"	"
+++	Univ./enterprise R&D	contract R&D	strong
+++	strong intramural	contract R&D	partial
+++	R&D		
++			
+++	Univ./enterprise R&D	project financing contract R&D	strong
+	rigid admin. STS	contract R&D	weak/ traditional
−	Uni.-AoS-relations	contract R&D	partial
	own 'socialist' model, no AoS/ branch R&D	contract R&D	strong
++	"		"
++	"		"
?	"		"
++			"
−			
?	"		"
−	?	?	weak

*1 Key to symbols:
 − weak/none
 + minor
 ++ partial
 +++ marked
*3 AIR = Academy-Industry-Relations

Table 2.1 continuing

Countries	Societal Framework at the Outset of Transformation	
	Situation of country	Orientation*2 (model/ influence)
I USSR	dissolution	–
– Russia	inner conflicts	unclear/ open (large country)
– Ukraine	newly independent/ Russian influence	unclear/ (middle sized) (Russia/ EU)
– Belarus	newly independent	unclear/ open (Russ.)
– Moldova	" /armed inner conflicts	unclear/ open (?)
– Baltic Rep.	newly independent	
. Latvia	newly ind./ Russ. infl.	(EU) Scandinavia
. Estonia	newly ind./ strong Russ. infl.	(EU) Scandinavia
. Lithuania	newly ind./ Russ. infl.	(EU) Scandinavia
II CEC		
– Poland	stable/ evolution '80s	EU (USA)
– CSSR	stable/ division of country	
. Czech Rep.		EU
. Slovakia		? (EU)
– Hungary	stable/ evolution '80s	EU
– Romania	military coup	(EU)
– Bulgaria	politically unstable	? (EU)
III Yugoslavia (SFRY)	dissolution	
– Slovenia	stable	EU(Austria/Germany)
– Croatia	stable/ war vs. Serbia	(EU)
– Bosn.-Herz.	civil war	?
– FRY . Serbia . Montenegro	relatively stable, but war/ intl. isolation	? (own tradition)
– Macedonia	stable / intl. isolation	?
IV Albania	civil war	?

*2 EU: candidate for membership;
(EU): interested in membership

REFERENCES

Braun, A., and T. Schubert (1990) "International collaboration in the sciences 1981–1985." *Scientometrics* 19, no. 1–2: 3–10.

Büschenfeld, H. (1997) "Der Transformationsprozeß in den Nachfolgestaaten Jugoslawiens." *osteuropa* 5: 488–502.

Bzhilianskaya, L. (1996) *The Transformation of Technological Capabilities in Russian Defence Enterprises, with special reference to dual-use technology*. STEEP Discussion Paper No 31. Brighton: Science Policy Research Unit (SPRU), ESRC Centre on Science, Technology, Energy and Environment Policy.

Czerwon, H.-J. (1998) "International scientific cooperation of EIT countries: a bibliometric study." Unpublished paper. Berlin: Wissenschaftszentrum Berlin für Sozialforschung (WZB).

Dagyte, I., J. Kristapsons, and H. Martinson (2000) *Baltic R&D Systems in Transition*. Stockholm: Sodertorns Hoghskola.

EBRD (1998) *Transition report update – April 1998*. London: European Bank for Reconstruction and Development (EBRD).

EBRD (2001) *Transition report update – April 2001*. London: EBRD.

EC (1998–2001) *Regular Reports on Progress towards Accession*. European Commission (EC). Internet: europa.eu.int/comm/enlargement.

Fedorko, A.(1995) *Finanzierung der Wissenschaft in Osteuropa Ende der 80er, Anfang der 90er Jahre – Zusammenfassende Auswertung von Forschungsberichten aus elf Ländern*. WZB-Paper P 95–402. Berlin: WZB.

Gaponenko, N. (1998) "Self-Organization and Politics in Russian S&T Transformation." *Transforming Science and Technology Systems – The Endless Transition? NATO Science Series 4: Science and Technology Policy – Vol. 23*. Ed. W. Meske, et al. Amsterdam: IOS Press, 118–128.

Gläser, J., and W. Meske (1996) *Anwendungsorientierung von Grundlagenforschung? Erfahrungen der Akademie der Wissenschaften der DDR*. Series by the 'Max-Planck-Institut für Gesellschaftsforschung', Cologne, Vol. 25. Frankfurt a.M.: Campus Verlag.

Kocka, J., and R. Mayntz (Eds.) (1998) *Wissenschaft und Wiedervereinigung – Disziplinen im Umbruch*. Berlin: Akademie-Verlag.

Kramarenko, V.G. (1998) "What is the fate of S&T in the Republic of Moldova?" *Transforming Science and Technology Systems – The Endless Transition? NATO Science Series 4: Science and Technology Policy – Vol. 23*. Ed. W. Meske, et al. Amsterdam: IOS Press, 150–152.

Kutlaca, D. (1996) "The transformation of the S&T system in Yugoslavia directed to a new innovation system." Unpublished paper. Berlin: WZB.

Mindeli, L., and L. Pipiia (1998) "Financing Russian R&D: Crisis and Possible Solutions." *Transforming Science and Technology Systems – The Endless Transition? NATO Science Series 4: Science and Technology Policy – Vol. 23*. Ed. W. Meske, et al. Amsterdam: IOS Press, 27–39.

Nadiraschwili, A. (1994) *Die Transformation der Wissenschaft in den Ländern der ehemaligen UdSSR*. WZB-Paper P 94–401. Berlin: WZB.

Nauka Rossii (1992) *Nauka Rossii segodnja i zavtra* (Science in Russia today and tomorrow). Moscow: Russian Academy of Sciences, Analytical center for problems of socioeconomic and scientific-technological development.

Rouvez, A. (1995) *Industrial Restructuring and Defence Conversion in Russia. Targets and Impact of European Technical Assistance*. A BONIFICA-Grouping publication (Commission of the European Communities) Moscow: Delegation of the European Commission in Moscow.

Stern, N. (1997) *The Transition in Eastern Europe and the Former Soviet Union: Some strategic lessons from the experience of 25 countries over six years*. Working Paper No. 18. London: EBRD.

Weber, M., W. Meske, and K. Ducatel (1999) *The Wider Picture: Enlargement and Cohesion in Europe*. The Futures Project, EC JRC, EUR 19035 EN. Seville: Institute for Prospective Technological Studies (IPTS).

3. ANALYZING THE TRANSFORMATION OF S&T IN CEECS – THE THEORETICAL AND METHODOLOGICAL APPROACH

Werner MESKE

There is no clear definition of what constitutes a science and technology (S&T) system. For our present purposes, an S&T system is understood to be the network of public and private institutions that undertake scientific and technological activities and in which the process of scientific and technological innovation is carried out. According to UNESCO and OECD definitions, scientific and technological activities are "... systematic activities which are closely concerned with the generation, advancement, dissemination and application of scientific and technical knowledge in all fields of science and technology. These include such activities as R&D, scientific and technical education and training and the scientific and technological services... Scientific and technological innovation may be considered as the transformation of an idea into a new or improved product introduced on the market, into a new or improved operational process used in industry and commerce, or into a new approach to a social service" (OECD, 1994, pp. 18 – 19).

Thus scientific organizations and research and technological development (R&D) activities are the core of the S&T system (STS) in all countries. In any developed society, S&T is both distinguished from other social sectors and linked to them via functionally and organizationally specialized entities. Thus the STS constitutes a particular social subsystem alongside and connected to other subsystems, which constitute its 'environment'. Linkage occurs both in the performance of its functions, that is through its inter-institutional output, and in the utilization of services provided by other social subsystems, especially through inter-institutional input into the STS. As a complex system, the STS in developed societies is also differentiated internally, having intra-institutional specialization, exchange and cooperation. The most important parts of an STS can be divided into three categories. The most important *functional elements* are teaching, basic and applied research, informational and other transfer and/or scientific services and the preparation of innovations through the development and testing of new products and technologies. The most important *organizational elements* are universities and other tertiary institutions (higher education), R&D institutes in the public sector, R&D institutes and departments in the industrial (private) sector (especially in companies) and the suppliers of scientific information and other service providers, while the most important *financial elements* are institutions or programs/projects, whether with public, private or mixed funding, along with the relevant sources of funding or support institutions.

There is sometimes a lack of complete congruence between activities and organizations, both of which may be attributed to the STS. As a result, both duplication and overlapping occur within the STS (universities with their dual activities of teaching and research, for example), as well as between it and other social subsystems (in the case of industrial R&D in particular, as this belongs to both the STS and the economic system). At the same time, national structures and processes are being increasingly influenced by international structures and processes, above all through the integration of scientists into international scientific communities, but also indirectly, especially through technology transfer (cf. Figure 3.1).

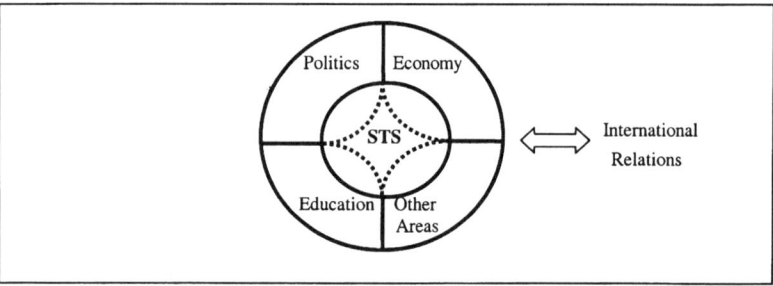

Fig. 3.1 S&T system (STS) in society

In most member countries of the Council for Mutual Economic Assistance (CMEA/COMECON), this sector was not clearly defined and hence there was no single definition of organizations and functions that applied to all the countries (Meske, 1990). However, in line with the UNESCO definition of scientific and technological activities (UNESCO, 1978), it was generally understood – beside higher education (HE) teaching – to encompass basic research, applied research and technological development all the way through to the introduction of new products and technologies. In the GDR this included, for example, the development of prototypes of new products, the construction and testing of experimental large-scale systems, expenditure for patents and licenses as well as work in the area of standardization (Definitions, 1975). On the other hand, in the GDR the entire sector of social sciences and humanities was excluded from R&D statistics. This is merely one example of how the structures and functions of STS in the socialist countries, both statistically and in reality, differed markedly not only from those in other countries, in particular OECD nations, but also from one another, as a comparison of the AoS of the GDR and the CSSR shows (Meske and Müller, 1978).

This lack of an adequate statistical definition of the STS has its counterpart in the absence of a comprehensive and applicable theory of scientific and technological development; there are currently only a number of different conceptual approaches. On the one hand, these different approaches reflect the very diverse development of science and technology in different types of political systems (Kröber and Krüger, 1987; Kröber, 1988; Felt et al., 1995) and in countries with different levels of industrialization (cf. Nelson, 1993). On the other hand, however, their diversity also reflects the many different aspects of science and technology that can be considered and analyzed. This is particularly evident in the fact that the transformation of S&T in the former socialist countries has not hitherto been studied in its full complexity but rather for the most part in terms of specific aspects.

Account has to be taken of the fact that the systemic transformation in formerly socialist countries has led to changes not only in the environment in which S&T operates but also in S&T itself. It cannot simply be assumed that this is merely the S&T system's reaction to changes in its environment, that is, in the words of Schimank and Stucke (1994), 'coping with trouble'. Political changes were undoubtedly a *trigger* for activities within the STS; however, the concepts of the self-organization of science (Krohn and Küppers, 1989) and of actor-centered institutionalism (Mayntz and Scharpf, 1995) suggest that active participation in STS reorganization and activities influencing its environment can also emanate from the STS itself.

It is not surprising, therefore, that although countless analyses of transformation processes have already been undertaken (see Schwarz, 1995), no single theory has been formulated that might serve as a basis for the analytical study of the transformation of science and technology in the formerly socialist countries. This applies, firstly, to the process of 'transformation' itself, since no consistent and viable theory of the course of (economic) transformation processes is either currently available or, in all likelihood, to be expected (Götz, 1998, p. 8). Furthermore, experiences in East Germany, which are surely the most pronounced, suggest that the 'institutional transfer' (Lehmbruch, 1992) from West to East that took place by no means represents the full extent of the transformation or the entire range of interactions that occurred in reality. An understanding of the complexity and long-term nature of these processes is only gradually beginning to emerge (Rudolph, 1995; see also the various papers in the special issue 'Transformationsanalysen' in *BISS public*, 1999, No. 27 and 28). At present, the various elements or 'building blocks' of a theory that can be used to understand and explain transformation processes in transition situations are being put in place in various areas (on business management organizations, see Alt et al., 1996; on research organizations, see in particular Wolf, 1996, p. 74–97).

Studies of the transformation of S&T tended to focus initially on the transformation of individual sectors and organizations, on quantitative changes in the resources utilized, on individual fates or political and other decisions (cf. Wolf, 1996, p. 23; Kocka, 1998, p. 10). More recently, cognitive changes have also begun to be considered (Kocka and Mayntz, 1998; Gläser, 1998). This approach can be explained by the fact that the initial concern was to understand and explain a rapid and unexpected historical process and its impact, in other words to answer the questions 'what exactly happened?' and 'why?' (Mayntz, 1994, p. 24). The result was an increasing number of precise analyses of (partial) processes in the transformation of science and technology, first in East Germany and then also in other Central and East European Countries (CEECs) (see in particular Meske, 1993; Mayntz et al., 1995; *Social Studies of Science*, 1995; Webster, 1996; Mindeli and Nadiraschwili, 1997; Dyker, 1997; Meske et al., 1998; Meske, 2000). These analyses showed that the transformation of science systems embedded in economic and political transformation processes is undoubtedly much more problematic than the restructuring of research systems in economically and politically stable societies, no matter how comprehensive such a restructuring may be (Schimank, 1995, p. 10).

However, the growing number of specific studies undertaken has, however, not only increased and consolidated the body of detailed knowledge but has also contributed to a perception of the general, that is the cross-national, problems, factors and phases in the transformation of STS. Our own experiences and analyses, together with an evaluation of the relevant literature, reveal that the following aspects are central to an understanding of the transformation of the STS in CEECs:

– The state of the STS at *the beginning* of the transformation

 This is an important point of reference for comparative assessments of changes in STS. At the same time it also determines the internal preconditions for the ensuing changes and is thus of considerable significance for the process of transformation itself. This concerns not only descriptions of the conditions at the outset, but above all their *evaluation*. Research on scientific capabilities has shown that a given capacity always has a dual character, since it is both the *result* of the preceding processes and the *precondition* and prerequisite for new, subsequent research processes (cf. Meske, 1988); this finding is currently being confirmed through the debate on whether the

STS inherited from socialism, or rather its constituent elements, are to be regarded as 'assets' or 'liabilities' (Meske, 1999 and 2002). The fundamental differences between East and West can lead to misunderstandings and also to very different assessments of the existing STS. It is known that in East Germany this contributed to errors in the modification of the STS, in particular as regards the alleged impossibility of transferring R&D capacities from the Academy of Sciences (AoS) to the universities, since the West wrongly assumed that the latter conducted no R&D in the GDR. These errors were later acknowledged also by those involved but could no longer be corrected (cf. Simon, 1995; Laitko, 1997). Even today, the focus of interest often lies in "a common structural heritage in the research system of Central and Eastern Europe rooted in the shared past" (Balazs et al., 1995, p. 615) and too little attention is paid to the differences between the countries.

– The STS '*environment*' and changes therein

The significance of the social 'environment' for the development of science systems has been shown by Krohn and Küppers (1989). System transformation makes the non-scientific environment especially important in the CEECs.

+ The system transformation began at more or less the same time throughout the entire socialist bloc. In many cases the collapse of this bloc, which often went as far as the dissolution of former states and the formation of new independent states, involved the transformation of the state framework for STS and thus also of its systemic structure. In addition, all S&T organizations and activities in the formerly socialist nations now had to function in the context of new *international* conditions in which old links were loosened or broken altogether and the opening up to the West created new possibilities (and needs) for international relations.

+ The social systems of the socialist states differed fundamentally from those of the OECD nations; this also affected the distinctions and links between the STS and other societal subsystems. Politics in particular had a strong influence on the structures and dynamics of the STS, both directly and indirectly – via the economy. (At the same time, however, this influence was not as strong as was claimed, both in the socialist bloc and often also in the West, as mentioned by Kocka, 1998a). The (intended) transition to a market economy and to democracy brings with it fundamental changes in the politics and economies of CEECs and thus in the relations between those subsystems and the S&T system.

+ The starting point of the system transformation lay in the political system but it then unfolded more or less simultaneously in the various subsystems. This gave rise less to simple cause and effect relationships than to a diversity of interrelationships. Since these changes are still going on in

many areas, in particular in the economy, there is great insecurity, especially in the restructuring of the STS, due to its considerable (not only economic) dependency upon politics and the economy.
– The *structure* and *mode of operation* of the STS and changes therein
The entire system of S&T in the individual CEECs underwent significant changes under the influence of the endogenous preconditions and internal activities arising from the situation at the outset as well as from the effects of the (national and international) environment. These changes occurred both top-down – especially as a result of relevant political and economic decisions – and bottom-up – through the activities of S&T actors, which were linked in particular to the democratization process and the achievement of greater individual autonomy. As a result, the previous system relations dissolved and the STS began to fragment; moreover, greater differences began to emerge as its individual organizations and activities were restructured. This led to substantial differentiation in the restructuring of the STS both within the individual countries and ultimately between them. The industrial R&D sector constituted a particular problem in all countries as it probably underwent the greatest changes but has, for a long time, received scant attention from politicians and researchers.

The analyses presented in Chapters 1 and 2 took as their starting point the assumption that the S&T systems in the socialist countries were consonant with the essential features of the Soviet model, which provided the baseline from which deviations in the individual countries and other aspects of the differences and differentiations could be determined. In Part II, on the other hand, the individual countries with their quite different starting conditions and STS transformation processes to date form the starting point.

For this next step, the largely country-based analysis of the course of transformation so far undergone by the STS and its sectors in most of the CEECs, a preliminary basic model was developed (see Figure 3.2), which served as a framework for the subsequent research work. Taking as a starting point the central aspects of STS transformation alluded to above, a differentiated analytical procedure was selected and carried out by national experts from relevant countries. Thus the account of the S&T transformation processes in the individual CEECs begins with their conditions and particularities at the outset. This is followed by a characterization of the transformation processes in society and S&T during the 1990s, focusing in particular on relevant political and economic changes as well as changes in the sectors engaged in S&T. In conclusion, the current situation as well as future tasks and problems are briefly assessed. We consider these aspects to be fundamental points of reference for

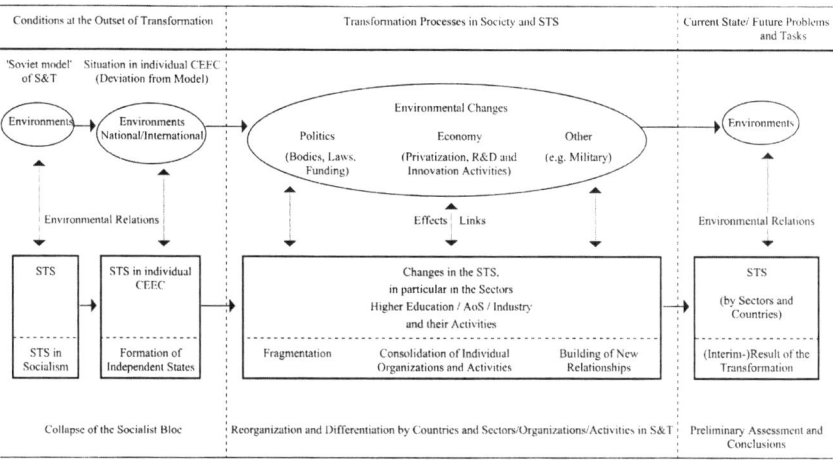

Fig. 3.2 Model of the institutional transformation of the S&T system in CEECs

ascertaining and assessing the measures undertaken to restructure S&T as well as their results in the 1990s.

In the first stage, the essential features of the STS and its 'environment' in each country prior to the system transformation were gathered together. The characterization of the environment and of the relations between its various spheres included in this model are based on the concept developed by Krohn and Küppers (1989). However, for pragmatic reasons and on the assumption that not all the spheres of the environments are equally relevant, we have focused on the areas of politics and the economy while excluding others, such as the public sphere, or including them only indirectly (via the political sphere, for example). Since the basic type of STS in the CEECs is outlined in Chapter 2, the emphasis in this stage was on the deviations from the Soviet model in the individual CEECs. In this way, the specific conditions in each country at the outset of the transformation could be more clearly characterized. Both aspects are essential for explaining both the general and the specific effects of the collapse of the socialist system on the individual CEECs. An understanding of these effects is, in turn, a necessary precondition for explaining the similarities and differences between the conditions in the various countries at the beginning of the transformation process.

The next stage entailed studying the fundamental changes that took place in the S&T systems and their environments in the CEECs during the 1990s. In accordance with the analytical model described above, this involved investigating the changes in the S&T environments and in the S&T systems themselves, as well as in the nature of the relations between the two. Particular attention was paid to the relations between S&T and the political and economic spheres.

Attention was focused here on the *institutions engaged in S&T activities*, in particular the processes of scientific and technical innovation. The core of these institutions is made up of the former sectors and organizations of higher education, the AoS and industrial R&D or their successors. Activity in the industrial R&D sector is closely associated with the changes in enterprises, so that both can be understood only in terms of their interaction. The connections and interactions between different S&T organizations, as well as between them and their 'non-scientific' environment, are also of importance here. Following the extensive fragmentation of the socialist STS, the formation of new relationships assumes a major significance. Politics plays a decisive role here, since it had directly or indirectly controlled the activities of all S&T facilities in the socialist regimes via the centralized management hierarchies.

The rebuilding of the *political system* has also led to changes in the environment in which all S&T institutions operate. What this environment is and whether or how it is specified and laid down in legal regulations depends primarily on political decisions. Such decisions can directly affect the future of organizations, such as, for example, the AoS. Further, stipulations on the general relationship between politics and S&T, S&T legislation, the establishment of political bodies to observe or implement S&T policy and their respective responsibilities influence the very basis of the livelihood and behavior of all institutions relevant for S&T. Because the political decisions taken during a process of systemic transformation not only establish new basic conditions but also often trigger a reorganization of S&T, these decisions are mostly considered first in the analysis. One important factor here is that, in many countries, it was the activities of scientists themselves and of their representatives in professional associations in reaction to the fragmentation of the socialist STS that led to the initial changes in science systems and compelled politicians to act. The economic crisis had a similar effect on politics as well as on science.

These analyses should result in a preliminary assessment of the current state of S&T in each country and, where possible, in conclusions or a summary of future tasks and problems.

The individual country analyses were carried out initially between 1996 and 1998 as part of a project entitled 'Institutional Transformation of S&T Systems and S&T Policy in Economies in Transition' (coordinator: W. Meske; cf. Meske, 1999), which was part of a larger project entitled 'Restructuring and Reintegration of Science and Technology Systems in Economies in Transition' funded by DG XII of the European Commission under the Targeted Socio-Economic Research (TSER) Program (coordinator: S. Radosevic; cf. SPRU, 1999). These country analyses, which included the period up to 1996/97, confirmed the existence of certain basic patterns in the transformation process and also revealed considerable differences in the changes that had taken place in the individual countries (cf. Meske et al., 1998; Meske, 2000). It had to be assumed, however, that the transformation processes had in no country led to the emergence of a qualitatively and quantitatively new R&D and innovation system, characterized by efficacy, effectiveness, efficiency. This remained therefore the main task and the dominant aspect in all countries in the second half of the 1990s.

In order to incorporate the changes that had taken place in the second half of the 1990s and to enrich the primarily retrospective analysis of already completed processes with a more forward-looking assessment of the prospects for meeting the challenges of the 21st century, the authors of the present book carried out a second round of analyses in 2000 and 2001. The resultant analyses of the changes in the STS of 15 CEECs are presented in Part II.

REFERENCES

Alt, R., R. Lang, and E. Weik (1996) "Auf dem Weg zur Theorie? Sammelrezension von Literatur zum betrieblichen Wandel in Ostdeutschland." *Die Betriebswirtschaft* no. 1: 85–109.
Balazs, K., W. Faulkner, and U. Schimank (1995) "Transformation of the Research Systems of Post-Communist Central and Eastern Europe: An Introduction." *Social Studies of Science* 25, no. 4: 613–632.
BISS public (1999) no. 27 and 28. (Special issue 'Transformationsanalysen').
Definitions (1975) *Definitionen wichtiger Kennziffern und Begriffe für Planung und Statistik. 1. Ergänzungsausgabe zur Originalausgabe von 1967.* Ed. Ministerrat der Deutschen Demokratischen Republik, Staatliche Zentralverwaltung für Statistik. Berlin: Staatsverlag.
Dyker, D.A. (Ed.) (1997) *The Technology of Transition. Science and Technology Policies for Transition Countries.* Budapest: Central European University Press..
Felt, U., H. Nowotny, and K. Taschwer (1995) *Wissenschaftsforschung.* Frankfurt a.M.: Campus Verlag.
Gläser, J. (1998) *Kognitive Neuorientierung der ostdeutschen außeruniversitären Grundlagenforschung als Folge des Institutionentransfers.* WZB-Paper P 98–402. Berlin: Wissenschaftszentrum Berlin für Sozialforschung (WZB).
Götz, R. (1998) "Theorien der ökonomischen Transformation." *osteuropa* no. 4: 339–354.
Kocka, J. (1998) "Einleitung." *Wissenschaft und Wiedervereinigung – Disziplinen im Umbruch.* Ed. J. Kocka and R. Mayntz. Berlin: Akademie-Verlag, 7–19.
Kocka, J. (1998a) "Wissenschaft und Politik in der DDR." *Wissenschaft und Wiedervereinigung – Disziplinen im Umbruch.* Ed. J. Kocka and R. Mayntz. Berlin: Akademie-Verlag, 435–459.
Kocka, J., and R. Mayntz (Eds.) (1998) *Wissenschaft und Wiedervereinigung – Disziplinen im Umbruch.* Berlin: Akademie-Verlag.
Kröber, G. (Ed.) (1988) *Wissenschaft – Das Problem ihrer Entwicklung. Vol. 2: Kom-*

plementäre Studien zur marxistisch-leninistischen Wissenschaftstheorie. Berlin: Akademie-Verlag.

Kröber, G., and H.P. Krüger (Eds.) (1987) *Wissenschaft – Das Problem ihrer Entwicklung. Vol. 1: Kritische Studien zu bürgerlichen Konzeptionen der Wissenschaftsentwicklung.* Berlin: Akademie-Verlag.

Krohn, W., and G. Küppers (1989) *Die Selbstorganisation der Wissenschaft.* Frankfurt a.M.: Suhrkamp Verlag.

Laitko, H. (1997) "Abwicklungsreminiszenzen. Nachdenken über das Ende einer Akademie." *hochschule ost* 1: 55–81.

Lehmbruch, G. (1992) "Institutionentransfer im Prozeß der Vereinigung: Zur politischen Logik der Verwaltungsintegration in Deutschland." *Verwaltungsintegration und Verwaltungspolitik im Prozeß der deutschen Einigung.* Ed. W. Seibel, et al. Baden-Baden: Nomos Verlag, 41–46.

Mayntz, R. (1994) *Deutsche Forschung im Einigungsprozeß. Die Transformation der Akademie der Wissenschaften der DDR 1989 bis 1992.* Frankfurt a.M.: Campus Verlag.

Mayntz, R., and F.W. Scharpf (1995) "Der Ansatz des akteurzentrierten Institutionalismus." *Gesellschaftliche Selbstregelung und politische Steuerung.* Ed. R. Mayntz and F.W. Scharpf. Frankfurt a.M.: Campus Verlag, 39–72.

Mayntz, R., U. Schimank, and P. Weingart (Eds.) (1995) *Transformation mittel- und osteuropäischer Wissenschaftssysteme – Länderberichte.* Opladen: Leske+Budrich.

Meske, W. (1988) "Wissenschaftsentwicklung als Prozeß der Reproduktion des Wissenschaftspotentials." *Wissenschaft – Das Problem ihrer Entwicklung. Vol. 2: Komplementäre Studien zur marxistisch-leninistischen Wissenschaftstheorie.* Ed. G. Kröber. Berlin: Akademie-Verlag, 245–265.

Meske, W. (Ed.) (1990) *Wissenschaft der RGW-Länder. Länderberichte zur Situation am Ende der 80er Jahre aus der DDR, Polen, der Tschechoslowakei, Ungarn, Bulgarien, der Sowjetunion, der Mongolischen VR, Vietnam und Kuba. Studien und Forschungsberichte No. 30.* Berlin: Akademie der Wissenschaften der DDR, Institut für Theorie, Geschichte und Organisation der Wissenschaft.

Meske, W. (1993) "The Restructuring of the East German Research System – a Provisional Appraisal." *Science and Public Policy* 20, no. 5: 298–312.

Meske, W. (1999) "Transformation of R&D in the Post-Socialist Countries: Asset or Liability?" *Innovation and Structural Change in Post-Socialist Countries: A Quantitative Approach.* Ed. D.A. Dyker and S. Radosevic. Dordrecht: Kluwer Academic Publishers, 137–152.

Meske, W. (2000) "Three-phase model – Changes in the innovation system in economies in transition: basic patterns, sectoral and national particularities." *Science and Public Policy* 27, no. 4: 253–264.

Meske, W. (2002) *Science in Formerly Socialist Countries – Asset or Liability within the New Societal Conditions?.* WZB-Paper P 02–401. Berlin: WZB.

Meske, W., and K. Müller (1978) *Vergleichende Untersuchung zu Entwicklung und Struktur der Potentiale der Akademie der Wissenschaften der DDR (AdW der DDR) und der Tschechoslowakischen Akademie der Wissenschaften (CSAV).* Studien und Forschungsberichte No. 8. Berlin: Akademie der Wissenschaften der DDR, Institut für Theorie, Geschichte und Organisation der Wissenschaft.

Meske, W., J. Mosoni-Fried, H. Etzkowitz, and G.A. Nesvetailov (Eds.) (1998) *Transforming Science and Technology Systems – The Endless Transition? NATO Science Series 4: Science and Technology Policy – Vol. 23*. Amsterdam: IOS Press.

Mindeli, L., and A. Nadiraschwili (Eds.) (1997) *Akademiceskie Instituty v Uslovijach Transformizii* (Institutes of the Academy under the Conditions of Transformation). Moscow: Zentr Issledovanij i Statistiki Nauki (Centre for Science Research and Statistics).

Nelson, R. (Ed.) (1993) *National Innovation Systems. A Comparative Analysis*. New York: Oxford University Press.

OECD (1994) *Proposed standard practice for surveys of research and experimental development – Frascati Manual 1993*. Paris: OECD.

Rudolph, H. (Ed.) (1995) *Geplanter Wandel, ungeplante Wirkungen. WZB-Jahrbuch*. Berlin: Edition Sigma.

Schimank, U. (1995) "Die Transformation der Forschungssysteme der mittel- und osteuropäischen Länder: Gemeinsamkeiten von Problemlagen und Problembearbeitung." *Transformation mittel- und osteuropäischer Wissenschaftssysteme – Länderberichte*. Ed. R. Mayntz, et al. Opladen: Leske+Budrich, 10–39.

Schimank, U., and A. Stucke (Eds.) (1994) *Coping with Trouble: How Science Reacts to Political Disturbances of Research Conditions*. Frankfurt a.M.: Campus Verlag.

Schwarz, R. (1995) *Chaos oder Ordnung? Einsichten in die ökonomische Literatur zur Transformationsforschung*. Marburg: Metropolis-Verlag.

Simon, D. (1995) "Westliche Theorie – Östliche Realität. Drei Szenen aus der deutsch/deutschen Wissenschaft." *Transit: Europäische Revue* 9: 159–168.

Social Studies of Science (1995) 25, no. 4. (EASST [The European Association for the Study of Science and Technology] Special Issue).

SPRU (1999) *Science, Technology and Growth: Issues for Central and Eastern Europe*. Brighton: Science and Technology Policy Research Unit (SPRU).

UNESCO (1978) *Recommendation concerning the International Standardization of Statistics on Science and Technology*. Paris: United Nations Educational, Scientific, and Cultural Organization (UNESCO).

Webster, A. (Ed.) (1996) *Building New Bases for Innovation. The Transformation of the R&D System in Post-Socialist States*. Cambridge: Anglia Polytechnic University.

Wolf, H. G. (1996) *Organisationsschicksale im deutschen Vereinigungsprozeß. Die Entwicklungswege der Institute der Akademie der Wissenschaften der DDR*. Frankfurt a.M.: Campus Verlag.

Part II

The transformation of S&T systems in individual CEECs in the 1990s

4. RUSSIA: TOWARDS A NATIONAL INNOVATION SYSTEM – INSTITUTIONAL CHANGES AND FUNDING MECHANISMS

Nadezhda GAPONENKO [1]

This chapter focuses on the institutional issues and changes in science and technology (S&T) funding mechanisms, which are of special importance to the national innovation system (NIS) in Russia (Breschi and Malerba, 1995; Edquist, 1997; Freeman, 1987; Freeman and Perez, 1988; Gaponenko et al., 1997; Lundval, 1992; Nelson, 1993; Niosi et al., 1993).

1. PERESTROIKA AND AFTER: GRADUAL TRANSITION GIVES WAY TO CHAOS

The initial period of S&T system (STS) transformation in Russia can be divided into two main phases: a gradual transition to a new order, followed by a chaotic period of change leading to the emergence of a new STS.

The transformation of the Russian STS began with the introduction of perestroika (rebuilding, reconstruction) in 1986, although some initiatives were already under way in the early 1980s. A gradual transition to a new order occurred under conditions of socio-political and socio-economic stability. The main elements of perestroika were as follows: conversion (from military to civil), the relaxation of state, party and departmental control over enterprises' activities, the growth of democratic freedoms and glasnost (openness, accountability), the partial elimination of economic and scientific autarchy and the establishment and development of an alternative, non-state sector of the economy (Gaponenko et al., 1993).

Several key institutional changes took place during this period. First, the status of various sectors of science began to change. The Academy of Sciences (AoS) became an independent organization that had a chance to transform the

---

[1] This paper is an interim product of a three-year project entitled 'National innovation system: the problems of transformation and the mechanisms of regulation', supported by the Ministry of Industry, Science and Technology of the Russian Federation, carried out by the Russian Institute of Economics, Policy and Law in S&T (RIEPL), Moscow, and coordinated by the author.

'Academy model' of science. Unfortunately, it took little advantage of this opportunity.

The branch institutes also gained opportunities to apply the results of their research in industry more actively than before. They now had the right to determine the scope of their research topics and to enter into contracts with other ministries and departments and with enterprises of various kinds. Many research institutes took an interest in utilizing their R&D in industry and began actively to seek contracts with enterprises. Those institutes that developed closer relationships with industry helped their scientists to survive during the next stage of transformation, in which the planning system was done away with completely (Dyker, 1997).

A second key institutional change associated with the processes of transformation started at the micro level. The status of researcher was beginning to rise relative of that of science administrator. At the same time, departments and research groups began to be differentiated on the basis of research quality and the demand for their output in the technology market. The processes already outlined above led to a trend towards hiving off small institutional structures from the institutes. The transformation of large institutes into associations of small research centers formed the next stage of transition. The formation of a non-state sector in the STS also began, but its development was inhibited by restrictions on small, innovative businesses and joint ventures.

The equilibrium of the previous STS was destroyed as a result of these changes, but the gradual transition to a new order had only just begun. The early reform policy known as perestroika had unexpected negative consequences for the economic and social situation. Attempts to undo the ill effects of inadequate reforms only accelerated the crisis, since the hidden defects of the Soviet STS remained concealed. The changes made to the economic system and the STS during the Gorbachev era caused social instability and made movement toward a new order chaotic. The greatest impetus came from the reckless destruction of the planned economy at that time, as institutions were dismantled without anything being put in their place. The demise of centralized management totally disrupted the interaction between science and industry, which had traditionally been channeled through the ministries. Moreover, the accumulation of financial resources in ministries and departments was also eliminated. As the old system of coordination disappeared, no provision was made for the creation of market mechanisms to organize the tasks of productive accumulation, financing of industrial research, provision of credit and insurance, etc. Since market institutions were only just beginning to emerge, the loss of the former system left scientists without suppliers of the necessary scientific equipment and materials. The perestroika period impaired innovative activities by redirecting resources to immediate consumption needs. Legal chaos aggravated the situation even further. The disintegration of the USSR led to the breakdown of economic, technological and informational ties.

Since relationships between central government and the regions were not transformed at the same time as central planning was done away with, the problems resulting from the inability to manage the economy that initially affected the center soon spread outward. Even though command-style organizations were eliminated, the mental habits associated with them persisted. The cumulative effect of these factors resulted in a chaotic transitional process. The main features of this chaotic transition were instability, high uncertainty and a diminishing capacity to make decisions as the issues became more complex and risky. This combination of factors diminished the ability of the Russian government to formulate and implement an active policy. Given that its capacities were less than those of other countries, self-organization became the main driving force behind the transition to a new order.

At the beginning of 1994, the impossibility of successfully reforming the economy without government taking an active role in formulating national policy was acknowledged. Although there was still considerable institutional instability at the political level, there was a shift towards sharing functions between various political institutions and putting in place the mechanisms required for coordinated action. At the same time, the transformation of Russia's innovation system was taking place against the background of a protracted economic crisis.

In 1995, GDP had fallen by 37.8 per cent compared with 1990 and industrial production to less than half; total economic investment amounted to just 30 per cent of that made in 1990. In 1995, the real money income of the population fell by 42 per cent compared with 1991; the state budget deficit in 1990 was equal to 1.3 per cent and in 1995 to 3.2 per cent of GDP. Total unemployment reached 6.7 million in 1995. The number of those whose income was below the official poverty line was 36.6 million, 10 per cent more than in 1994 (Gaponenko, 1996).

The major economic changes that substantially affected the STS were the high rate of inflation, the sharp decline in GDP, industrial output and investment, the rapid fall in domestic market demand and the transformation of its structure and the loss of market positions in the markets of Western and post-communist countries. Structural shifts in the economy in favor of the fuel and energy sector and strong business rivalries among domestic companies in the national market also played a role.

All of this underlines the necessity for a new institutional framework for Russia's STS following the breakdown of the USSR, one more in keeping with the transition from a planned to a market economy. Since the mid-1990s, a new stage of NIS transformation has been under way; it is marked by a gradually increasing role for government in the regulation of the NIS, supporting institutional changes and the transformation of financial mechanisms. By the mid-1990s, the minimum legislative base had been adopted, although monitoring of its implementation remained a key problem even at the beginning of 2001. Since the mid-1990s, competition in the home market has started to grow as a result of foreign technology flows into the Russian market. As a response to these new conditions, a set of presidential decrees and governmental resolutions was adopted in order to improve the competitiveness of national products and processes on the domestic and global markets; however, most of these measures were implemented only with difficulty or remained 'paper promises'. The government gradually went on to develop medium-term strategies to replace the short-term strategies that dominated at the beginning of the reforms. All these attempts took place under conditions of political and financial instability, which hampered progress.

2. NEW INSTITUTIONAL PROFILES SINCE THE MID-1990s

Outlining the NIS institutional environment will help to clarify the role of the various governmental agencies, the weight of public and private sectors in funding and carrying out R&D and the role of R&D-funding institutions and of those that perform related functions (such as technology transfer or human resources mobility, etc.). The lack of institutions for technology transfer and of networking among different NIS institutions hampered knowledge and technology transfer, and consequently, thwarted any attempts to improve the competitiveness of national firms and to aid their recovery from economic crisis. It became evident that an active policy of institutional change was required.

An overview of the set-up of those institutions in place in the Russian Federation (RF) at the beginning of 2002 is provided in Figure 4.1. The institutional map consists of seven layers, each with different functions.

The top layer comprises the general policy-making bodies: the Russian government, the Council of Federation and the State Duma, which play a key role in setting broad policy directions. The second layer represents institutions that formulate and implement science, innovation and technology policy. The third layer comprises the public and semi-public foundations and private investors, which, together with federal and regional authorities, finance and support the production and implementation of innovation; it also includes standard-setting agencies and institutions, which facilitate and support innovation. The fourth layer comprises the organizations that carry out R&D. The fifth layer includes those institutions facilitating technology diffusion, while the sixth comprises the service-oriented organizations in the NIS. The final

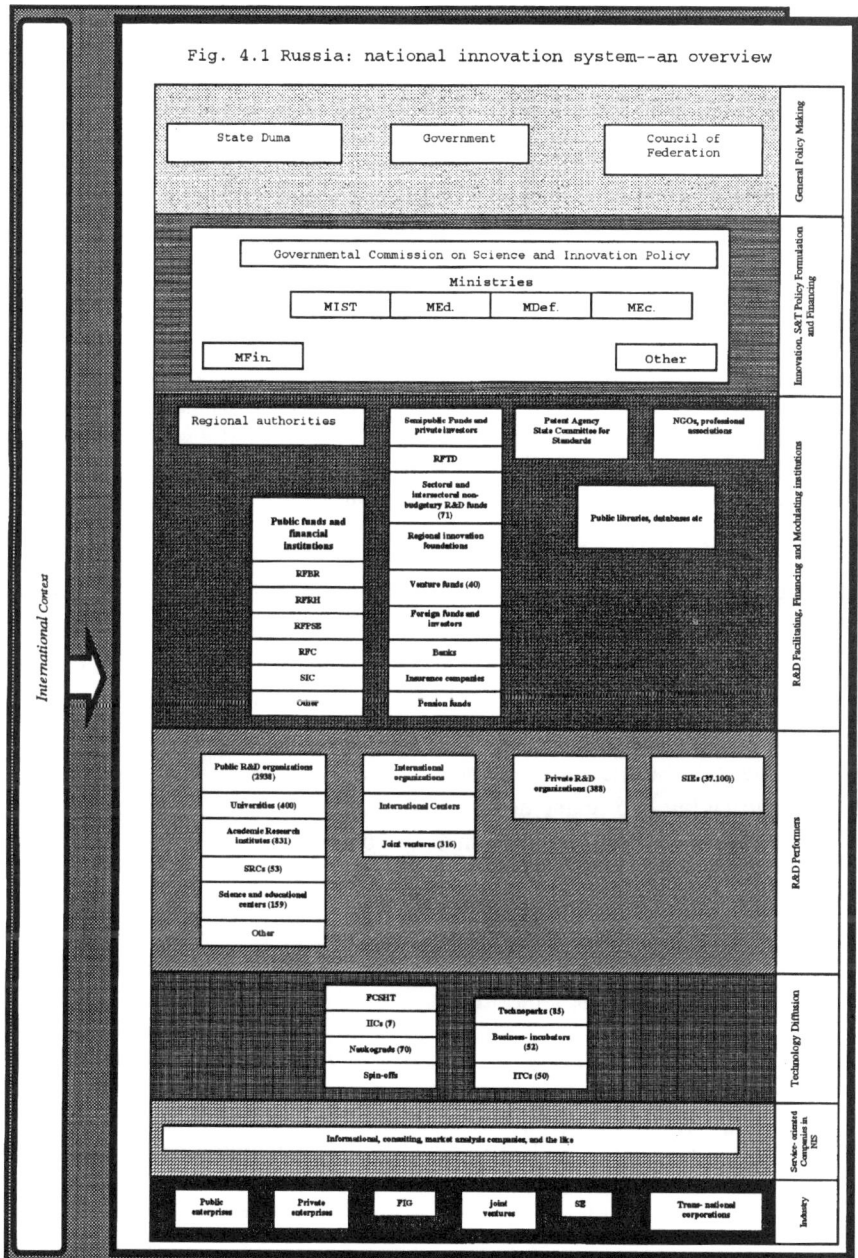
Fig. 4.1 Russia: national innovation system—an overview

Fig. 4.1
Abbreviations:

MIST	Ministry of Industry, Science and Technology
MEd.	Ministry of Education
MDef.	Ministry of Defense
MEc.	Ministry of Economy
MFin.	Ministry of Finance
RFBR	Russian Foundation for Basic Research
RFRH	Russian Foundation for Research in Humanities
RFPSE	Russian Foundation for Promotion of Small Innovation Enterprises in S&T
RFC	Russian Financial Corporation
SIC	State Investment Corporation
RFTD	Russian Foundation for Technological Development
SRCs	State research centers
FCSHT	Federal centers of science and high technology
IICs	Innovation-industrial complexes

level contains public and private industrial enterprises, corporations and joint ventures.

2.1. TECHNOLOGY AND INNOVATION POLICY FORMULATION AND IMPLEMENTATION

Government ministries (the second layer in Figure 4.1) are the main institutions responsible for policy-making. Many ministries are involved in science, innovation and technology policy formulation and implementation. The Ministry of Industry, Science and Technology (MIST) is the key body responsible for policy-making and financing in Russia. There are four key problem areas in the performance of these policy-making bodies:

i) coordination of the various ministries' actions;

ii) the coordination of functions, responsibilities and measures between federal and regional authorities;

iii) the transition to a participatory approach in policy development and implementation;

iv) policy transparency and accountability in NIS institutions.

The first steps towards encouraging the various ministries to enter into dialogue with each other and to coordinate their activities have already been taken as part of the development of national programs, although some ministries, such as the Ministry of National Resources and the Ministry of Agriculture and Food Production, are still excluded from this process. Steps are also being taken to involve non-governmental organizations (NGOs) and the private sector in policy development and the monitoring of newly introduced measures.

However, the status of these actors is still not clearly defined and their voices are not heard very often.

The Governmental Commission on Science and Innovation Policy (GC-SIP) was established in 1998 for the purpose of coordinating policy formulation across ministries as well as with regional authorities. It has also been proposed that the private sector, academics and NGOs should be involved in the Commission's activities. This is a promising step forward on the way towards involving the various stakeholders in policy coordination and the development of a genuinely participatory approach. However, various problems could emerge as a new policy-making culture emerges. The establishment of the institute responsible for policy coordination and clarification of the status of the various stakeholders are only the first steps.

2.2. THE INSTITUTIONS CARRYING OUT R&D

The institutions actually engaged in R&D constitute the fourth layer of Figure 4.1. At the beginning of 2002 there were 4 134 such organizations in the Russian NIS. The period of transition was marked by a decrease in the number of R&D organizations and the number of R&D personnel (although this trend had changed by 1998/1999) (see Figure 4.2).

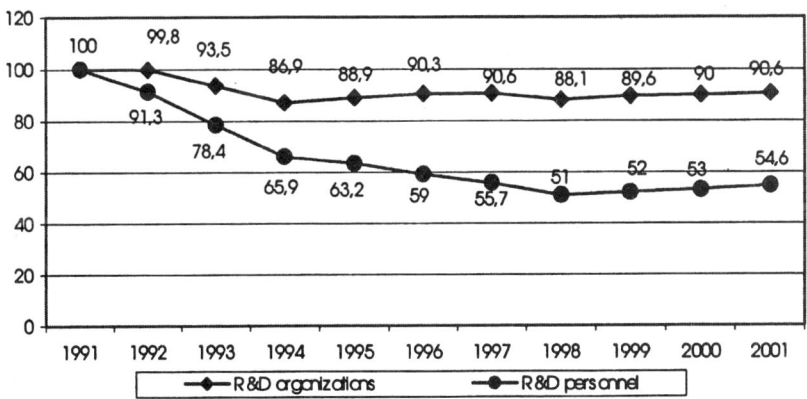

Fig. 4.2 Russia: R&D organizations and R&D personnel (in %, 1991=100)
Source: Gokhberg and Mindeli, 1996, p. 8, 22; 2001, p. 10, 28; SCSRF, 2000; 2001.

The R&D organizations have become more flexible. The number of personnel per research institute averaged 673 in 1990; by 2001, this figure had decreased to 216. Another remarkable feature of the 1990s was the increasing significance of private institutions. The public R&D organizations still dominate, with a share of 71.7 per cent of the total number of R&D organizations in 2000. However, the number of private institutions increased by a factor of 3.3

between 1993 and 2000. The number of personnel per private-sector research organization averaged 140 in 2000. The performance of research organizations has changed too. Relationships within research groups and among departments and collaboration among institutions and their relationships with the authorities have undergone substantial changes as well (cf. Gaponenko, 1995).

Academy of Sciences (AoS)

The research institutes of the Russian AoS enjoyed a privileged position during the Soviet period. Their number increased by a factor of 1.5 between 1990 and 2000. Although the number of R&D personnel decreased by 24 per cent, these institutes retained the capabilities to generate knowledge. Their share of basic research increased from 60 per cent in 1991 to more than 70 per cent in 2000, while the share of professors in the total number of R&D personnel increased from 8.2 per cent in 1990 to about 15 per cent in 2000.

The model of academy science changed substantially during the period of transition (Balazs, 1997; Gaponenko et al., 1995; Mayntz, 1994; Mayntz et al., 1998). The Russian AoS was granted the status of self-governing organization. Many academy scientists became integrated into the world scientific community. Some research institutes set up their own international centers. During the period of transition, the links between academy science and the universities were rather poor. The 'State Program for Supporting the Integration of Academy Science and Higher Education' was launched in the 1990s with the aim of strengthening the links between these sectors. As a result, 157 research and educational centers (R&ECs) were set up in 39 of the 89 regions of the RF. Some institutes established educational centers for the provision of advanced training in some areas.

In spite of the substantial changes that took place in academy science during the 1990s, many issues have still not been resolved and hamper the integration of academy research institutes into the NIS. The key problems faced by the Russian AoS research institutes are an aging personnel and poor links with industry. It should be noted that AoS research institutes, and indeed the universities as well, enjoy a considerable degree of autonomy; researchers need not ask for permission when they wish take up temporary outside appointments or conduct cooperative research with industry, that is there are no legislative barriers to there doing so. What is more, the sharp decrease in R&D expenditure has encouraged researchers to strengthen links with industry. However, the interaction of various factors has hampered these processes. A survey of 500 industrial companies in various branches and regions, carried out in 1996–1997, shed some light on the relationships between academy research institutes and some other NIS institutions and industry (Survey, 1997). The industrial enterprises had closer links with research organizations related to their branch of industry and seldom had dealings with AoS and state research centers (SRCs) (see Figure 4.3).

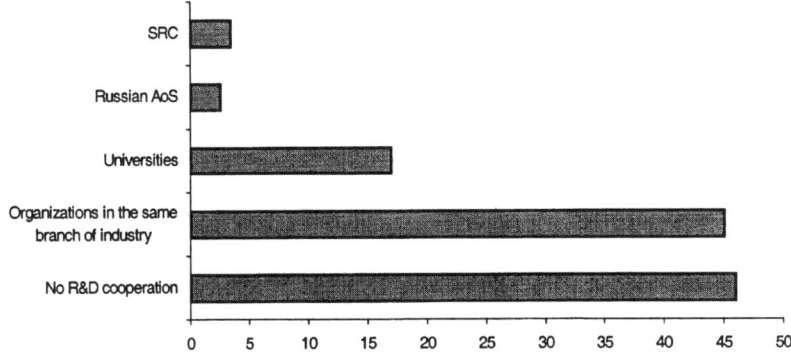

Fig. 4.3 Russia: R&D cooperation by industrial enterprises (percentage of enterprises)
Source: Survey, 1997.

Institutional and cultural barriers, coupled with the innovation climate, conditioned this trend. However, it should be noted that academy and university research institutes are already helping to improve the competitiveness of national companies. The survey also revealed that export-oriented companies call on academy and university scientists 2.5 times more frequently than those producing for the domestic market. Their key motivation is to acquire specific products that could improve their competitiveness in the world market.

State research centers (SRCs)
The MIST established SRCs in response to the dramatic decline in R&D expenditure at the beginning of the market reforms.

SRC status is granted to those research institutes that have qualified personnel, are well equipped and engage in R&D in areas related to the S&T priorities laid down (Gaponenko, 1998).

SRC status is granted for a two-year period, after which it is reviewed in accordance with defined criteria. This status means the institutes get priority treatment when it comes to government support. In 1993, the MIST launched the SRC support program. From 1995 to 1999, the share of budget allocations for the SRCs in receipt of government support decreased by 13 per cent (see Table 4.1). The number of SRCs decreased from 62 in 1993 to 58 in 2001.

As a result, the SRCs attempted to find private customers. Some were successful in adapting their activities to the demands of foreign customers. The share of foreign finance in non-budgetary funding sources increased from 20.3 per cent in 1995 to 38.9 per cent in 1999. Another remarkable feature of the second half of the 1990s was the decrease in commercial activity. At the beginning of market reforms, many research organizations as well as SRCs diversified their activities in order to survive the sharp decrease in R&D expenditure. Commercial activity brought in a remarkable amount of income for research institutes and SRCs, although at the end of the 1990s incomes from commercial activity fell considerably.

The activities of SRCs vary from center to center as well as from department to department within centers. Some SRCs have found their niche in the domestic market and have even broken through to the world market, creating new products and processes at their experimental base. For many SRCs, however, the forging of links with industry remains a problem.

In 1995, the Association of SRCs was founded in order to coordinate strategy and protect SRCs' interests.

Table 4.1 Russia: R&D expenditure in state research centers by source of funds (in %)

	1995	1999
1. Total	100.00	100.00
2. Budget Allocations	57.78	44.50
of which (as % of total budget allocation)		
2.1 MIST	89.50	72.40
2.2 Other	10.50	27.60
3. Non-budgetary sources	42.22	55.50
of which (as % of total non-budgetary sources)		
3.1 Non-budgetary foundations	5.70	2.70
3.2 Scientific and technical services	50.10	43.50
3.3 Foreign financing	20.30	38.70
3.4 Commercial activity	17.40	7.30
3.5 Other	12.20	7.80

Source: RIEPL, 2000a.

Small innovative enterprises (SIEs)

SIEs emerged in the R&D sphere in Russia in 1987 (Gaponenko, 1998). In 1991, there were only 10 600 SIEs in the RF; by 1995, their number had increased to 73 000. However, by 2000, the number of SIEs had decreased to 30 900. In 1995, there were 362 000 persons involved full-time or part-time in small innovative business; by 2000, that number had fallen to 191 900. The strongest SIEs have survived both fierce competition and the shift of domestic demand (before the financial crisis and default of August 1998) to foreign investment and consumer goods.

To sum up, organizations engaged in R&D have undergone substantial changes. However, AoS research institutes and SRCs are still characterized by considerable inertia. Although most directors of research institutes have already adapted to new conditions and changed their management methods, inter-generational conflict still persists, with a top-down approach to management being adopted in some institutes, to the detriment of individual and group initiative.

2.3. TECHNOLOGY DIFFUSION

Technology diffusion is a particular problem for Russia (Gaponenko, 1998b; Radosevic, 1997). Historically, the unequal distribution of R&D organizations and industrial enterprises across the country has led to the concentration of research institutes in some big cities. Science happened to be concentrated in the two large cities of Moscow and St. Petersburg. In 1989, 30.8 per cent of the total number of those employed in science, 50.6 per cent of professors and 43 per cent of 'doctors of science' worked in Moscow. And if the Moscow suburbs are included in this calculation, these indicators rise to 59, 58.8 and 52.4 per cent respectively; in other words, more than half of all scientists working in the country were based in or near Moscow. The corresponding indicators in St. Petersburg were 16.2, 16.2 and 16.1 per cent. Thus three quarters of the total number of those employed in science were concentrated in those two cities. The centrally planned system of knowledge production and transfer aggravated the problem. Thus Russia entered the world of the market economy with a lack of institutions responsible for the transfer of knowledge and technology and with the heavy burden of an innovation culture that was a hangover from the past. From the very beginning of the market reforms, support for institutions involved in technology diffusion was one of the priorities of science and innovation policy (see Figure 4.1, fifth layer). These institutions contribute to the diffusion of knowledge and technologies, to the formation of a new business culture and business ethics, to the creation of a learning environment for SIEs and to the rapid dissemination of the best experiences across the country. However, when the reforms began, very limited financial resources were allocated to support these institutions. Only since the end of the 1990s have the authorities begun to implement a more active policy.

Science parks (KNOWN AS TECHNOPARKS IN THE RF)

In 2000, there were 85 science parks in the RF. About 70 were university-based and they were home to more than 1 100 SIEs and about 200 service firms. The Ministry of Education launched a program to support these university-based science parks. Two trends were evident at the end of the 1990s, namely an increase in numbers and changes in the science park model. Science parks were being established even in remote regions of the RF (in the far east and in eastern Siberia) and some were being set up by associations of innovative enterprises, for example, rather than around universities. These changes were driven by the need of innovative enterprises to create a learning environment in which common strategies could be developed and experiences shared. New science parks emerged as a result of self-organization on the part of the leading innovators.

Innovation and technology centers (ITCs)
The MIST announced the establishment of ITCs in the industrialized cities in 1997. The establishment of 18 ITCs was supported by the federal authorities. The regional authorities picked up the idea, and supported the foundation of a further 17 ITCs, so that by the year 2000 there was a total of 35 ITCs in Russia, and 50 by the year 2002. The award of ITC status gives the facility priority treatment in the allocation of federal and regional government support. More than 500 SIEs were involved in ITCs in the year 2002. A survey of spin-offs located in ITCs showed that ITCs are very attractive to SIEs (Survey, 2000; see Figure 4.4).

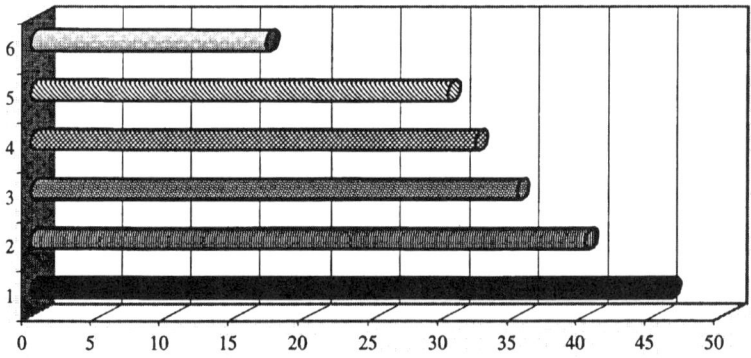

Fig. 4.4 Russia: the benefits of ITC for spin-offs (percentage of firms)
Key:
6. Opportunities for the advanced training of personnel
5. Location of space for firms
4. Good environment for sharing experiences and diffusion of best practice
3. Access to information
2. Financial support
1. Greater opportunities for forming networks
Source: Survey, 2000.

Business incubators

The transitional period also saw an acceleration of the development of business incubators. These institutions have no formal status yet, although regional and local authorities have begun to support them. In some regions of the RF, the support of business incubators is included in the small business support strategy developed by the regional authorities. Some regional and local authorities have started to coordinate measures for the regulation and support of small businesses with those running business incubators.

The National Partnership of Business Incubators (NPBI) was established in 1996 and had 52 members by the year 2000. The NPBI provides information, legal support, advanced training for SIE staff, etc.

Innovation-industrial complexes (IICs)

In 1999, the MIST, together with other ministries, announced the establishment of IICs, whose purpose would be to strengthen the links between science and industry, to accelerate knowledge and technology diffusion, to reduce the time required to get new products and processes to market and to improve competitiveness. The idea was to put in place structures in which both industrial enterprises and research institutes would be involved. Four IICs were set up in 1999/2000, in St. Petersburg, Moscow (2), and Zelenograd. Preparations for establishing three more IICs were being made in the year 2001. MIST documents stated that up to 50 per cent of R&D expenditure geared to the development of high-tech products and processes would be financed by this ministry.

Naukograds

There are about 70 '*naukograds*', or 'science cities', in Russia. A *naukograd* is a city that grew up around a big research institute and whose infrastructure developed around the research institute's activities. *Naukograds* differ from science parks (or technoparks). A big research institute acts as the kernel of a *naukograd*. Industrial companies located in different parts of the RF can draw on the knowledge produced by these institutes; in the Soviet period, these processes were carefully regulated. The *naukograds* developed during the Soviet period were oriented towards knowledge production, primarily for the defense sector but also in areas related to other S&T priorities. The dramatic decline in R&D expenditure in the 1990s led to a sharp drop in R&D activity and a brain drain, as well as high unemployment and social instability in the *naukograds* (RIEPL, 2000). It was more difficult for these cities to adapt to new conditions than it was for the R&D organizations in industrialized centers. It was obvious that measures had to be taken to transform these cities and to adapt them to the new conditions. However, it was not until the end of the 1990s that the government gave some indication of the direction the changes would take. The proposals included the development of high-tech industrial enterprises and/or SIEs and the development of educational centers and an innovation infrastructure for knowledge transfer. A federal law 'On the status of *naukograds* in the RF' and related government decrees were adopted to regulate these processes. The criteria and procedures for granting *naukograd* status, as well as the arrangements for federal and local authority support, were laid down in this legislation. So far only three cities (Obninsk, Korolev, Dubna) have been granted *naukograd* status.

Federal centers of science and high technology (FCSHTs)

The establishment of FCSHTs was announced in the 'Draft plan for the reformation of Russian science for 1998 – 2000' and confirmed by a government decree in 1999. The legislative base for the regulation of procedures and the criteria for granting FCSHT status were established by the MIST in the same year. FCSHT status was to be granted to those research institutes (1) that had capacities for providing competitive high-tech products and processes in the 21st century, (2) that were experienced in the training and advanced training of R&D personnel and (3) that were already cooperating with corporations, financial institutions and universities. The applicants were required to develop a program designed to solve certain key high-tech issues, to present a list of organizations that would contribute to the program's implementation and to set up a network to implement the program. It was also envisaged that the objective of the program should not be to develop basic technological innovations but rather to produce and implement innovations. The idea behind the establishment of FCSHTs is to forge links between research institutes, corporations and other institutions in order to make breakthroughs in certain key areas, to accelerate knowledge transfer and to improve competitiveness in high-tech fields (Kitova and Tcherkasov, 2000). The ITCs and the FCSHTs share the objective of improving competitiveness in the high-tech sector. However, the two organizations differ in that the FCSHTs are supposed to be predominantly forward-looking, focusing on the most significant issues of the future.

In the year 2000, the commission charged with granting FCSHT status evaluated 40 applications and concluded that three applicants met all the requirements laid down in the legislative base. The commission recommended that the MIST should ask the Government Commission for Science and Innovation Policy to grant FCSHT status to three research institutes. However, these institutes have not yet been given this status.

In fact, the legislative base is weak, in particular with regard to financial support. There is also no consensus among policy-makers on many issues related to the organization and performance of FCSHTs or on the key priorities.

Research spin-offs
The transitional period saw the emergence of research spin-offs. Although the State Committee for Statistics does not provide data on spin-offs, formation rates across the country or their numbers and activities, the survey of spin-offs (Survey, 2000) highlights the innovation activities of these firms and their links with other NIS institutions. They tend to be innovative: about 40 per cent of those surveyed introduced radical product innovations, more than 48 per cent introduced product improvements and about 37 per cent introduced radical process innovations. Forty per cent of respondents were seeking to break through to the world market and about 20 per cent were already operating in that market.

Spin-offs form links with public research institutes and industry. About 85 per cent of the firms surveyed had cooperated with various organizations in R&D (see Figure 4.5). Their partners were usually SIEs, co-founders (that is

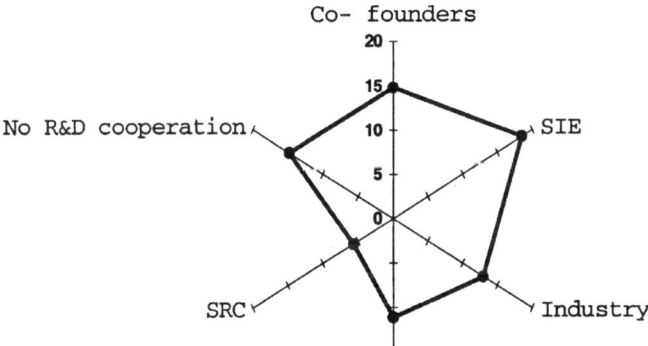

Fig. 4.5 Russia: R&D cooperation by spin-offs (percentage of spin-offs)
Source: Survey, 2000.

individuals or organizations, usually academics or research institutes, involved in the founding of spin-offs) or industrial enterprises. They rarely cooperated for R&D purposes with institutes in the Russian AoS or with SRCs. About 50 per cent of firms stated that they were already experienced in cooperative R&D and already had good partners and links with research organizations. Eleven per cent of firms indicated that is was difficult to find partners and about 17 per cent of spin-off directors noted that, while they had no experience as yet, they did realize the need for cooperative R&D.

The survey helped to identify the main factors driving R&D cooperation as well as the barriers to such cooperation. The lack of qualified personnel, equipment and financial resources, coupled with the growth of interdisciplinary R&D and the need to accelerate the development of new products and new processes, were key factors in encouraging cooperation between spin-offs and other research organizations and industry. The weakness of the legislation governing intellectual property rights, divergent research interests and cultural barriers were identified by spin-off directors as the main impediments to cooperation. It can be concluded that R&D cooperation has encouraged the establishment of networks and contributed to the development of a new innovation culture.

Labor mobility between firms is the most powerful mechanism for knowledge transfer, for the dissemination of best practice and for the transmission of tacit knowledge. The survey showed that spin-offs were very active in involving highly skilled personnel from both the private and public sectors as contract or temporary workers (part-time employment). Only three of the 100 firms investigated did not employ personnel formerly with other NIS organizations. Universities, SIEs and industry were the main sources of highly skilled personnel for the spin-offs. Academic research institutes, SRCs and research institute co-founders were less active in collaborative research (see Figure 4.6).

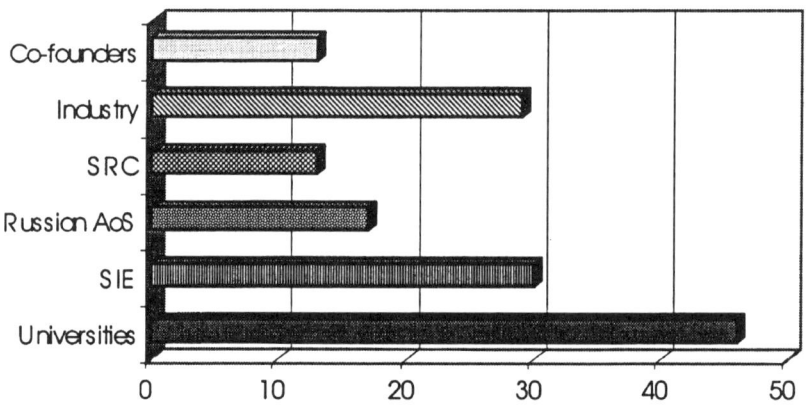

Fig. 4.6 Russia: labor mobility between firms (involvement of personnel per spin-off, in %, by source)
Source: Survey, 2000.

It is likely that the concentration of academic research institutes in a small number of major cities, the lack of jointly managed laboratories and inadequate experience in the use of IT to link researchers from various regions of the RF working on collaborative projects, together with cultural barriers, were all factors contributing to the low level of involvement among AoS researchers in the activities of spin-off firms.

The relative lack of collaboration with individuals and organizations involved in setting up spin-offs was surprising. It may be that spin-offs regard co-founders as competitors or that they are more likely to be set up by the most advanced departments or groups of researchers and that the personnel employed by co-founders are not highly skilled enough to meet their requirements. The survey showed that the growth of interdisciplinary R&D encouraged labor mobility between firms. It should be also noted 50 per cent of respondents declared it was difficult to find qualified personnel, although this might be because the more highly skilled researchers are now serving the needs of foreign customers.

To sum up, it can be concluded that spin-offs make a special contribution to knowledge and technology transfer, to network-building and to linkages within the innovation system mediated through human resources.

2.4. SERVICE FIRMS

The period of transition also saw the emergence of firms providing services in various fields, such as consultancy, product certification, marketing, information and legal services, etc. (see Figure 4.1, sixth layer). Most of these firms developed in response to domestic demand. Although that demand was very sluggish, increasing competition in the domestic market and gradual integration into the world market caused it to rise somewhat. Many service firms are located on science parks or operate as part of business incubators or ITCs. Working in close cooperation with SIEs, universities and corporations, they are responsive to changes in demand while at the same time helping to shape that demand and contributing to the development of a new business culture. These service firms are playing an increasingly important role in the innovation process. It is apparent that the demand for information services in the Russian market is growing rather faster than that for other services. Investment companies and banks, research institutes and science parks are developing databases in response to the demand. Support for an information infrastructure is also one of the declared priorities of science and innovation policy.

2.5. TOWARDS THE STRENGTHENING OF DOMESTIC NETWORKS IN THE NIS

The institutional map of the Russian NIS has changed substantially during the period of transition. New institutions have been established or have emerged out of the process of self-organization. Institutions established during the Soviet period have undergone fundamental changes. However, the distinctive features of the RF's NIS continue to be fragmentation, a lack of technology transfer and service-oriented firms, poor links between science and technology and inadequate inter-institutional cooperation.

In 1997, the MIST, together with the Ministry of Education, the Russian Foundation for the Promotion of Small Enterprises in S&T (RFPSE) and the Russian Foundation for Technological Development (RFTD) launched the 'Interdepartmental Program for the Support of Innovation Activity in S&T'. This was the first program to have as its strategic target the strengthening of domestic NIS networks, as well as measures to overcome institutional, regional and informational fragmentation within the NIS. There was a consensus among those involved in setting up the program that public and private finance, federal and regional budgets as well as foreign loans should be combined to achieve this goal.

In 1997, program priorities included the development of an innovation infrastructure in the regions of the RF, the establishment of ITCs and support for R&D commercialization by SIEs. The establishment of eight ITCs was supported by the MIST (33.1 per cent of total support), by the Ministry of Education (7.9 per cent), by the RFPSE (36.2 per cent) and by regional authorities (22.8 per cent).

In 1998, many other ministries and agencies joined the program. In the process of developing measures for the period 1998–2000, the various agencies began to enter into dialogue with each other and to coordinate their activities. Four priorities were established for the two-year period: support for institutional changes and information infrastructure development; support for high-tech projects implemented by SIEs; the training and advanced training of personnel; the development of a legislative base.

In 1999, the development of information networks linking science parks, ITCs, SRCs and other structures within the NIS was announced. As part of this scheme, the program provided technical support and initiated the development of problem-oriented databases and software, which were made available to the various structures in NIS. This was the first stage in the process of strengthening links between all the institutions in the Russian NIS. However, it is clear that this is not enough. Financial mechanisms must be changed and should also be used to strengthen networks.

3. THE FINANCING MECHANISMS

At the beginning of the market reforms, R&D expenditure fell sharply in real terms; the share of GERD in GDP declined from 2.03 per cent in 1990 to 0.74 per cent in 1992. This downward trend in R&D expenditure continued up to 1995, when it gave way to a gradual increase (excluding the drop caused by the 1998 financial crisis – see Figure 4.7).

The main changes in the shares of the various sources of R&D financing are as follows:

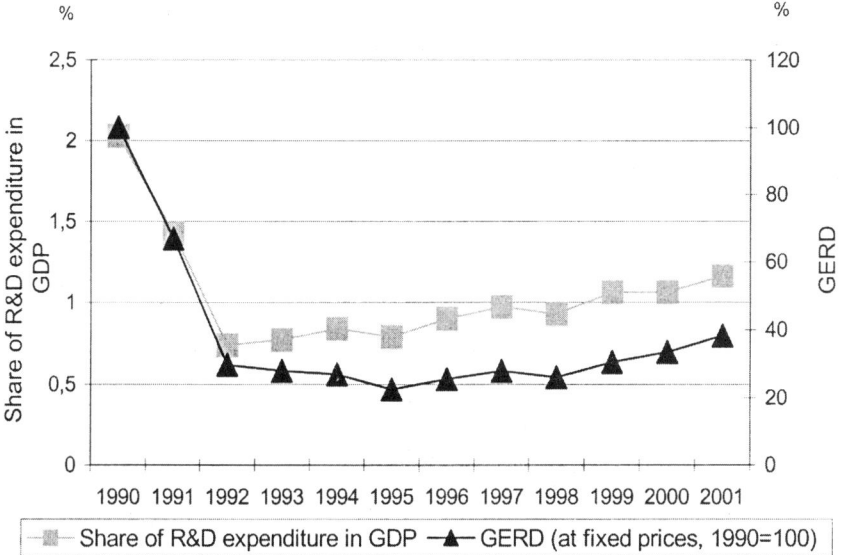

Fig. 4.7 Russia: R&D expenditure
Source: SCSRF, 2000; 2001.

Table 4.2 Russia: trends in R&D spending by source of funds (percentage share)

	1994	1995	1996	1997	1998	1999	2000	2001
Budget allocation for R&D	61.0	60.5	60.7	59.6	52.2	49.9	53.7	53.9
Industrial sector	19.9	17.4	15.3	15.5	17.3	15.7	18.7	16.5
Non-budgetary R&D funds	6.3	6.7	6.2	6.0	5.5	6.9	6.5	7.3
Foreign financing	2.0	4.6	5.6	7.4	10.3	16.9	12.0	13.3
Other sources	10.8	10.8	12.2	11.5	14.7	10.6	10.1	9.0

Source: SCSRF, 2000; 2001.

- budget allocations for R&D as a percentage of GERD fell from 92 per cent in 1992 to 53.9 per cent in 2001;
- the share of the industrial sector was 16.5 per cent in 2001;
- the role of foreign financing in total R&D expenditures changed, with its share increasing from 2.0 per cent in 1994 to 13.3 per cent in 2001;

Table 4.3 Russia: national R&D expenditure

	GERD as % of GDP (1995; Russia: 2001)	Government financing of R&D as % of total R&D (1995; Russia: 2001)
Russia	1.16	53.9
USA	2.6	34.6
Norway	1.7	43.5
Switzerland	2.7	28.4
Japan	2.8	20.9
Denmark	1.8	39.2
Canada	1.7	33.7
Austria	1.5	47.6
Australia	1.6	47.5
Germany	2.3	37.0
Netherlands	2.0	42.1
France	2.3	42.3
United Kingdom	2.1	33.3
Spain	0.9	43.6
Korea	2.7	19.0
Portugal	0.6	65.2
Greece	0.5	46.9
Czech Republic	1.2	35.5
Hungary	0.8	47.9
Mexico	0.3	66.2
Poland	0.7	64.7
Turkey	0.4	64.5

Source: NSF, 1998; OECD, 1998; SCSRF, 2001.

– the share of goal-oriented funds in the budget allocations for civil R&D increased from 2.1 per cent in 1993 to 8.0 per cent in 2000;
– the share of extra-budgetary R&D funds (sectoral and intersectoral) rose to 7.3 per cent in 2001 (see Table 4.2).

The public sector's share in R&D funding remains rather high compared with some industrialized and transitional countries (see Table 4.3). However, it was not realistic, in conditions of crisis, to wait for the private sector to expand.

The initial responses of policy-makers to the drastic decline in R&D expenditures were to introduce the goal-oriented budgetary and extra-budgetary funds to finance basic research, to encourage industry to support applied research and to channel some budgetary financial resources towards the SIEs (Gaponenko et al., 1995). The establishment of these funds also led to changes in funding mechanisms, primarily an increase in competitive funding for S&T projects, a strengthening of the expertise available for projects, the channeling of financial resources in applied science to S&T priorities and the formation of public-private partnerships (see Figure 4.1, third layer). The RFTD and the RFPSE experimented with providing loans to fund innovation and technology projects, which turned out to be very successful. These loans were very attractive, since the interest rates were significantly lower than those offered by commercial banks. The repayments in turn were used to fund further innovation and technological projects.

In the mid-1990s, the government attempted to change the way budgetary funds for civil R&D were allocated. It was decided to concentrate financial resources on priorities, to finance S&T projects on a competitive basis and to gradually introduce a contract system in the R&D sector (Gaponenko, 1998a). These changes were enshrined in the Law on Science and the State Science and Technology Policy, adopted in 1996. It was also envisaged that the federal budget funds allocated for civil R&D should amount to no less than 4 per cent of total federal budget expenditure. It was established in law that basic research would be mainly supported from the federal budget and a strategy to strengthen public-private partnership in applied science was adopted. This legislation represents a step forward in the process of dividing functions and responsibilities in the area of S&T support and NIS regulation between the federal and regional authorities.

After this law was passed, some progress was made in the transformation of financial mechanisms. However, political and financial instability, coupled with weak criteria for the allocation of financial resources, prevented financial resources from really being concentrated on priorities. For many years, the R&D system was under-funded, despite the amounts set aside in the federal budget (Mindeli and Pipiia, 1998).

The private sector was in no hurry to increase its funding of applied science. Since their capital assets were old and worn-out and their financial opportunities restricted, industrial enterprises spent most of their money on the purchase of new equipment. In 1999, the share of R&D expenditures in innovation expenditures was about 14.2 per cent, which was much lower than in other European countries (see Table 4.4); this might be explained in part by the lower salaries paid to academics in the RF. The private sector had to tread a path between the two forces determining its innovation strategy: increasing competition in the domestic market and its ambitions to break through to the world market. This latter aspiration increased interest in raising R&D expenditure, but the high risks and uncertainties, financial and political instability and worn-out capital assets hampered the growth of R&D expenditure. Inadequate financial, credit and tax policy also dampened down the private sector's interest in increasing its R&D expenditure.

Table 4.4 Russia: innovation expenditure (percentage share)

Countries	R&D	Patents and licenses	Market analysis
Russia	14.2	0.4	0.7
Australia	35.1	1.5	7.6
Belgium	44.7	1.5	6.6
Denmark	40.1	5.3	8.2
Germany	27.1	3.4	13.2
Greece	50.6	6.4	13.2
Italy	35.8	1.2	1.6
Netherlands	45.6	6.1	19.8
Norway	32.8	4.2	5.5
Portugal	22.9	4.1	5.4
Spain	36.4	8.0	8.8
United Kingdom	33.5	4.6	6.6

Note: Data do not total 100% as "other expenditures" are not included in the table.
Sources: ABS, 1994; Bosworth et al., 1996; OECD, 1998a; SCSRF, 2001.

4. SUMMARY AND OUTLOOK

4.1. FROM CHAOTIC TRANSITION TO NEW FORMS OF GOVERNANCE

The Russian Federation's national innovation system has traveled a long way in a short time. After a chaotic transitional period, it is now accepted that government should play a major role in regulating the systems' transformation and development. Financial resources for R&D are no longer allocated in accordance with the numbers of staff employed in the research organizations; a competitive, project-based system of funding has been introduced, together with a contract system. Huge institutes with hundreds of employees have given way to a more flexible and adaptable institutional set-up.

One remarkable feature of the transitional period was the growing complexity of the NIS. New institutions such as SRCs, international centers, joint ventures, IICs, ITCs, SIEs, private institutions, science parks, business incubators and research spin-offs were established or emerged as a result of self-organization. SIEs, and in particular research spin-offs, played a special role in the transformation of the NIS, encouraging the development of networks and contributing to the formation of a new innovation culture.

Several years ago, the main issue on the agenda was making good the deficit in technology diffusion institutions and building a balanced institutional set-up for the NIS. Some measures have been introduced to facilitate the institutional changes required, although the problems are far from being resolved.

However, the rapidly changing economic and social conditions [2] have already brought another issue to the fore, namely the need to strengthen networks and to put in place an adaptive and learning innovation system.

Small research centers and SIEs have brought flexibility to the Russian NIS. However, this is not enough. The Russian AoS and the state research centers are two institutions that have changed little. It would appear that joint laboratories could bring flexibility to basic research and applied science, making these fields more productive and responsive to rapid changes. This should be placed on the agenda.

NIS networks in Russia developed through self-organization. The government underestimated the synergy that could emerge if such networks were strengthened and focused its measures rather on the organizational set-up. It is now time to develop the financial schemes that could contribute to the linking of all NIS institutions.

4.2. Changes in policy

In the 1990s, innovation policy underwent substantial changes (Gaponenko, 1997). At the beginning of the market reforms, the main objectives were to distribute functions among different policy-making bodies, to save those organizations with the most valuable capabilities, to provide support for SIEs, to decentralize financing mechanisms and to develop a legislative base for the protection of intellectual property rights. In the institutional context, SRCs were established and the program supporting them was launched. Many institutions, such as research spin-offs, financial and industrial groups, small research centers, international research centers, etc., emerged as a result of self-organization. Parallel to this, the MIST initiated the establishment of goal-oriented, extra-budgetary funds, which actually turned out to be very successful. During this period, the government reduced its role in NIS regulation. There was a consensus that market forces would drive the changes and play the major part in creating a balanced innovation system.

---

[2] Statistical reference for economic and social indicators: in 2000 GDP was 3.3 per cent higher than in 1995; in the year 2000, total annual GDP growth was 7.7 per cent; in the year 2001, compared to 1995, industrial output had increased by 9.6 per cent over 1995; in 2001 the annual growth was 5.0 per cent; in the same year, the real money income of the population had increased by 9.1 per cent over 1999. Total unemployment in the year 2001 was 6.4 million (SCSRF, 2000; 2001).

From 1995 onwards, there was a gradual shift of policy towards better coordination of activities, both among the various policy-making bodies and between the regional and federal authorities, support for SIEs and the development of an innovation infrastructure. The government was also concerned to concentrate resources on S&T priorities, to distribute those resources on a competitive basis and to build up public-private partnerships. Implementation of these measures was affected by political and financial instability, a high level of risk and inadequate macro-economic policy, as well as by the rapid influx of foreign technologies on to the domestic market, which weakened the national industrial sector and R&D system.

From 1997 onwards, the emphasis was on the development of institutions responsible for technology diffusion and the linking of science and technology in order to improve competitiveness and reduce the time required to bring products and processes to market, in particular high-tech products and processes. Financial mechanisms were also put in place in order to concentrate resources on S&T priorities, to encourage high-tech breakthroughs and to promote public-private partnerships. It was decided that the government should formulate an active policy and that the private sector and NGOs should be involved in policy-making. At this stage, an inadequate legislative base, an underdeveloped financial infrastructure and stock market, the lack of technology transfer institutions and poor links between NIS institutions, coupled with high taxes and inadequate macroeconomic policy, were the main obstacles to successful implementation of these policies.

Table 4.5 Russia: the main directions of Russian innovation policy (1998–2000)

Development of a legislative base
Support of institutional change
Intellectual property rights protection
Support of critical technologies
Development of innovation infrastructure, including expertise, certification, financial infrastructure, training and advanced training of personnel, information infrastructure
Support of SIE
Involvement of all stakeholders in policy development and implementation
Support of environmentally friendly technology

Source: Concept of innovation policy of the RF for 1998–2000, Moscow, 1998.

At last, in 1998, the Russian government for the first time adopted the concept of innovation policy for 1998–2000, and the related measures for 1999–2000. The broad directions of innovation policy are outlined in the draft

proposal (see Table 4.5), as are the respective responsibilities of the various policy-making bodies and agencies. In institutional terms, the draft proposal provides for:
- the development of the private sector and the setting up of regional innovation centers;
- support for science parks, business incubators and ITCs;
- the forging of links between SIEs and corporations;
- the development of an innovation infrastructure;
- the establishment of FCSHTs and of a University of Innovation, Technology and Business Undertakings;
- measures for personnel training and advanced training, and the promotion of an information infrastructure.

Thus policy and the mechanisms put in place to implement it have changed substantially during the last decade. There has been a gradual shift away from uncoordinated attempts to support some organizations or to change only some of the funding mechanisms towards the creation of a whole new institutional framework and a system of financing suited to the new conditions.

4.3. FINANCING AS AN IMPORTANT TOOL

The failures and successes of the 1990s were evaluated and a strategy for the future outlined in the 'Draft proposal for reform of R&D system financing', adopted in 2000. The following measures were envisaged in this proposal:
- the introduction of programmed and targeted methods of government support for R&D;
- the allocation of budget expenditure to S&T priorities;
- the support of basic research mainly from the federal budget;
- the strengthening of public-private partnership;
- the introduction of the repayment principle in applied science in conjunction with an extension of competitive funding for S&T projects;
- the expansion of extra-budgetary sources for R&D support.

In 1998, the 'Draft proposal on innovation policy' was adopted by the Russian government. Some additional measures were introduced, namely the allocation of financial resources from foreign loans for the development of an innovation infrastructure, the purchasing of equipment, licenses and know-how in order to introduce new products and processes, support for venture funds and equipment leasing.

At the end of 2000, 40 venture funds were registered in the RF, although only 15 of them were actually operational. The European Bank for Reconstruction and Development was the main source of venture capital in the Russian market. The main features of venture capital activity are briefly outlined below.

Venture fund support for projects in Russia usually kicks in when they have reached an advanced phase rather than at the outset. The main clients of venture funds are the export-oriented companies or companies dependent on the import of component parts. Only about 1 per cent of venture capital is invested to support high-tech products and processes. The political, economic and financial instability, eroded legislative base, underdeveloped stock market and insurance system, together with the lack of motivation for long-term investment, have hampered the growth of venture capital. The banks, private pension funds, corporations, insurance companies and private investors have not been prepared to 'invest in risk'. In the light of this situation, the MIST initiated the establishment of the Venture Innovation Foundation (VIF) in 2000. The RFTD and the RFPSE provided finance for the VIF during the start-up phase. Regional and sectoral venture funds are planed, and it is proposed that the VIF, ministries and regional authorities should contribute to their development in the initial stages.

Some regional authorities have already established, or plan to establish, regional venture funds. In Moscow, the city government supported the foundation of the 'Rossinnovation' venture fund and the regional small business support program provides for the establishment of investment companies to support SIEs engaged in high-risk projects.

The future development of venture capital will depend to a great extent on the development of the legislative base, on market and corporate transparency, on the development of an insurance system, on advanced training of personnel and on changes in business ethics and culture.

To sum up, it is clear that the Soviet system of financial resource distribution is being replaced. However, the new mechanisms are not yet working properly. It would appear that the eroded legislative base, underdeveloped ethical codes both in the business community and among decision-makers, coupled with a lack of market transparency and inadequate accountability in all NIS institutions, are some of the main factors hampering the transition to a new mode of financing.

REFERENCES:

ABS (1994) *Innovation in Australian Manufacturing*. ABS bulletin 8116.0. Sydney: Australian Bureau of Statistics (ABS).

Balazs, K. (1997) "Is there any Future for the Academies of Sciences?" *The Technology of Transition: Science and Technology Policy for Transition Countries*. Ed. D. Dyker. Budapest: Central European University Press, 161–184.

Breschi, S., and F. Malerba (1995) "Sectoral Innovation Systems: Technological regimes, Schumpeterian Dynamics and Spatial Boundaries." Paper prepared for

the Conference of the 'Systems of Innovation Research network' coordinated by C. Edquist, Soderkoping, September 7–10, 1995.
Bosworth, K., P. Stoneman, and U. Sinha (1996) *Technology Transfer, Information Flow and Collaboration: An Analysis of the CIS*. EIMS Publication no. 36. Luxembourg: European Commission.
Dyker, D. (Ed.) (1997) *The Technology of Transition*. Budapest: Central European University Press.
Edquist, C. (Ed.) (1997) *Systems of Innovation: Technologies, Institutions and Organizations*. London: Pinter Publishers.
Freeman, C. (1987) *Technology Policy and Economic Performance – Lessons from Japan*. London: Pinter Publishers.
Freeman, C., and C. Perez (1988) "Structural crises of adjustment: business cycles and investment behaviour." *Technical Change and Economic Theory*. Ed. G. Dosi, et al. London: Pinter Publishers, 38–66.
Gaponenko, N. (1995) "Transformation of the Research System in a Transitional Society: The Case of Russia." *Social Studies of Science* 25, no. 4: 685–703.
Gaponenko, N. (1996) "Transformation of the System for Innovation in a Society in Transition: The Case of the Russian Federation." Warsaw: UN/ECE Seminar on State Policy in Economics in Transition Aimed at Promoting Innovation in Industry. June 24–25.
Gaponenko, N. (1997) "Innovacii i innovacionnaya politika v perehodni period k novomu tehnologitcheskomu poryadku" (Innovations and innovation policy in the transitional period to a new technological order). *Voprosi Ekonomiki* (Issues of Economy) no. 9: 84–98.
Gaponenko, N. (1998) "Self-Organization and Politics in Russian S&T Transformation." *Transforming Science and Technology Systems – the Endless Transition? NATO Science Series 4: Science and Technology Policy-Vol. 23*. Ed. W. Meske, et al. Amsterdam: IOS Press, 118–129.
Gaponenko, N. (1998a) "Finansovie mehanizmi regulirovaniya i podderzhki innovacionnoi aktivnosti" (Financial mechanisms of regulating and supporting of innovation activity). *Nauka i innovacii v perehodni period k obchestvu osnovannomu na znaniyah* (Science and innovations in the period of transition to a knowledge-based society). Ed. Y. Yakovetc. Moscow: International Kondratieff Foundation Press, 72–85.
Gaponenko, N. (1998b) "Transfer Tehnologi i innovacionnaya strategiya v promishlennosti" (Technology transfer and innovation strategy of industry). *Belorusski Ekonomitcheski Zhurnal* (Issues of Economy of Belarus) no. 3: 45–55.
Gaponenko, N., A. Polonsky, and O. Vjugin (1993) *Russia: Economy and Science on the Way of Reforms*. Moscow: Centre for Science Research and Statistics (CSRS).
Gaponenko, N., L. Gokhberg, and L. Mindeli (1995) "Transformation der Wissenschaft Russlands.'' *Transformation mittel-und osteuropäischer Wissenschaftssysteme*. Ed. R. Mayntz, et al. Opladen: Leske+Budrich, 382–569.
Gaponenko, N., J. Yakovetc, and V. Kushlin (1997) *Teoriya Innovaci v Rinochnoi Ekonomike* (Theory of Innovations in the Market Economy). Moscow: IKF Press.
Gokhberg, L., and L. Mindeli (Eds.) (1996) *Russian Science and Technology at a Glance: 1995*. Data Book. Moscow: CSRS.

Gokhberg, L., and L. Mindeli (Eds.) (2001) *Russian Science and Technology at a Glance: 2000*. Data Book. Moscow: CSRS.
Kitova, G., and V. Tcherkasov (2000) "Federalni Centri Nauki i Visokih Tehnologi: Problemi Organizacii" (Federal Centers of Science and High Technology: the Problems of Organization). *Regonomika* (Regional Economy) vol. 2: 223–233.
Lundval, B.-A. (1992) *National Systems of Innovation: Towards a Theory of Innovation and Interactive Learning*. London: Pinter Publishers.
Mayntz, R. (1994) *Deutsche Forschung im Einigungsprozess. Die Transformation der Akademie der Wissenschaften der DDR 1989 bis 1992*. Frankfurt a.M.: Campus Verlag.
Mayntz, R., U. Schimank, and P. Weingart (1998) *East European Academies in Transition*. Dordrecht: Kluwer Academic Publishers.
Mindeli, L., and L. Pipiia (1998) "Financing Russian R&D: Crisis and Possible Solutions." *Transforming Science and Technology Systems – the Endless Transition? NATO Science Series 4: Science and Technology Policy – Vol. 23*. Ed. W. Meske, et al. Amsterdam: IOS Press, 27–40.
Nelson, R. (Ed.) (1993) *National Innovation Systems: A Comparative Analysis*. Oxford: Oxford University Press.
Niosi, J., P. Saviotti, B. Bellon, and M. Crow (1993) "National Systems of Innovation: In Search of a Workable Concept." *Technology in Society* 15: 207–227.
NSF (1998) *Science and Engineering Indicators*. Washington DC: National Science Foundation (NSF).
OECD (1998) *Science, Technology and Industry Outlook*. Paris: OECD.
OECD (1998a) *National Innovation Systems: Analytical Findings*. Paris: OECD/Working Group on Innovation and Technology Policy.
Radosevic, S. (1997) "Technology transfer in global competition: The case of economies in transition." *The Technology of Transition*. Ed. D. Dyker. Budapest: Central European University Press, 126–161.
RIEPL (2000) *Analiz socialnih i ekonomitcheskih uslovi v naukogradah – Godovoi Otchet* (The analyses of social and economic conditions in Naukograds – Annual Report). Moscow: Russian Institute of Economics, Policy and Law in Science and Technology (RIEPL).
RIEPL (2000a) *Razrabotka bazi dannih dlya analiza deyatelnosti GNC – Godovoi Otchet* (The development of a database for the analysis of activity by SRCs – Annual Report). Moscow: RIEPL.
SCSRF (2000; 2001) *Social and Economic development of Russia*. (Yearbooks) Moscow: State Committee for Statistics of the RF (SCSRF).
Survey (1997) Unpublished survey of 500 industrial enterprises of various branches of industry, located in different parts of the Russian Federation, carried out in 1996–1997 within the framework of the project "Innovations and Technology Strategy of Industry", supported by the Ministry of Industry of the Russian Federation, and coordinated by the author. The questionnaire consisted of 30 questions, which were designed to highlight the market and innovation strategy of industrial enterprises, the issues of technology transfer via the domestic market, the relationships between all NIS institutions, barriers to the introduction of basic and modified innovations, etc.

Survey (2000) Unpublished results of a survey of 100 spin-offs was carried out within the framework of the project "National Innovation System: the Problems of Transformation and the Mechanisms of Regulation", supported by the Ministry of Industry, Science and Technology of the Russian Federation, carried out by the Russian Institute for the Economy, Policy and Law, and coordinated by the author. The questionnaire consisted of 35 questions, which were designed to highlight the market and innovation strategy of spin-off firms, their relationships with organization co-founders and other NIS institutions, the issues of labor mobility, etc.

5. UKRAINE: INSTITUTIONAL CHANGES IN S&T IN A PERIOD OF ECONOMIC DECLINE

Lidiya KAVUNENKO

The years of independence have turned out to be a difficult and controversial period of transformation for Ukraine. There are now many opportunities to assert newly won political and economic freedoms, which is one of the main achievements of Ukrainian independence. At the same time, however, Ukraine's dependency on exports meant that the economic crisis of 1997– 1998 considerably worsened the situation in such industries as iron and steel and chemicals.

1. ECONOMIC BACKGROUND AND S&T TRANSFORMATION

In terms of its industrial capabilities, Ukraine had been one of the most developed republics within the USSR. In 1991, before the disintegration of the USSR in August of that year, the population was 51.9 million. The economy was based on the powerful iron and steel, heavy machinery and agricultural equipment industries. The Ukraine was the world's second largest producer of coal, cast-iron, steel and rolled iron. However, the abrupt breakdown in the economic links within the USSR in 1991 had an enormous impact on the economic situation of the newly emerged countries (Dobrov, 1990; Malitsky and Nadiraschwili, 1995; Meske and Nadiraschwili, 1994).

The population of Ukraine decreased by seven per cent between 1990 and 2000. Over the same period, the economically active population decreased by more than 22 per cent. The state budget revenue was collected with great difficulty and the budgets of most enterprises and institutions were not balanced. Economic problems persist to this day because of the lack of structural change and the stagnation of deregulation and privatization processes, which limits the prospects for private entrepreneurship. Loss-making enterprises still enjoy a wide range of privileges and subsidies. In contrast, successful local enterprises are under considerable tax pressure, which pushes them into operating in the black economy, which accounts for between 40 and 60 per cent of GDP, depending on the estimate (Malitsky, 2000). Considerable difficulties are also being experienced in reaching budgetary goals. The foreign trade situation remains extremely complicated. The volume of foreign trade has continued to decline, though imports have declined much faster than exports.

The gravity of the economic crisis is reflected in the continuing decline of all social development and economic performance indicators (Table 5.1). Erratic stabilization processes in the manufacturing sector are further aggravated by problems in the financial area. The most serious problem is the flow of funds out of manufacturing industry in order to service state debt, which amounted to 6.7 per cent of GDP in 1997.

The restructuring of the economy is faltering because of inadequate innovation. According to expert estimates, the scientific intensiveness of industrial production does not exceed 0.3 per cent, which is 10–20 per cent of the world level. About 90 per cent of Ukrainian products today lack the relevant S&T provision. The low level of technology and the lack of motivation to produce high-quality products add to the commercial non-viability and non-competitiveness of industrial products.

The Ukraine had a 13 to 15 per cent share of the USSR's total research capability. Some of the more significant characteristics of scientific research in the Ukraine were the high concentration of experimental bases operated by the USSR research organizations, a highly developed regional structure, the dominance of industrial research (accounting for about 50 per cent of total R&D personnel), a high share of applied research, even in the Academy of Sciences (AoS), and an active search for new forms of research organization. In terms of its R&D potential relative to population, Ukraine was better prepared for independent development than the other (mostly very small) Soviet republics, which played a more specialized role as 'suppliers' for Russia.

The Ukrainian science and technology system (STS) had been a leader in many areas of research, in particular in space studies, theoretical physics, mathematics, the welding industry, protective and reinforcing coatings and biotechnologies. By many criteria, the Ukrainian S&T potential matches that of most developed countries. It must, however, be noted that Ukrainian science has for many years been used only partially for the benefit of the Ukrainian nation. The current crisis in Ukrainian science is caused by many negative economic and political factors. The core economic factor is the critical decline in R&D funding. Funding per researcher in Ukraine is considerably less (monthly salary is about 75 USD) than in Germany, France or Great Britain. Such financing is not only inadequate but makes it impossible to conduct essential scientific research. In the year 2000, the number of researchers had fallen by more than 50 per cent compared to 1990; there are 25 researchers per 10 000 people in Ukraine, one third less than in developed countries. In addition, there is considerable hidden unemployment because of the economic crisis.

Table 5.1 Ukraine: trends in economic activity

Indicators	1991	1992	1993	1994	1995	1996	1997	1998	1999	2000
Population (mill.)	51.9	52.1	51.9	51.8	51.5	51.1	50.7	50.2	49.9	49.5
Employment (mill.)	25.0	24.5	23.9	23.0	23.7	23.2	22.6	22.3	21.8	21.6
GDP per capita (as % of the previous year)	90.6	89.7	85.7	77.4	88.5	90.7	97.5	99.0	100.0	100.4
Real GDP annual growth, %	-8.7	-9.9	-14.2	-22.9	-12.2	-10.0	-3.2	-1.9	-0.4	-0.4
Real GDP growth per capita, %	-9.4	-10.3	-14.3	-22.6	-11.5	-9.3	-2.5	-1.2	0.4	0.3
Annual rate of inflation, %	390	2 100	10 256	501	281.7	139.7	110.1	120.0	119.2	125.8
Overall budget deficit (as % of GDP)	7.1	13.8	5.1	8.9	6.6	4.9	6.7	2.2	2.0	
Total science expenditures (as % of GDP)	2.5	1.6	0.7	0.6	0.6	0.5	0.5	0.4	0.4	0.3

Source: Dergkomstat, 1998 – 2000; Dergkomstat, 2001.

The unfavorable economic and social environment in which the new independent science system is developing is not conducive to innovation or to the creation of organizational links between scientific progress and market reforms. Moreover, Ukrainian science, which developed out of a centralized, state-funded system, was poorly prepared for dealing with market conditions and a democratic society. Consequently, fundamental changes were needed in S&T organization in order to keep pace with those in the economy, education, management and other spheres. These changes have been taking place in Ukraine since its independence, partially under government guidance, although the transformation of the science system largely eludes state control.

The Ukrainian S&T system has changed radically since the dissolution of the Soviet Union in 1991. The transformation has been affected by major changes in the external and internal environments. The most significant of these changes is the *disintegration of the international socialist science system and, with it, the disappearance of the links between scientific centers*. This system was established under the umbrella of the 'Council for Mutual Economic Assistance' (CMEA) and took the form of bilateral and multilateral cooperation between the research institutions of the former socialist countries.

Another factor that affected transformation of Ukrainian science was the *collapse of the Soviet Union's internal system of labor deployment and scientific cooperation*. The consequences of the disintegration of economic relations between the republics of the Soviet Union have proved to be even graver for science than for other social institutions. The breakdown of the Soviet S&T system ruptured long-standing links between research organizations and deprived scientific organizations of very substantial financial support from the government.

The grave situation demanded drastic measures if the Ukraine's science system was to be transformed. The key problems to be addressed were as follows:

– transformation of the STS from a regional to a national system;
– developing a new S&T policy and creating a legislative base for it;
– the establishment of new, democratic decision-making bodies;
– the restructuring of the system of research organizations.

However, the only measures that have been implemented are the establishment of national scientific administration bodies and the introduction of legislation governing their functioning.

2. S&T POLICY IN INDEPENDENT UKRAINE

At the beginning of S&T reform, lengthy policy debates were held at the various levels of management. Before the collapse of the USSR in 1991, Ukraine had no national bodies responsible for S&T management and organization.

The first steps towards the restructuring the S&T system, taken in 1992, were aimed at creating the basic structures needed to implement a new higher education (HE) and science policy and to effect institutional changes within the S&T system. The State Committee on Science and Technology was established and given responsibility for formulating and supervising government S&T policy. Two foundations were established under its auspices: the State Foundation for Basic Research (SFBR, covering the natural sciences) and the State Innovative Foundation (SIF). In the year 2000, SIF became the State Innovative Company of Ukraine.

From 1992 through 2000, the government agencies in charge of S&T policy were gradually integrated into a single system. Uncertainty about S&T goals and the means of achieving them was the most important negative factor affecting S&T transformation. It took several years to prepare the documents in which the governments of the newly independent states of the former USSR formulated their positions on national science. In Ukraine, the document 'Main approaches to S&T policy and to the improvement of the science management system in Ukraine' (Main approaches, 1996) claimed that the general aims of socio-economic development made S&T a high priority for the state and for society as a whole. The activities of the various agencies are regulated by the Law on the Principles of Government Policy in the Sphere of S&T and S&T Activities. The Law on Scientific and Technological Activity was passed in 1998. The draft plan for scientific, technological and innovative development was adopted in 1999.

Two S&T committees were established in Parliament, together with a body for the management of S&T policy attached to the President's office. In May 1996, the State Committee on Science and Technology was reorganized into the Ministry of Science. In the year 2000 the Ministry of Science became the Ministry of Education and Science (cf. Figure 5.1).

The purpose of the State Foundation for Basic Research (SFBR) is to supplement the systems of goal-oriented, competitive funding for grant-based investigations. The main task of the State Innovative Foundation is to provide financial support for innovation in firms seeking to implement modern technologies and know-how and to introduce new and competitive products. These foundations fund goal-oriented research projects through open competition. In the seven years of the Foundation's existence, six calls for S&T projects have been announced and over 2 000 experts working for 300 organizations have

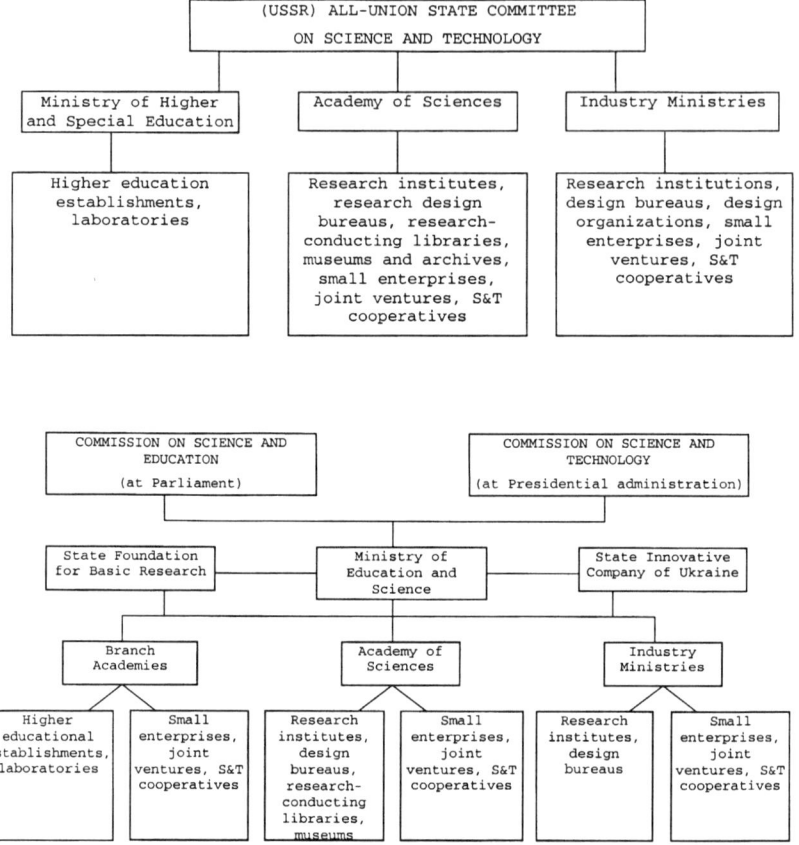

Fig. 5.1 Ukraine: the organization of the S&T system in 1990 and 2000
a) in 1990
b) in 2000

been involved in developing projects. As a result, grants have been awarded to about 10 000 participants from all regions of Ukraine.

The National Academy of Sciences (AoS) has traditionally received more than 60 per cent of the total volume of funding from the SFBR (see Table 5.2).

The projects thus funded have proved to be sufficiently well founded for the overwhelming majority of them to have produced technological solutions that can be put into practice immediately. Thus, while focusing its attention on the grant-supported development of the basic sciences, it is also very important for the Foundation to make prospective users aware of completed projects. In addition, the Foundation is also concerned to promote the further development and practical implementation of projects by state S&T institutions. The areas

Table 5.2 Ukraine: SFBR grants by organization, in %

Organizations subordinated to	1993	1995	1997
National Academy of Sciences	67.3	66.9	64.0
Ukrainian Academy of Agricultural Sciences	2.9	1.7	1.3
Ministry of Education	15.0	20.8	21.6
Ministry of Health	1.6	1.5	1.4
Other ministries	6.6		3.8
Other organizations	6.5	9.0	7.7
Total	100.0	100.0	100.0

Source: Kiyk, 1997.

Table 5.3 Ukraine: SFBR funding and results, 1997

Fields of funding	Share of funding (%)	Projects carried out	Publications	Reports
Informatics, mathematics, mechanics	11.5	170	1 483	1 060
Physics and astronomy	23.2	256	1 378	965
Chemistry	13.2	195	470	403
Scientific fundamentals of advanced technologies	18.6	209	576	336
Biology	15.7	204	131	310
Earth and space science and environmental problems	11.8	240	693	225
Fundamental aspects of the humanities	2.2	34	172	116
Fundamental social-economic problems	2.3	79	237	162
Forecasting of science and education development	1.4	48	135	98

Source: Kiyk, 1997.

funded by the SFBR and the main results are shown in Table 5.3. The Foundation's activity is based on the notion that fundamental research must be made one of the priorities of state S&T policy as it is the country's attitude to the basic sciences that defines its position in the world community.

The State Innovative Foundation supports the following main scientific fields: agricultural technologies, medicine, energy saving, new materials and other applied sciences (see Table 5.4). During the period 1993–1998, most of the projects were in the fields of energy saving and agricultural technologies. However, the largest volume of funds was allocated to projects in the fields of energy saving and new materials.

After the breakdown of the USSR, scientific links between the research institutions collapsed. Ukrainian S&T policy has focused particularly on strengthening cooperative links with the other newly independent states (NIS) of the former USSR. One of the most important steps taken in the reconstruction of scientific relations within the NIS was the establishment of the International Association of Academies of Sciences (IAAS) in 1993. IAAS is an international, non-governmental scientific organization. It includes the AoS of Armenia, Azerbaijan, Belarus, Georgia, Kazakhstan, Kyrgyzstan, Moldova, Russia, Tajikistan, Turkmenistan, Ukraine, and Uzbekistan. The main task of the IAAS is to unite national AoS in the search for solutions to the most important scientific problems. It has supported a wide range of research in such areas as physics, astronomy, mathematics, chemistry, mechanics, informatics, biology, economics, the humanities and sociology. IAAS has held international conferences, been involved in the training of young scientists in the scientific centers of the NIS, technology transfer, etc. The AoS of Ukraine is the leader in many areas of both basic and applied research. Between 1994 and 1997, the number of international agreements concluded with foreign partners increased from 25 to 46, while the number of joint projects involving foreign partners increased and those involving the former Soviet republics decreased (see Table 5.5).

3. THE UKRAINIAN SCIENCE SYSTEM

Like all the other republics of the former USSR, Ukraine inherited the key features of the Soviet STS, namely extreme compartmentalization, funding from a single governmental budget and centralized management and control structures. The prevalence of large organizations, insensitive to any change or adjustment of their institutional forms, and a lack of communication with the world community created additional problems (Kaukonen, 1994; Kavunenko, 1996; Klochko, 1994; Meske and Nadiraschwili, 1994; Nesvetailov, 1990).

3.1. CHANGES IN ORGANIZATIONAL STRUCTURE

Although the number of R&D personnel decreased by more than 50 per cent between 1990 and 2000, the number of S&T organizations increased between 1992 and 2000, for various reasons. In line with the Soviet model, the STS in

Table 5.4 Ukraine: State Innovative Foundation (SIF) projects

a) Structure of the projects carried out

	1993		1994		1995		1996		1997		1998	
	No.	%	No.	%	No.	%	No.	%	No.	%	No.	%
Technologies in agriculture	24	21.4	150	23.1	127	27.3	60	27.9	81	31.4	23	26.4
Medicine	13	11.6	74	11.4	38	8.2	25	11.6	20	7.8	9	10.3
Energy saving	34	30.4	219	33.7	120	25.8	74	34.4	82	31.8	30	34.5
New materials	13	11.6	113	17.4	133	28.5	39	18.1	66	25.6	12	13.8
Other	28	25.0	93	14.3	48	10.3	17	7.9	9	3.5	13	14.9
Total	112	100	649	100	466	100	215	100	258	100	87	100

b) Structure of financing, in %

	1993	1994	1995	1996	1997	1998
Technologies in agriculture	8.1	20.2	15.8	16.8	24.4	17.5
Medicine	34.5	9.2	9.0	10.3	11.7	9.8
Energy-saving	29.1	38.5	31.8	45.4	29.2	29.1
New materials	17.1	16.3	24.6	16.8	32.5	20.4
Other	11.2	15.8	18.9	10.7	2.2	23.2
Total	100.0	100.0	100.0	100.0	100.0	100.0

Source: Minnauki, 1999.

Table 5.5 Ukraine: the AoS and international cooperation

	1994	1995	1996	1997
International agreements	25	37	44	46
Joint projects with foreign partners–total:	473	583	424	495
Joint projects with NIS	95	93	54	70
Grants	131	222	228	243
Training	14	47	159	215
Business trips abroad	3 200	3 120	2 877	2 756
Visits by scientists from abroad	2 484	1 664	1 389	1 297
Participation in international conferences, symposia, workshops, exhibitions	700	577	572	697
Organized international conferences, symposia, workshops, exhibitions	93	77	159	105
Published abroad:				
– monographs	25	63	16	12
– articles	1 039	1 346	1 220	1 630
Memberships in international organizations, associations, committees	248	257	261	264
Carried-out contracts (total)	6 387	5 316	4 512	2 912
– for NIS	641	426	282	208
– for other countries	491	613	464	445

Source: NANU, 1998.

the Ukraine historically comprised three sectors: the AoS, the universities and industries. S&T distribution among the various sectors is shown in Table 5.6.

Table 5.6 Ukraine: S&T potential by science sector, 1999 (%)

Sector	S&T personnel	Finance	S&T organizations
Academy	24	20	20
HE	12	6	12
Industry	52	62	59
Other (in-house R&D)	12	12	8

Source: Dergkomstat, 2000a.

The AoS sector is the leader in many areas of both basic and applied research. In addition, this sector has its own longstanding tradition of training highly qualified researchers with doctorates. Many AoS institutes contain subdivisions of university departments, where graduates can prepare for research degrees. Academy institutes prioritized basic research. The availability of equipment for producing innovations in accordance with the linear model (new idea – basic research – applied research – industrial application) has been one of the principal attributes of AoS institutes in the various engineering fields.

The higher education (HE) sector in Ukraine includes universities and colleges specializing in various areas. The central mission of this sector is the training of highly qualified personnel for all economic sectors. Universities are also active in R&D, although to a lesser extent than the AoS sector.

In terms of the number of personnel involved and of S&T entities, the industrial sector is the major player in R&D. Its main task is to conduct applied R&D and to introduce the results into the production cycle.

From the beginning of 1989, new organizational forms began to be established in Ukraine, namely, S&T cooperatives, engineering centers, small innovative firms, joint high-tech enterprises, cooperatives, innovative entities, associations and consortia. The variety of names given to such organizations reflected the broad spectrum of services they offered. Services varied from the development of software, documentation and methods to technology introduction and product and plant design.

Under the centralized system of funding, the management and control of S&T activities was, in theory, a simple matter but turned out, in practice, to be ineffective. The following factors contributed to the failure to put scientific results into practice:
– the absence of an up-to-date policy on stimulating S&T development;
– the lack of a competitive environment and market for high-tech products;
– a lack of effective mechanisms for transferring research findings to production;
– the monopolistic position of the large enterprises and the lack of organizations with sufficient flexibility to implement scientific findings and inventions.

Financial problems accelerated the process of reform within the STS. The number of S&T organizations increased between 1990 and the year 2000 (see Table 5.7 and 5.8). The so-called branch institutes, which were subordinated to all-union ministries, became separate Ukrainian research institutes. Technological units in the various enterprises changed both their areas of specialization and their organizational forms. New institutional forms were established during this period; one of them is the 'Science and Technology Center of Ukraine' (STCU), established in 1993.

The STCU is an intergovernmental organization dedicated to preventing the proliferation of technologies and expertise that can be applied to the production of weapons of mass destruction, including nuclear, biological and chemical weapons and missile delivery systems. The STCU was founded un-

Table 5.7 Ukraine: distribution of R&D organizations by type (number of units)

	1990	1991	1992	1994	1996	1997	1998	1999	2000
Total	1400	1344	1350	1463	1435	1450	1518	1541	1506
Research institutes	479	502	553	672	682	724	784	805	787
Design bureaus	358	358	341	323	292	275	274	274	245
Pilot production plants	24	17	15	17	17	12	11	11	13
Design and design prospecting organizations	117	88	77	57	58	53	53	53	60
Higher educational establishments	145	146	144	148	153	153	158	163	160
R&D and design divisions in industrial enterprises	138	104	102	126	109	97	97	97	93
Others	139	129	118	120	125	136	141	143	131

Source: Dergkomstat, 2000a.

Table 5.8 Ukraine: distribution of R&D organizations by science sector (number of units)

	1991	1994	1996	1997	1998	1999*
Total	1 344	1 463	1 435	1 450	1 518	1 541
Academy of Sciences	290	268	280	285	294	300
Industry	804	921	893	915	969	981
Higher education	146	148	153	153	158	163
Others (in-house R&D)	104	126	109	97	97	97

* Estimation
Source: Dergkomstat, 2000a.

der the terms of an agreement signed by Canada, Sweden, Ukraine and the USA. Its headquarters are in Kiev, with an international staff of scientific, financial and administrative experts. The STCU develops, approves, finances and monitors S&T projects that engage the former Soviet weapons science community in Ukraine in peaceful activities. Work began soon after Ukrainian independence, when the governments of Canada, Sweden, Ukraine and the USA recognized that Ukraine was in a critical phase of its transition to a market economy. Disarmament and the conversion of industrial-technical potential from military to peaceful endeavors were seen as crucial aims. The STCU governing board held its first meeting in December 1995 and approved the first 13 projects, worth a total of 1.7 million USD. By December 1996, the STCU had approved funding for a total of 85 projects valued at 10.4 million USD. The projects support 1 650 highly qualified scientists and engineers, who are using their skills, originally developed for military purposes, in peaceful, civilian R&D projects. They are interacting with the international scientific community, becoming well versed in international standards for proposal writing and forming partnerships with Western commercial organizations in order to develop projects with practical applications in Ukraine and international markets.

Since independence, HE institutions in Ukraine have experienced considerable changes (although some of the changes were initially merely formal). Five former HE institutes have become universities (Kavunenko, 1996 and 1996a). Some large institutes have been divided into smaller institutes and new institutions have been established. Some technical colleges have become state HE establishments empowered to confer degrees. Private HE organizations have also been set up.

3.2. THE FINANCING OF S&T

Problems related to science budgeting as well as to sources of funding are at the core of the science reform process, since these factors influence both the direction and pace of reform. The underlying principle for distributing funds from all-union structures to the republics was 'institutional' financing. Attempts to reform the procedures and principles of budgeting in the S&T sphere had already been made in the USSR. In the late 1980s, it was proposed that the principles and methods of cost calculation (pricing) should be applied at all stages of the innovation cycle. However, lack of competition and worsening economic performance at the national level caused a decline in demand for S&T products.

Long-term investments proved to be ineffective. Conversion led to a decrease in the funds earmarked for R&D, mostly in natural sciences, mathematics and engineering. Although the progress of market reforms revealed the need for other reforms, the socio-economic crisis restricted the number of possible alternative development policies for S&T. According to official statistics, the science budget in 1991 prices (estimated) declined in every scientific sector. In 1995, the total budget was only 26 per cent of the 1991 budget (see Table 5.9). In 1996, a new Ukrainian currency (Hryvna/UAH) was introduced.

The research institutions had been funded from two main sources: government and enterprises (on the basis of contracts). Their finances were seriously undermined by the collapse of all-union S&T programs and of projects funded by the former socialist countries or through contracts with government ministries and agencies. Defense research started to decline as early as the late 1980s, and the cuts went on until the break-up of the USSR. Attempts were made to reduce the effects of such drastic shrinkage by involving the institutes in conversion programs, but this was insufficient to resolve the problem. In 1992, the funding of all-union programs ceased completely.

Between 1990 and 2000, the most fundamental change in institutes' funding was the decline in the amount of money brought in by contracts with enterprises. Before the crisis, the institutes had received the majority of their funds (about 55 per cent; cf. Table 5.10) from such contracts. In the mid-1990s, this share fell to about 40 per cent. The economic crisis led to a fall-off in external R&D contracts, making it difficult to carry out not only applied research but basic research as well. Expenditure on science has decreased year on year. Expenditure in the year 2000 was only 30 per cent of that in 1991 (in 1991 prices calculated on the basis of a GDP deflator).

As well as the changes in the distribution of funding sources, the distribution of costs has also changed. The greatest share is accounted for by labor costs, which in the year 2000 represented more than 60 per cent of most institutes' costs, and the trend was upwards. The second largest cost category is overheads (rent, water, heat, electricity, etc.), accounting for more then 30 per cent in 2000; again, the trend is upwards. As a result, spending on R&D projects decreased to extremely low levels. The institutes have no funds to ac-

Table 5.9 Ukraine: S&T budgeting by science sector

a) Total volume

	million krb (1991 prices)				million UAH (current prices)		
	1991	1994	1995	% of 1991	1996	1997	1998
Total	7 297.7	2 403.7	1 933.7	26.5	943.6	1 318.6	1 209.6
AoS	1 320.2	549.4	420.6	31.9	226.1	258.4	241.9
HE	483.3	188.9	125.1	25.9	58.2	73.4	58.3
Industry	5 138.9	1 536.7	1 248.2	24.3	591.7	905.3	880.5
Others (in-house R&D)	355.3	128.7	139.8	39.3	67.5	81.5	80.2

b) Structure (%)

	1991	1994	1995	1996	1997	1998
Total	100.0	100.0	100.0	100.0	100.0	100.0
AoS	18.1	22.9	21.8	24.0	19.6	19.2
HE	6.6	7.9	6.5	6.2	5.6	4.6
Industry	70.4	63.9	64.5	62.7	68.7	69.8
Others (in-house R&D)	4.9	5.3	7.2	7.2	6.2	6.4

Source: Dergkomstat, 2000a.

quire equipment and renew facilities. The reduction of funds is manifested in worsening support for R&D (including IT support) and the disruption of communication flows and therefore of relations with users of S&T products.

The involvement of research teams in various competitive programs and projects can be regarded as a positive aspect of the reorganization of research. Such competitions have undoubtedly provided new opportunities for research teams to display initiative and to become involved in new areas of research. Besides active involvement in national and international competitions, institutes and researchers have started to look for S&T programs and potential users of R&D products. Sometimes they initiate new programs and submit them for the consideration of ministries, agencies or enterprises. In conditions of crisis, such efforts often do not reach the desired targets but they are at least evidence of new behavior.

The hidden side of the process is the dramatically increasing number of minor projects being carried out in R&D departments on a contract basis. The expansion and overlap of research topics is unlikely to improve performance or produce important new findings. The need constantly to submit project applications results in researchers' spending an increasing share of their total working time on activities that have nothing directly to do with scientific endeavor.

The need to search for new sources of funding seems to have been heightened by the changes to the budget system and procedures. The principle of institutional fund distribution still predominates within the STS. The only alternative seems to be competitive bidding for research funding. Such a system could serve as a real basis for reform within the science system, leading to the effective prioritization of research activities, better organization of research teams, a more innovation-oriented approach, the closure of unproductive institutes and the cessation of research in irrelevant areas.

3.3. CHANGES IN SCIENTIFIC PERSONNEL

The reduction of funding has led to outflows of scientific personnel. There have been no concerted efforts to reduce staff numbers; rather, people have left of their own volition, typically because of the uncertain future of science and low salaries. The flows are into business and other, more financially attractive areas of activity, as well as to other countries. In Ukraine, the 'brain drain' has meant that the number of specialists involved in R&D has decreased by half since 1990. Approximately 7 to 8 per cent of all researchers have left for abroad. Moreover, more than 30 per cent of scientists still working within their original workgroups are now conducting research for various foreign customers. Within the science infrastructure, tacit unemployment has risen to an incredible level. A career in scientific research has become less and less attractive. According to official statistics, fewer than half of all scholars with

Table 5.10 Ukraine: sources of funds for R&D (in %; total = 100)

	1990	1991	1994	1996	1998	1999
Governmental budget	45.0	11.9	47.2	36.0	35.0	34.0
Non-budget reserves			5.9	8.0	9.0	11.0
Funds of foreign companies		0.3	1.9	6.0	4.5	5.2
Stockholders			1.7	2.0	3.0	3.5
Funds from contractors, total	55.0	80.7	42.5	42.0	40.5	40.0
Contractors located on the territory of:						
– Ukraine	*20*	*32.8*	*36.4*	*37.0*	*36.0*	*37.0*
– other CIS countries	*35*	*47.9*	*6.1*	*5.0*	*4.5*	*3.0*
Other sources		7.1	0.8	6.0	7.5	6.3

Source: Dergkomstat, 2000a.

Table 5.11 Ukraine: S&T personnel by function (1 000 persons)

	1989	1990	1991	1992	1993	1994	1995	1996	1997	1998	1999	2000
Total	594.4	494.2	449.8	380.8	345.8	323.9	293.1	262.5	233.3	215.0	205.0	191.5
Scientists and engineers, of whom:	379.4	313.3	295.0	248.5	222.1	207.4	179.8	162.5	142.5	133.4	124.0	119.0
– Doctors (DoS)		3.2	3.4	3.8	4.0	4.1	4.1	4.1	4.3	4.5	4.1	4.2
– Candidates (CoS)		29.3	27.8	27.4	26.0	24.3	22.9	22.0	20.6	19.9	19.0	18.0
Auxiliary personnel			103.1			76.3		55.7	51.3	45.4	41.0	
HE teaching staff involved in R&D on a part-time basis			36.1			40.0		45.4	46.8	48.8	52.0	

Source: Dergkomstat, 2000a.

scientific and academic degrees and resident in Ukraine are involved in R&D. This internal brain drain was caused to a considerable extent by the demand for highly qualified specialists that arose during the period of transition to a market economy. Such an economy requires specialists in economics, computers and other spheres. This is especially true for newly created private firms involved in consulting, engineering and other services. As a result, many highly qualified specialists able to change their sphere of activity without additional training or retraining courses, are eager to find employment in private business, which is much more lucrative and offers better promotion opportunities. Such tendencies are clearly having a negative effect on the science system. In the absence of any concerted attempt to reverse them, the brain drain is likely to continue.

Since 1992, the Ministry of Statistics has been collecting data on the emigration of people with the degrees of doctor and candidate of science. The figures show that, in 1992, 300 such individuals (57 doctors of science and 243 candidates of science) left the country. The figures for 1994 and 1995 are in excess of 500 and 600 respectively (among them 95 doctors of science). Of the doctors of science, three per cent worked in engineering, 15 per cent in medical sciences and 16 per cent in physics and mathematics. However, the brain drain in Ukraine remains on a smaller scale than in Russia (Kavunenko, 1997). The number of highly promising researchers and the inflow of young people have both declined year on year. In recent years, the average age of research institute staff has begun to rise. The total number of scientists and engineers decreased by 62 per cent, while the number of candidates of science decreased by 39 per cent between 1990 and 2000 (see Tables 5.11, 5.12 and 5.13). These (comparatively low) rates of decrease in the numbers employed in the STS can be explained by that fact that research institutes have introduced enforced unpaid vacations, often lasting several months. This means that a large number of scientific workers are registered in the official statistics but are not actually engaged in scientific activity or being paid. This practice was accepted in order to keep the institutes' research potential intact. In some cases, entire departments and research teams, even whole institutes, are working part-time in order to be able to cover labor costs with reduced budgets. Taken as a whole, these measures constitute a 'science survival strategy' designed to maintain personnel. The downside, however, is the increase in the average age of scientific staff which, combined with the declining social prestige of science, will lead to an overall decline in the volume and quality of teaching and research in the sciences. Soviet science policy, which sought to achieve a constant increase in the number of scientists and in the involvement of young people, is now out of favor.

The present conditions call for a selective approach to staffing in each in-

Table 5.12 Ukraine: scientists and engineers by R&D sector (1 000 persons)

	1991	1994	1996	1997	1998	1999*
Total	295.0	207.4	160.1	142.5	134.4	124.0
AoS	54.6	42.5	36.5	34.9	34.5	34.0
HE	26.1	16.0	13.8	13.0	12.5	13.0
Industry	188.8	129.1	92.7	82.2	75.4	65.0
Others (in-house R&D)	25.5	19.6	17.1	12.4	12.0	12.0

* Estimation.
Source: Dergkomstat, 2000a.

Table 5.13 Ukraine: doctors and candidates of science by R&D sector (persons)

	1991	1994	1996	1997	1998	1999*
Doctors total	3 432	3 995	4 151	4 309	4 510	4 682
– AoS	2 287	2 680	2 917	2 981	3 184	3 320
– HE	118	200	229	312	286	315
– Industry	1 018	1 107	981	1 011	1 033	1 040
– Others (in-house R&D)	9	8	24	5	7	7
Candidates total	27 843	24 277	21 357	20 637	19 829	19 285
– AoS	13 089	12 083	11 029	10 719	10 271	10 050
– HE	3 888	3 094	2 674	2 704	2 629	2 595
– Industry	10 604	8 893	7 434	7 062	6 777	6 500
– Others (in-house R&D)	262	203	220	152	147	140

* Estimation. Source: Dergkomstat, 2000a.

stitute, the exploitation of international contacts and migration in order to improve professional training and a flexible approach to attracting young people into scientific careers.

4. CONCLUSIONS

In the ten years that have elapsed since the S&T reform was launched, the balance of power in research policy decision-making has gradually shifted. Between 1998 and the year 2000, the government adopted a number of proposals on S&T policy and strategy. They included legislation on scientific and technological activity, scientific and technological expertise and innovative activity.

In 1999 and early 2000, important measures were taken to change state scientific and technical policy. In particular, the head of state officially declared Ukraine's transition to the innovation model of economic development. The Supreme Council of Ukraine voted in favor of a new approach to the country's scientific and technological development and to innovation that provides for the state to play a more active role in S&T development and the application of new knowledge in economic and other activities. Amendments were introduced to the new legislation on science and S&T activities. Academic pension were substantially increased and the wages of several categories of academics are to be raised by presidential decree as from 2001.

The new approach to scientific and technological development envisages the following strategic and tactical measures, which are intended to promote S&T and innovation activities:

Funding:
- budget policy to be geared exclusively towards supporting basic innovations in a limited number of priority areas chosen because they are vital to the country's security, its economic and technological position and its competitiveness in domestic and foreign markets;
- functional, goal-oriented mobilization and concentration of financial resources in special funds for innovation development;
- transition to a plurality of sources of finance for innovation through the establishment of non-budget funds at the federal, regional and branch levels;
- formation of a normative legal basis and mechanisms for government stimulation of private and foreign investment in the sphere of innovation.

Organizational structure:
- formation of a new organizational structure for the science system that combines disciplinary and problem-oriented approaches with precise structural delimitation of commercial and non-commercial sectors of scientific potential;
- establishment of a new network of scientific organizations articulated around research centers of various status (national, branch and regional);
- foundation of various types of R&D units in firms;
- establishment of technopolises and science parks (known as technoparks in the Ukraine) with the aim of putting in place structures that, under market conditions, would bring research much closer to the production process;
- development of an innovatory infrastructure in order to speed up technology transfer. This must be seen against the background of a steady quantitative and qualitative decline in domestic experimental science (cf. Table 5.14) and similarly negative tendencies in the development of technical standards and the application of inventions.

Table 5.14 Ukraine: finished experimental developments (in 1 000)

	1991	1994	1996	1997	1998	1999*
Total	82.0	60.4	44.1	42.6	41.8	41.5
New products and processes (technologies)	33.7	24.2	14.9	11.5	10.2	10.0
Developments with incorporated inventions	6.8	3.9	2.3	1.7	1.3	1.3

* Estimation
Source: Dergkomstat, 2000a.

These and other actions are to be taken as Ukraine's scientific system enters the next stage of its transformation, a stage to be characterized by purposeful and managed changes in the science and innovation systems, with an emphasis on qualitative changes.

REFERENCES

Dergkomstat (1998–2000) *Narodne Hospodarstvo Ukraini* (National Economy of Ukraine – Statistical Yearbooks). Kiev: Dergkomstat.
Dergkomstat (2000a) *Nauka Ukraini* (Science of Ukraine – Statistical Yearbook). Kiev: Dergkomstat.
Dergkomstat (2001) *Ukraina v zifrach u 2000* (Ukraine in figures in 2000). Kiev: Technika.
Dobrov, G.M. (1990) *Nauka o nauke* (Science of science). Kiev: Naukova Dumka.
Kaukonen, E. (1994) "Science and Technology in Russia: Collapse or new dynamics?" *Science Studies* 7, no. 2: 23–36.
Kavunenko, L. (1996) "The Problems of S&T relations in Academy – University – Industry during the Transition Period in Ukraine." *Book of Abstracts, International Conference 'Universities and the Global Knowledge Economy: A Triple Helix of University – Industry – Government Relations'. University of Amsterdam, January 3–6. Amsterdam: University of Amsterdam, 59–60.*
Kavunenko, L. (1996a) "The Transformation Processes in Academy Institutes of Ukraine." *The Development of Science and Technological Potential in Ukraine and Abroad. Supplement to Nauka i naukovedenie (Science and Science of Science).* Kiev: STEPS Centre, 21–28.
Kavunenko, L. (1997) "Scientific and Technological Potential of Ukraine during the Transition Period." *People and Technology in the Process of Transition – Vol. 2: The Aftermath of 'Real Existing Socialism' in Eastern Europe.* London: Macmillan Press LTD, 249–262.
Kiyk, B.R. (1997) "Fond fundamentalnich doslidgen yak element dergavnoi pidtrimki doslidgen" (Foundation of fundamental research as element of state support of

science). *Nauka i naukovedenie* (Science and Science of Science) no. 1 – 2: 141 – 148.

Klochko, U.A. (1994) "Utechka spesialistov is nauchnech Organizacij Ukraini" (Brain drain in scientific organizations of Ukraine). *Nauka i naukovedenie* (Science and Science of Science) no. 11 – 12: 173 – 180.

Main approaches (1996) "Main approaches to S&T policy and to the improvement of the science management system in Ukraine." Kiev: Visti Verchovnoi Radi Ukraine *Informatsiini zbirnik* 37: 125 – 152.

Malitsky, B.A. (2000) *Nauchno-technologicheski potencial Ukraini: problemi i perspectivi razvitiya* (Scientific and technological potential of the Ukraine: problems and prospects of its development). Kiev: STEPS Centre.

Malitsky, B.A., and A. Nadiraschwili (1995) "Preobrazovanie nauchnoy sistemi Ukraine v usloviyach radikalnoi transformazii obschestva" (Changing the science system of Ukraine under conditions of the radical transformation of society). *Nauka i Naukovedenie* (Science and Science of Science) no. 3: 1 – 62.

Meske, W., and A. Nadiraschwili (1994) "Umbruch der Wissenschaft in Mittel- und Osteuropa." *Institutionenvergleich und Institutionendynamik – WZB-Jahrbuch 1994*. Ed. W. Zapf, et al. Berlin: Sigma, 349 – 376.

Minnauki (1999) *Zvit Dergavnogo Innovaciynogo fonda* (Report of the State Innovative Foundation). Kiev: Minnauki (Ministry of Science).

NANU (Nacionalnoi Academii Nauk Ukraini) (1998) *Zvit Nacionalnoi Academii Nauk Ukraini* (Report of the National Academy of Sciences of Ukraine). Kiev: Naukova Dumka.

Nesvetailov, G. (1990) "Bolnaya nauka v bolnom obschestve" (Sick science in a sick society). *Sotsiologicheskie issledovaniya* no. 11: 43 – 54.

6. BELARUS: TRANSFORMATION OF THE S&T SYSTEM

Gennady A. NESVETAILOV/Anton A. SLONIMSKI [1]

The Belarussian Socialist Soviet Republic had a population of ten million and, with 4.6 per cent of all research organizations and 3 per cent of all researchers, held third place for S&T among the republics of the USSR. The Academy of Sciences (AoS) of the Belarussian SSR (with 15 per cent of all the republic's researchers) was the center of science, running leading scientific schools and conducting research in optics and spectroscopy, theoretical physics and mathematics. Industrial R&D was characterized by a high share of design offices and organizations engaged in development for the metal-working industries. The last two decades of the USSR were characterized by the development of applied defense R&D, which was financed mainly through contracts from outside the republic.

1. ECONOMIC BACKGROUND

When Belarus became independent in 1991, it was the richest of the twelve republics of the Commonwealth of Independent States (CIS) in terms of per capita income – a status reflecting the republic's steady growth during the 1970s and early 1980s. It had developed an industrial sector that, in terms of its share of total GDP, made it one of the most heavily industrialized countries in the world. The agricultural sector was modernized and came to depend not only on heavy equipment, much of which was manufactured in Belarus, but also on imported fuel. Industrial and agricultural development was based on oil and gas from Russia that was priced at less than ten per cent of world prices in 1990. Enterprises in Belarus also enjoyed access to material inputs for processing and final assembly at internal transfer prices, as well as heavy injections of capital investment and preferential access to the markets of the former Soviet Union (FSU) and Eastern Europe, due to artificially low transport costs and the absence of customs barriers. As a result, Belarus became one of the most trade-reliant nations in the world, with imports and exports each constituting 50 to 60 per cent of GDP (World Bank, 1997). It was also one of

---

[1] The first draft of this chapter was written by G.A. Nesvetailov in 1997/1998. Sadly, he died in 1999 and A.A. Slonimski kindly agreed to update the chapter.

the FSU republics most deeply dependent on intra-regional trade, which was equivalent to 70 per cent of total trade.

Belarus lost its privileged position with the breakup of the FSU. Traditional markets collapsed, and the cost of critical inputs (especially energy) increased sharply in real terms. The terms of trade loss in 1991–92 has been estimated at 11 per cent of GDP. The resulting decline in production and increase in inflation were primary factors contributing to the subsequent economic crisis.

The failure of policies to restore sustainable growth has hurt the people of Belarus. The average poverty rate has risen from about 5 to about 25 per cent (World Bank, 1997). Only about 5 percentage points of this increase in poverty have been the result of higher income inequality. The rest has been the direct result of economic decline.

The potential for restoring growth in Belarus is excellent. The nation has a highly educated population, a disciplined work force, close proximity not only to traditional Russian markets but to the rich European markets as well and a history of heavy reliance on external trade. By Soviet standards, its manufactured products were highly sophisticated.

Specialization within the FSU led Belarus to develop such civil industries as synthetic fibers, agricultural machinery, electronics and electrical engineering, mechanical engineering and oil extraction. Unfortunately, the technological base of production has fallen substantially behind that of OECD countries – some analysts estimate a lag of 15 to 20 years – because of a high level (more than 40 per cent) of wear and tear on production facilities.

Belarus was not a blank sheet in the world of education and science. The education level in the republic – 627 specialists per 10 000 inhabitants received a higher education – was the fourth highest in the FSU after Armenia, Estonia and Georgia (EBRD, 1991, p. 216–223). The Belarussian SSR also ranked first in the FSU in both the number of patent applications and inventions registered per 100 000 inhabitants and per 1 000 researchers (Belgospatent, 1994, p. 6). The capital, Minsk, ranked sixteenth in the world (and fourth in the FSU) in scientific output. The most scientifically advanced fields were physics and mathematics, while engineering sciences enjoyed the largest proportion of resources (Nesvetailov, 1991, p. 81).

Thus the newly independent Belarus inherited several features of a developed country. Its per capita consumption of energy and raw materials was very high and similar to that of Sweden, Iceland and the United Kingdom. The energy expenditure per unit of national product was similar to that of OPEC countries.

According to various estimates, the military industrial complex (MIC) once constituted between 40 and 60 per cent of Belarussian industry (BP, 1995, p. 45). Many large enterprises known for their civilian products also produced defense products, which accounted for between 20 and 80 per cent of their production plan. The number of people directly employed in the MIC or fulfilling its orders was at least 1 million. The regional structure of employment in the USSR was characterized by the extreme concentration of the arms industry, which produced complete assembled weapons on Russian territory. The industry of Belarus differed from that of Russia in that it had no such enterprises. In terms of the number of MIC enterprises per 10 million inhabitants, Belarus (16.9) ranked ahead of the Ukraine (13.7) but far behind Russia (25.8) (Paukert and Richards, 1991, p. 166–169). After the collapse of the FSU, all of the republics except Russia inherited the fragmented parts of the single and indivisible MIC. They were able to produce only separate components of military equipment, however sophisticated they may have been, rather than complete pieces of equipment. Russia was and still is the only country to consume these products and assemble them into modern weapons systems. Belarus' share in the MIC of the FSU (estimated to be 3.3 per cent) was on a par with many other indices of the country's scientific-industrial development. The major branches of the Belarussian MIC were electronics, radio and telecommunications equipment. The republic produced 30 per cent of the FSU's computers, 39 per cent of its precision optical equipment and 23 per cent of its radio equipment (Alesin, 1996, p. 46).

Many of the FSU's defense industries had their own research institutes and design offices on Belarussian territory. A number of scientific organizations run by the Belarussian AoS and in the higher education sector were not part of the defense industry, but they too carried out a large amount of defense R&D. In some years, revenue from state-funded and contract defense work accounted for between 30 and 70 per cent of their total income (Dernovoj, 1994, p. 14). The long-term orientation of research institutes and design offices toward servicing the MIC meant that many of their activities bore little relation to the urgent needs of the Belarussian economy.

2. THE FIRST PHASE OF S&T TRANSFORMATION: 1991–1994

The first phase of the transformation of the S&T system started in Belarus before the dissolution of the FSU, primarily in connection with the conversion of military production. The funds allocated to science were reduced and the number of workers in the S&T system declined. During this period, state control of S&T was weak, with the government failing to adopt a socio-political development strategy. The absence of a coherent plan for economic reform caused Belarus to lag far behind other transitional economies.

The socioeconomic crisis substantially restricted the choice of alternative development paths for S&T and created a situation of uncertainty. On the one hand, nobody stated openly that the republic did not need science. On the other hand, science was no longer considered a priority field of social development. The reduction in science funding as a share of GDP is an indicator of science's lowlier status: from 2.27 per cent of GDP in 1990 it fell to 1.43 per cent in 1991 and to 0.78 per cent in 1993. After a rise to 1.09 per cent of GDP in 1999, funding fell again to 0.82 per cent of GDP in the year 2000, 34 per cent of the 1990 level (Table 6.1). There was no notion of using science to drive the reforms, merely of protecting the S&T system from complete breakdown.

Since the state was weak and there were virtually no market mechanisms yet in place, scientific organizations and even their structural subunits had to deal with their problems on their own if they were to survive. Their differing willingness and ability to adapt to the changing social environment meant that basic science teams varied considerably in the extent to which they were able preserve their potential.

Legislation should have played an important role in the transformation of science. It took the parliament more than a year after the dissolution of the FSU to draft and adopt a series of laws regulating S&T. The most important one was the Law on Fundamental State Science and Technology Policy adopted on January 19, 1993. It provided for the institutional separation of science into three sectors, namely the academy, university and industrial sectors.

The academy sector's share of Belarus' scientific potential at that time amounted to about 14 per cent of all R&D carried out, 30 per cent of scientific organizations, 36 per cent of research equipment and 43 per cent of doctors of sciences involved in R&D. In the higher education sector, state institutions predominated (37 out of a total of 48), most of them operating under the aegis of the Ministry of Education (Nesvetailov, 1995). The association of non-state higher-education institutions included only educational institutions in the humanities and social sciences, in which research activity was not very highly developed.

As of January 1, 1994, the industrial sector was made up of 174 organizations, in which about 70 per cent of all specialists with higher education in the country involved in R&D were employed. Most of the organizations formerly under the control of the USSR central authority were transferred to the Ministry of Industry, which thereafter had the highest S&T potential.

The government took the view that an evolutionary approach to transformation was not only reasonable from a pragmatic standpoint but also well suited to the mentality of the Belarussian people. Reforms in science became part of general state policy. Among the most notable changes in the 1991–1994 period were the conversion of defense R&D, the transfer of research institutes and design offices formerly under the control of the USSR central authorities to Belarus, a sharp reduction in the funds allocated to R&D and the liberalization and intensive expansion of international scientific cooperation. Belarus founded the Basic Research Foundation in 1991, the first former Soviet republic to establish such an institution. This organization, a new one in post-Soviet science, was set up to administer a system of grants for financing small projects on a competitive basis. This was a realistic mechanism for the democratization and decentralization of science. However, these changes were introduced without there being a coherent state strategy for the reform of science. Consequently, many of the reforms in S&T were initiated 'from the bottom up'.

In Belarus there was no one body responsible for managing innovation activity. This inhibited the formation and implementation of state S&T policy. Repeated attempts to obtain parliamentary approval for such an organizational unit failed. In 1993, finally, the government bypassed parliament and adopted a resolution to establish the Committee for Science and Technology, which was answerable to the Council of Ministers of the Republic of Belarus. In September 1994, after Belarus had been transformed from a parliamentary to a presidential republic, the Committee became, as the State Committee for Science and Technology (SCST), a legally independent body within the Ministry of Education and Science.

Thus the first phase of science transformation in Belarus took place without strong state control, and its main influencing factors were the inertia of the accumulated S&T potential and the loss of traditional R&D customers.

3. THE SECOND PHASE OF TRANSFORMATION SINCE 1995

The second stage of science transformation in Belarus involved more fundamental changes, which were related to the establishment of presidential government in 1994. Following this, the influence of the state in all social spheres, including science, started to increase rapidly. Administrative rather than economic management methods were used more and more extensively.

Efforts to establish standards and introduce legislation were intensified. In late 1994, the government issued the 'Resolution on Measures to Preserve and Use Effectively the Science Potential of the Republic'. In accordance with this resolution, various measures strengthening state control over science were adopted in 1995, with policy being based on S&T priorities and the competitive selection of projects (Galinovski and Prokoshin, 1996). State S&T programs were adopted as the main means of implementing this policy.

Since 1996, state S&T programs and individual S&T projects have been set up in accordance with the S&T priorities approved by the government:
- the development of statehood in Belarus;
- health;
- dealing with the consequences of the Chernobyl nuclear accident;
- environmental protection and management of natural resources;
- the production, processing and preservation of agricultural products;
- power and transportation;
- computerization, telecommunications and communications;
- resource conservation technologies;
- new materials and technologies;
- increasing the competitiveness of mechanical engineering and (radio) electronic production.

The development of state S&T programs is a multi-stage process:
- identification of the main socioeconomic and S&T problems in the area of science and technology priorities;
- development of proposals for the implementation of the programs;
- proposals submitted for review to the coordination boards for priority trends at the State Committee for Science and Technology (SCST);
- approval of the list of programs by the Council of Ministers;
- competition for inclusion of R&D projects in the various programs held by the Academic Board of the programs;
- state examination of the S&T programs conducted by the Board of Experts at the SCST;
- approval of the programs by the Council of Ministers.

Organizational, scientific and methodological control over the development and implementation of the programs is exercised by the SCST.

It can be assumed that the state authorities consider the system of S&T programs to be the core of the national innovation system currently under development. In 1996, funding for these programs accounted for 54.3 per cent of the total funds allocated to science from the state budget (Gajsenok, 1997). In 1999, such funding accounted for 40.9 per cent of total science funding (Nedilko and Vojtovich, 2000, p. 17) and for 39.5 per cent in the year 2000 (Lesnikovich and Volotovski, 2001, p. 22). The new order should encourage ministries and enterprises to concentrate their very limited resources on solving the most urgent problems of applied S&T in Belarus. However, practical implementation of the new mechanism has revealed a sharp contradiction between the complex, multi-stage process and the extreme labor intensity of expert activity, on the one hand, and the small amount of funding allocated to each program, on the other. This is hardly surprising if we bear in mind that the 1996 – 2000 list of 48 state S&T programs was supplemented by 39 basic research programs. This proliferation of programs actually deprives the notion of 'priority' of any real significance.

Such a policy is typical of some of the transitional economies. It arises out of a desire to avoid any fundamental restructuring of the S&T potential for fear of the huge social problems that restructuring might bring in its wake.

The tightening of state control over science can be clearly demonstrated by the example of the status of the Academy of Sciences of Belarus. On December 19, 1990, during the period when restructuring and democratization were taking place within the FSU, the Parliament of Belarus adopted the 'Resolution on the Status of the Academy of Sciences of the BSSR'. The Resolution proclaimed the AoS of Belarus to be a self-managed organization of the Republic providing independently for the activities of its institutes in basic research and the training of researchers. This status was confirmed on January 10, 1993 by the Law on General State Science and Technology Policy: "The Academy of Sciences of Belarus is a self-managed scientific organization with appropriate status established by the laws of the Republic of Belarus. Funds for its activity are allocated annually from the budget". A similar wording was preserved in the Law on Scientific Activity prepared by the parliament and signed by the President of the Republic of Belarus on October 21, 1996. However, on May 15, 1997, the President issued the edict which transformed the Academy of Sciences of Belarus into the National Academy of Sciences (NAS) of Belarus with the status of a higher state scientific organization. As before, the Statute of the NAS of Belarus was approved by the Council of Ministers. However, it contained a fundamentally new provision stipulating that the President of the NAS of Belarus, after election by the General Meeting of the Academy, was to be confirmed in office by the President of the Republic of Belarus, have the status of minister and be a member of the Council of Ministers.

The increased state control over the development of S&T was also reflected in the material support given to researchers. The Edict of the President of the Republic of Belarus of September 27, 1996 provided for additional remuneration for those holding certain scientific titles and degrees. Since January 1, 1997, doctors and candidates of sciences have been paid a bonus of six and three times the minimum salary respectively, provided that they work in organizations and institutions financed out of the state budget. The stipend for postgraduate students was raised to the level of a junior researcher's salary and that for doctoral candidates to the level of a senior researcher's salary.

The Edict of the President of the Republic of Belarus of September 29, 1996, which granted the Belarussian State University (BSU) the status of top-ranking higher education institution in the national system of education, was even more radical. The salaries of professors and researchers at the BSU were increased on average by a factor of 1.7 (Galinovski and Prokoshin, 1997, p. 50). The ratio of full-time students to teachers at the BSU was decreased to 6:1 on August 1, 1997 (as opposed to 10:1 overall in higher education institutions in Belarus).

The Law on Scientific Activity provided for staged increases in the average monthly salaries of researchers and teachers in higher education such that by January 1, 1998 they were to reach a level of at least 1.5 times the average monthly wage in industry (Galinovski and Prokoshin, 1997, p. 20). The same law stipulated that state funding of scientific activity be a protected item of current budgetary appropriations, that is not subject to cuts during budget revisions.

The 1995 – 1997 period was characterized by the interaction of two factors: the state tightened its grip on the development of science and research institutes and design offices continued to adapt to new social conditions. Eventually, in the second phase of transformation, the drastic cutbacks in science resources, with their shock effects, gave way to more moderate reductions and even to growth in some indicators. In particular, after 1993, science funding as a share of GDP increased slightly and the share of S&T in total employment stabilized at 1.0 per cent (cf. Table 6.1). The number of workers employed in R&D fell by 36 per cent in 1992, by 9.6 per cent in 1995 and by only 3.2 per cent in 1996. Due to the Russian crisis, it continued to decline until 1999 and it was not until the year 2000 that the number began to rise (see Table 6.2).

As the economy plunged into crisis, salaries became a severe problem, with average monthly pay in R&D institutions lagging far behind the national average for industry (81 per cent in 1994, for instance). Hardly surprisingly, there was an exodus of R&D personnel and the share of young people began to fall. For two years, in 1995 and 1996, there was a tendency towards wage parity: in 1996 the average salary in science equaled that in industry (Goncharenko, 1997, p. 154), while in the year 2000 it grew slightly (1 per cent higher) (Lesnikovich and Volotovski, 2001, p. 44). During the 1990s, the number of researchers rose only in the humanities and in medical sciences, while the greatest reduction was in engineering sciences (Table 6.3). The number of younger researchers in the various areas of science differs considerably (Table 6.4). The share of postgraduate students in engineering has declined from 36.8 per cent in 1990 to 21.6 per cent in 2000, while the percentage of postgraduates in the humanities has risen from 18.7 to 24.7 per cent.

Some of the science manpower indicators were affected by general changes in the labor market. For example, the threat of unemployment among the young resulted in a constant increase in the number of full-time graduate students until 2000. Many of them continued to study as graduate students, though they did not intend to devote their whole lives to science.

Table 6.1 Belarus: changes in S&T resources, 1990–2000

Indicator	1990	1991	1992	1993	1994	1995	1996	1997	1998	1999	2000
GDP (%)	100	99	89	83	73	65	67	75	81	84	89
S&T allocations											
– as % of GDP	2.27	1.43	0.82	0.78	0.80	0.89	0.88	0.85	0.82	1.09	0.82
– as % of 1990 level	100	59	32	36	29	28	27	29	30	38	34
S&T personnel as % of total labor force	2.0	2.0	1.6	1.2	1.1	1.0	1.1	1.0	1.0	1.0	1.0

Source: MSA, 2001, pp. 51, 106; Gajsenok, 1997, p. 50; Lesnikovich and Volotovski , 2001, p. 8; Belorusskii Ekonomiceskii Shurnal, 2001, p. 147.

Table 6.2 Belarus: manpower in S&T (persons)

	1990	1991	1992	1993	1994	1995	1996	1997	1998	1999	2000
Total number of workers in S&T of which:	107 296	90 999	58 278	51 181	43 472	39 300	38 030	33 200	32 477	31 791	32 926
Scientists & engineers	59 400	50 463	33 685	30 474	26 141	23 771	23 324	19 598	19 153	18 817	19 707
– having doctoral degree	542	593	634	684	685	712	744	728	747	771	819
– having candidate degree	5 896	5 814	5 101	4 848	4 579	4 405	4 313	4 115	4 017	3 892	3 856
Service personnel	26 078	18 710	13 789	12 886	10 099	8 956	8 816	6 843	6 670	6 713	6 751

Source: Belorusskii Ekonomiceskii Shurnal, 2001, p. 145.

Table 6.3 Belarus: distribution of researchers by scientific field

Scientific Field	1993*		1995*		1997		2000	
	number	%	number	%	number	%	number	%
Natural	8 482	21.7	7 235	23.6	4 874	24.8	4 554	23.0
Medical	1 744	4.5	1 748	5.7	1 145	5.9	1 244	6.3
Agricultural	2 010	5.2	1 645	5.4	1 282	6.6	1 126	5.7
Engineering	20 401	52.4	14 566	47.8	10 285	52.5	10 730	54.6
Humanities	2 873	7.4	2 757	9.0	499	2.5	621	3.1
Social	2 695	6.9	2 221	7.3	1 288	6.6	1 194	6.1
Others	751	1.9	367	1.2	225	1.1	238	1.2
Total	38 956	100.0	30 539	100.0	19 598	100.0	19 707	100.0

* In 1993 and 1995 higher education institutions instructors carrying out S&T were taken into account.
Natural: physico-mathematical sciences, chemical sciences, biological sciences, geological and mineralogical sciences;
Medical: medical sciences, pharmaceutical sciences;
Agricultural: agricultural sciences, veterinary sciences;
Humanities: architectural sciences, art criticism, philology, historical sciences, pedagogical sciences;
Social: philosophy, economics, sociology, political science, juridical sciences.
Source: Belorusskii Ekonomiceskii Shurnal, 2001, p. 146.

Research funding, personnel structure and student admissions in higher education also stabilized. In 1996, the number of applicants increased sharply in almost all specialties, including paid training. Graduates began to be hit hard by unemployment. Many of them had to apply to the employment services and find work in areas for which higher education was not a requirement. However, it was realized that cutting graduate numbers was strategically unwise (Vetokhin et al., 1997) and 2.8 per cent of the entire population is now in higher education (in EU member states, this indicator is as high as 3 per cent).

4. QUALITATIVE CHANGES IN THE INNOVATION SYSTEM

A set of measures designed to create a national innovation system was outlined in the 'Program for the Development of Scientific Innovation Activity in the Republic of Belarus' drafted by the State Committee for Science and Technology (SCST) and approved by the government on February 26, 1996 and amended on July 10, 1997 (Galinovski and Prokoshin, 1996 and 1997). In accordance with this program, a review of 207 research institutes and design offices operating under the aegis of no fewer than 29 ministries was conducted in 1996. The first stage of the review was based on self-assessment by the organizations and the second on an analysis by the ministries of the results of this self-assessment and on proposals for reforming the network of research institutes and design offices.

Analysis of the questionnaires filled in by the research institutes shows that individual organizations have developed their own survival tactics. As a result, efforts to restructure and transform the science and technology potential have not always produced the desired results. In particular, the share of the science and innovation system directly targeted at production decreased from 53.3 per cent in 1990 to 41.3 per cent in 1996, as measured by the number of specialists employed by the organizations surveyed. For the National Academy of Sciences of Belarus, this indicator increased from 15.3 to 21.1 per cent. Such structural shifts might have been reasonable if priority had been given to basic research within a linear model of innovation. In a severe economic crisis, however, the available technical base can be updated without the need for intensive input from basic science. Between 1990 and 1996, the most significant change in the R&D institutes' budgets was the decline in revenue from contracts with enterprises.

The percentage of research in the entire R&D sector increased from 34.6 per cent in 1990 to 46.9 per cent in 1996 and remained at a similar level until the year 2000 (Table 6.5). This trend was in keeping with the traditions of Soviet S&T, in which the final stages of the innovation cycle were weak. More-

Table 6.4 Belarus: number of post-graduate students by scientific field

Scientific Field	1990 persons	1990 %	1995 persons	1995 %	1997 persons	1997 %	1999 persons	1999 %	2000 persons	2000 %	% to 1990
Natural	579	19.4	526	17.0	628	15.4	733	15.3	724	13.9	125
Medical	113	3.8	121	3.9	182	4.5	218	4.5	254	4.9	225
Agricultural	147	4.9	144	4.7	181	4.4	241	5.0	248	4.8	169
Engineering	1 097	36.8	842	27.3	1 060	26.0	1 127	23.5	1 126	21.6	103
Humanities	558	18.7	825	26.8	481	24.1	1 156	24.1	1 287	24.7	231
Social	434	14.5	403	13.1	738	18.1	1 034	21.5	1 225	23.6	282
Others	56	1.9	221	7.2	305	7.5	290	6.1	339	6.5	605
Total	2 984	100	3 082	100	4 075	100	4 799	100	5 203	100	174

Source: Belorusskii Ekonomiceskii Shurnal, 2001, p. 149; MSA, 2000, p. 176–177.

over, the transition period brought with it additional difficulties, as producers were unable to spend large amounts on pilot projects or on the production and testing of prototypes. This hindered the advance of science-intensive production for internal and external markets.

Table 6.5 Belarus: technological structure of S&T activity (% of expenditure)

Indicator	1990	1994	1996	1998	1999	2000
S&T tasks performed	100.0	100.0	100.0	100.0	100.0	100.0
of which:						
+ research	34.6	38.7	46.9	45.8	38.6	42.8
− basic	8.1	11.4	14.5	18.0	19.0	18.9
− applied	26.5	27.3	32.4	27.8	19.6	23.9
+ development	65.4	61.3	53.1	54.2	61.4	57.2

Source: Belorusskii Ekonomiceskii Shurnal, 2001, p. 147.

Statistics for 1999 suggest there have been some positive changes in the technological structure of S&T work, with a decrease in the percentage of design work and an increase in activity related to the manufacturing and testing of prototypes. It is to be hoped that these trends reflect an improvement in producers' finances and an increased interest in innovation. In any event, Belarus still faces an enormous task in restructuring S&T work in favor of the final stages of the innovation cycle in order to re-orient S&T activity towards the needs of the market.

It is well known that large scientific organizations were the linchpins of the Soviet model of science. This resulted in 'gigantomania' and the overwhelming bureaucratization of scientific life, just as it hindered the generation of new ideas and alternative approaches. The splitting of large scientific organizations into smaller units was an important element in the transformation of the Belarussian S&T system. The AoS, for example, had seven institutes with 500 or more employees; by 1996, only two such large institutes remained in the NAS, a further six institutes employed over 300 people, 17 between 100 and 200 and 15 institutes fewer than 100.

Under the terms of the Law on Scientific Activity, research can be carried out by ad hoc research teams as well as by scientific organizations. This reflects the increasing democratization and decentralization that characterizes post-dependence Belarus and has brought R&D management there closer to the organizational models that prevail throughout the world.

The analysis of the activities of research institutes and design offices included a self-assessment of the level of the R&D that they conducted. This analysis revealed that the market ideology that had taken hold in Belarus had

had no effect on organizations' ability realistically to assess their own output. As before, these self-assessments were exaggerated and of little value to science policy-makers. For example, according to the questionnaires filled out by research institutes and design offices, 4 per cent of their completed work was at a level higher than the world standard (19 per cent in the NAS) and 52 per cent was equal to the best world developments. Overall, as few as 25 per cent of the research institutes and design offices put their output in the intermediate category, while 65 per cent declared themselves to be operating at a high level. These assessments were confirmed by their supervising ministries.

Patent statistics show that patent activity in Belarus declined at the beginning of the 1990s and has increased again since 1996 (Table 6.6). The number of patent applications from foreign applicants fell drastically, from 1 005 in 1994 to 361 in 1996 and 204 in 2000. Some experts believe that this decline can be ascribed mainly to the collapse of the innovation system, since foreign competitors see that Belarussian scientists and producers have lost their positions in some currently important fields and therefore see little need to protect their designs in Belarus territory (Nedilko, 1996, p. 27).

The nature of the relationship between the academy and higher education sectors is an issue that remains unresolved. Little competition has developed between them, but equally there are few signs of any real integration. As before, the two sectors are operating largely independently of each other. For example, after the national basic research programs had been approved, the Ministry of Education put in place a parallel system of ministerial programs. And after the science development plan for the whole republic had been approved, the Ministry of Education decided to develop its own particular proposals for science development.

Higher education is ahead of the state research institutions in accommodating the private provision of services. In view of the high level of demand for educational services, non-state universities have been founded by various bodies. However, activity in this area seems to have tapered off. While 7 licenses were issued in 1993 and 1994 and 5 in 1995, the number of licenses was down to one in 1996 (Vetokhin et al., 1997). In that year, non-state universities employed no more than 4 per cent of full-time teaching staff in higher education.

Changes in social requirements and values are reflected in the structure of applications from prospective students. For example, in 1996, the highest number of applicants was in physical education and sport (eight applicants per place), followed by psychology (six), law (5.3) and economics and production management (5.6). The smallest number of applicants per place was in physics and radio physics (1.3) (Vetokhin et al., 1997, p. 53). Such an applicant structure certainly has a negative effect on the development of scientific innovation activity in Belarus and on the development and dissemination of advanced technologies.

Small scientific innovation enterprises (SSIEs) have helped to some extent to keep research personnel in the S&T system. These firms can be defined mainly as economic entities that develop and implement new advanced technologies. Such entities exist in almost all scientific organizations, in the form of cooperatives, small enterprises, joint stock and limited companies and joint ventures. Unfortunately, the legal base now being put in place continues to take little account of the peculiarities of innovation activity.

Privatization in the S&T sphere has not yet become an important part of state policy, and is unlikely to do so. For example, as of the end of the first quarter of 1997, as few as 14 organizations involved in scientific and technological activities had been privatized (Gajsenok, 1997, p. 33). Innovative small firms are also experiencing great difficulties. For instance, about one hundred firms taking various legal forms were set up within the AoS between 1990 and 1992, but by 1995 only 53 of them were still functioning (ibid., p. 34). In the republic as a whole, the share of SSIEs in all small enterprises does not exceed 5 per cent and in recent years has declined to less than 2 per cent. In 1992, they accounted for 8.3 per cent of total employment in small enterprises; by 1996, this share had fallen to 2 per cent and declined further to just 1.1 per cent in the year 2000 (Table 6.7). Thus conditions favorable to the development of small, innovative businesses do not yet exist in the country.

This is illustrated particularly clearly by the fate of efforts to establish science parks in Belarus, where they are known as technoparks. In January 1994, for example, the Minsk Scientific-Technological Park was founded as a closed joint-stock company. Its organization was based on experience gained in Russia and Germany, and in particular on the technological centers that had existed in the former GDR. The principal partner in Germany was the Warnemünde Technopark near Rostock (MON/MMI, 1995, p. 166). In the first year, four small enterprises were established and ten enterprises were incorporated into the park on the basis of cooperation agreements. They had a total of 100 employees. A similar science park was created in Mogilev.

However, recent years have shown that the absence in Belarus of an environment conducive to innovation has hindered the development of these enterprises. Organizational structures may have been borrowed, but the other elements required for a fully functioning innovation system are not yet in place. This is hardly surprising during a period of economic crisis, when struggling producers are unable to support new activities and the dissemination of advanced technologies. The state has to be the main source of funds and support for scientific innovation, primarily through budget allocations for scientific organizations.

Table 6.6 Belarus: patent activity indicators

Indicator	1993	1994	1995	1996	1997	1998	1999	2000
Received applications for inventions	1 494	1 688	1 039	1 059	1 162	1 209	1 189	1 198
of which:								
– from national applicants	828	683	624	698	752	910	993	994
– from foreign applicants	666	1 005	415	361	410	299	196	204
Patents for inventions and industrial designs	693	522	434	605	633	698	591	679

Source: Belorusskii Ekonomiceskii Shurnal, 2001, p. 148, BellSA, 2001, p. 56–58.

Table 6.7 Belarus: small scientific innovation enterprises

Indicator	1992	1993	1994	1995	1996	1997	1998	1999	2000
Number of small enterprises (SE)	11 055	11 428	14 165	14 813	20 077	22 754	24 061	26 787	25 706
of them with S&T activity	843	560	600	577	834	601	537	503	412
as % of total number of SE	7.6	4.9	4.2	3.4	4.2	2.8	2.2	1.9	1.6
Staff in innovative SE	21 368	7 544	6 944	6 700	3 400	6 200	6 600	5 680	3 160
as % of total number of employees in SE	8.3	3.7	3.0	3.8	2.0	2.7	2.4	1.7	1.1

Source: Lesnikovich and Volotovski, 2001, p. 97.

Between 1991 and 1994, virtually no new rules governing innovation were drawn up and there was felt to be an acute shortage of written documents regulating scientific and technological development. In the second transformation period (1995–1997), many of the elements of a national innovation policy were put in place and numerous methodological documents and standards were adopted. This brought a certain degree of order to S&T activities and extended the role of the state in the planning of R&D and the dissemination of research results. However, it is evident that the tight state budget and meager commercial support have led to the abandonment of many initiatives and a failure to implement much of the legislation. In particular, the creation of the Innovation Foundation has been delayed for many years, the 'Draft Plan for the Development of Science in the Republic of Belarus' approved by the Government in 1996 remains ineffective and an attempt to introduce regulations providing researchers with strong social guarantees has failed. These examples confirm that administrative efforts alone, without economic support from enterprises, are not enough to put in place a workable innovation strategy. These trends have continued up to the present. In 2000, an initial survey on innovation activities was conducted among enterprises with more than 100 employees in major industries (Lesnikovich and Volotovski, 2001, pp. 29–31). Of the 200 enterprises that were contacted, 176 responded and of those 79, or 42.4 per cent, had produced any innovations at all in the previous three years. Despite this, now fewer than 117 of them had their own R&D or project planning and technology departments. New or technologically modified products accounted for 9.7 per cent of total output. The bulk of these innovations were the product either of developments in the enterprises themselves or of cooperation with other domestic companies and institutes (74 per cent). The only noteworthy foreign cooperation that still existed was with Russia. The long-established practice of making scientific research findings freely available, rather than charging for them, is still the main means of technology transfer within the country. Enterprises generally lack the funds or the state support required for high-risk innovations, and in any case, as already noted, conditions conducive to innovation (international exchange of information, a technology market, cooperative networks, etc.) do not yet exist to any great extent in Belarus.

International S&T cooperation is characterized by the intensification of contacts in traditional regions and centers and the development of links outside these established areas, as well as by the use of new forms of cooperation, such as joint R&D projects, contracts, participation in international S&T events (seminars, conferences, exhibitions), S&T information exchange, exchange of scientists and specialists and technical assistance.

To date, the Republic of Belarus has concluded more than 30 bilateral agreements on cooperation with Russia, with other CIS countries, Germany and Middle East countries.

Belarussian scientists traditionally participate in the annual Hannover Trade Fair, which helps them to acquire experience of international fairs and to improve Belarus' performance in the scientific and economic spheres. The establishment of contacts with Western (primarily German) specialists is one of the most significant results of these activities. Belarus took part in EXPO 2000, also held in Hannover, Germany.

For the last 10 years, Belarus has cooperated with the UNO Development Program, which provides support for work on the scientific problems currently of greatest interest to Belarus, such as the commercialization of S&T results and the organization of innovation activity. Since 2000, UNIDO funds, 200 000 USD in all, have been used to set up an international network of technology transfer centers. Between 1996 and 2000, Belarussian scientists carried out projects supported by UNESCO, INCO-Copernikus, TEMPUS, TRANSFORM, CRDF, etc.

5. SITUATION AND PROSPECTS IN 2001

Perhaps paradoxically, the integrity of the Belarussian S&T system has been maintained rather than lost during the period of transformation that started in 1991. This outcome has been facilitated by the reassertion of state control over S&T, the establishment of a legal basis for scientific innovation activity and targeted support for research institutes and design offices that used to answer to the central authority in Moscow.

In fact, however, throughout the period of transformation, the S&T system in Belarus has been in 'survival mode', suffering as it has been from an acute shortage of demand for the results of scientific innovation activity as well as from the effects of budget deficits and the lack of priority attached to funding for S&T. The state did no more than establish the preconditions for radical reforms in science and technology, while the main transformation processes took place in primary research teams and individual organizations.

In the first phase of the transformation process, the preservation of personnel was the most urgent problem; in the second phase, however, the most pressing issue was the updating of the science base, in both material and technical terms. Initially, the S&T system in the post-Soviet period was poorly equipped by world standards. During the period of transformation, expenditure on research equipment and instruments was reduced to a minimum. For example, in 1995 at the Belarussian State Polytechnic Academy (the leading polytechnic college), expenditure on equipment and instruments accounted for as little as 0.5–0.7 per cent of total expenditure. In 1996, capital production assets per employee in agricultural science were half the level of those in agricultural production; in 1990, these same indicators had been equal (Sumonow, 1997, p. 22).

By 1995, the resources that could be drawn on in order to ensure the survival of the S&T system were almost exhausted. As a result of the protracted socioeconomic crisis, the content of Belarussian R&D was now outdated, its material and technical base impoverished and its human resources were seriously depleted. The country faced a choice: either it could support science as a resource of strategic importance for the country's economic development or it could allow its S&T and innovation systems to go to rack and ruin (Gajsenok, 1996, p. 42).

The system of S&T and innovation management that has been put in place in Belarus since 1995 forms a basis for dealing with the problem of activating state science and technology policy for the coming years. Its main thrust should be the formation of an innovative type of economy, one in which production is reordered both structurally and technologically. In this connection, the SCST considers that the state should not only regulate but also directly manage innovation processes (Gajsenok, 1996, p. 43). There are several reasons for this, the first being the fact that most of the country's research base and its experimental and production base as well are owned by the state. In an unfavorable economic environment, the profit-oriented commercial sector is unable to make a major contribution to innovation processes. When new technological systems are being developed, the state rather than the private sector always plays the leading role. Since GDP had been increasing slowly since 1995 and the share of S&T expenditures within the GDP had stuck at about 0.8–0.9 per cent, science funding began to increase in the second half of the 1990s (cf. Tab. 6.1). However, the scale of the increase has been insufficient to complete the necessary restructuring of the S&T system. The share of personnel costs in total expenditure increased from 53.4 per cent in 1998 to 55.6 per cent in the year 2000, whereas the share of investment in equipment decreased from 7.3 to 4.1 per cent (Lesnikovich and Volotovski, 2001, p. 42).

The transition from 'survival mode' to major reform may also have to include very painful processes such as the restructuring of the existing network of scientific organizations. The priority problem for the near future is how to maintain the level of state S&T funding and increase revenue from extra-budgetary sources. In the 1990s, the state had an approximately 50 per cent share in R&D funding.

Social policy in the S&T field has hitherto been 'soft', with as many workers as possible being retained. As a result, expenditure per employee was far below the critical level that allowed normal professional activity to be conducted. Were this trend to continue for a long time, the 'survival' tactics themselves would become pointless because the scientific teams would go into irreversible decline. The nation needs a competitive scientific innovation system.

Moreover, numerous alternative and independent sources of funding, to be acquired by attracting joint stock and private companies, banks, insurance and investment companies, should become a very important element of the national innovation system. In order to minimize the innovation risk, it is necessary to introduce and expand the practice of providing tax, credit, depreciation and other incentives. Among the means of increasing solvent demand for science-intensive production through the implementation of state policy, the following are of relevance for the near future:
- creation of an extensive consumer protection system;
- introduction of harsh sanctions for outdated and low-quality production;
- state orders for high-tech products;
- establishment of organizations that can provide technological expertise;
- prohibition on imports of low-quality or outdated products and technologies.

All these measures are indicators of the general situation in Belarus, which is characterized by the absence of a functioning market economy and innovative enterprises.

The 'Program of Social and Economic Development of the Republic of Belarus for the Period 2001 – 2005' was adopted in 2001. One of its aims is to double the share of GDP devoted to R&D. In October 2000, the President of the Republic of Belarus signed the decree 'On the improvement of science and the reforming of the National Academy of Sciences'. Under the terms of this decree, appropriate measures are currently being prepared with the aim of developing science and increasing its effectiveness and its role in the country's social and economic progress.

REFERENCES

Alesin, A. (1996) "Konversija cherez torgovliu oruzhijem?" (Conversion through weapon trade?) *Belorusskij rynok* (Belarussian Market) no. 4: 46–47.
Belgospatent (1994) *Godovoj Otchet 1993* (Annual Report 1993). Minsk: Gosudarstvennoje Patentnoje Vedomstvo – Belgospatent (State Patent Office of the Republic of Belarus).
BelISA (2001) *Nauka Republiki Belarus. Kratkii statisticheskii sbornik* (Science in the Republic of Belarus. Short statistical collection). Minsk: Belorusskij Institut Sistemnogo Analiza i Informacionnogo Obespechenija v Nauchno-technicheskoj Sfere (BelISA) (Belarussian Institute of System Analysis and S&T Information).
Belorusskii Ekonomièeskii Shurnal (2001) "Statistièeskije Materialy" (Statistical data). *Belorusskii Ekonomièeskii Shurnal* (Belarussian Economic Journal) no.3.
BP(1995) *Belarus turning to people – National Human Development Report 1995*. Minsk: Belarus Publishers (BP).

Dernovoj, V. (1994) "Perekowka mechej na orala: kak ona proishodit v Belarusi" (Beating swords into ploughshares: how it's going on in Belarus). *Nacional'naja Ekonomicheskaja Gazeta* (National Economic Gazette) no 5: 14.
EBRD (1991) *A Study of the Soviet Economy – Vol. 1.* Paris: European Bank for Reconstruction and Development (EBRD)/OECD.
Gajsenok, V.A. (Ed.) (1997) *Razvitije Nauki Belarusi v 1996 godu* (Science development in Belarus in 1996 – Analytical report). Minsk: Gosudarstvennyj Komitet po Nauke i Technologijam Respubliki Belarus (State Committee for Science and Technology of the Republic of Belarus – SCST).
Galinovski, O.I., and V.I. Prokoshin (Eds.) (1996) *Organizacija nauchno-tekhnichskoj dejatelnosti v Respublike Belarus – Sbornik normativno pravovych aktov* (Organization of S&T activities in the Republic of Belarus – Collection of legislative acts). Minsk: SCST/Belorusskij Institut Informacii i Prognozirovanija (Belarussian Institute of Information and Forecast).
Galinovski, O.I., and V.I. Prokoshin (Eds.) (1997) *Organizacija nauchno-tekhnichskoj dejatelnosti v Respublike Belarus – Sbornik normativno pravovych aktov, vtoroj vypusk* (Organization of S&T activities in the Republic of Belarus – Collection of legislative acts, second edition). Minsk: SCST/BelISA.
Goncharenko, A.M. (Ed.) (1997) *Kratkij ochet o dejatelnosti Akademii Nauk Belarusi 1992–1996* (Short report on the activity of the Academy of Sciences of Belarus in 1992–1996). Minsk: *Akademia Nauk Belarusi* (Academy of Sciences of Belarus).
Lesnikovich, A.I., and I.D. Volotowski (Eds.) (2001) *Razvitije nauki Belarusi v 2000 godu* (Science development of Belarus in 2000 – Analytical Report). Minsk: SCST.
MON/MMI (1995) *Innovacionnyje Centry Belarusi: Podhody, sostojanije, perspektivy* (Innovation centers of Belarus: general approaches, current situation, prospects of development). Proceedings of the International Conference in Mogilev, June 5–6. Mogilev: Ministerstvo Obrazovanija i Nauki Respubliki Belarus (MON) (Ministry of Education and Science of the Republic of Belarus)/ Mogilevskij masinostroitelnyj institut (MMI) (Mogilev Institute of Mechanical Engineering).
MSA (2000) *Statistical Yearbook of the Republic of Belarus 1999.* Minsk: Ministry of Statistics and Analysis of the Republic of Belarus (MSA).
MSA (2001) *Statistical Yearbook of the Republic of Belarus 2000.* Minsk: MSA.
Nedilko, V.I. (1996) *Sostojanie i problemi razvitia nauki Respubliki Belarus v 1995 godu – Analiticheskii doklad* (Current state and the problems of science development in the Republic of Belarus in 1995 – Analytical Report). Minsk: SCST/BelISA.
Nedilko, V.I., and A.P. Vojtovich (Eds.) (2000) *Razvitije Nauki Belarusi v 1999 godu – Analiticheskii doklad* (Science Development in Belarus in 1999 – Analytical report). Minsk: SCST.
Nesvetailov, G.A. (Ed.) (1991) *Nauchnyj Potencial Respubliki* (Scientific Potential of the Republic). Minsk: Nauka i Tekhnika.
Nesvetailov, G.A. (1995) "Transformacija nauki v Belarusi" (Science Transformation in Belarus). *Problemi prognozirovania* (Problems of Forecasting) no 3: 114–125.

Paukert, L., and P. Richards (Eds.) (1991) *Defence expenditure, industrial conversion and local employment.* Geneva: International Labour Office.

Sumonow, M. (1997) "Skupoj platit dwazhdy" (Miser pays twice). *Belorusskij Rynok* (Belarussian Market) no. 41: 22.

Vetokhin, S.S., S.W. Tsedrik, and A.V. Makarov (1997) *Sostojanie i perspektivi razvitija vysshei shkoli Respubliki Belarus 1995–1996* (Current state and the prospects for the development of higher education in Belarus 1995–1996). Minsk: Respublikanskij Institut Vysshej Shkoly Respubliki Belarus (Republican Institute of the Higher Education of the Republic of Belarus).

World Bank (1997) "Executive Summary." *Belarus: Prices, Markets, and Enterprise Reform.* A World Bank Country Study. Washington, DC: The World Bank, xiii–xiv.

7. ESTONIA: TRANSFORMATION OF THE R&D SYSTEM

Helle MARTINSON

The process of establishing political and economic independence and a market economy in Estonia has been under way for nearly fifteen years. Radical reforms started in Estonia in 1987–88, when the idea of economic autonomy was first debated and developed into proposals for economic and social reforms. During the period of political liberation and democratization, the raising of the Iron Curtain and a rapid increase in international cooperation, Estonia's fundamental political need was to preserve and develop the state and the nation as a part of democratic Europe. In 1991, the goal of political independence was attained and Estonia re-established its sovereign statehood.

The transformation from a centrally planned to a market economy caused serious economic decline and a structural crisis in Estonia at the beginning of 1990s. The radical changes in the economic environment that created the necessary preconditions for economic stabilization and revival took place mainly between 1992 and 1994. The introduction of a national currency (EEK – Estonian kroons) and escape from the ruble zone were the most significant factors in curbing inflation and creating the basis for gross domestic product (GDP) and production growth.

The reform of R&D in Estonia was launched from the 'bottom up' as the result of an initiative started by the Union of Scientists in 1989. In the course of devising a new S&T policy and restructuring the R&D system, a complicated set of problems had to be solved. Estonia inherited quite a large and capable pool of intellectual expertise from the former USSR as well as a broad network of research institutions. However, this system had been structured in such a way as to be an integral part of the Soviet R&D establishment, in which the center determined the thrust of R&D policy to meet the needs of a large, homogeneous country. Such a format was inappropriate for a small independent republic. This was one of the reasons why the structural reforms were more radical in Estonia (and the other Baltic States) than in other former socialist countries. The main R&D policy problems in need of rapid solutions were the following:

1. A new R&D policy suited to a small independent state had to be outlined and a legislative base for it had to be created;
2. New democratic decision-making bodies, e.g. R&D councils, had to be set up;
3. The R&D funding system had to be reorganized, with project-based financing replacing institution-based financing;
4. The system of research institutes had to be restructured and the Soviet-type Academy of Sciences (AoS) reorganized into Western-type associations of the academic elite;
5. University education and research had to be reformed;
6. Research institutes had to be integrated into the universities;
7. Estonia needed to introduce its own academic degrees;
8. New R&D data collection and analysis systems had to be put in place.

Three clearly defined phases in the formation of R&D policy in the sovereign Estonian Republic can be discerned:

1. the dramatic years of the collapse of the USSR and the initial period of independence;
2. the period between 1994 and the year 2000, characterized by a certain degree of economic stabilization and revival;
3. the period from the year 2000 onwards, when a sustainable, balanced R&D and innovation system began to emerge.

1. THE DISSOLUTION OF THE SOVIET R&D SYSTEM (UP TO 1994)

In the early years of emerging political liberalism and rapid economic change, when Soviet military forces were still present in Estonia, only the broad outlines of the course to be taken by the country's economic and social development were settled. The country's principal objectives were to return to the European fold and to establish a market economy. The economy was liberalized even more quickly than private ownership and competition were developed. During this period, economic policy in Estonia was driven largely by short-term macroeconomic considerations. Concrete national policies in areas such as industrial development, defense, agriculture and health had yet to be defined.

The main objectives in the reform of the R&D system were the abolition of hierarchical management and planning, the removal of barriers within the system and the establishment of free exchanges of information with the Western world and the freedom of movement and cooperation. In 1990, several important steps planned by a group of scientists within the Estonian Union of Scientists were approved by Estonia's first post-Soviet government. However, as the political situation was extremely volatile, implementation of these 'bottom-up' changes was quite inconsistent. As a result, the only changes to the R&D system during this phase were organizational and structural ones. With the benefit of hindsight, it is evident that rapid changes in the political and economic environment, including withdrawal from the former Soviet economic sphere, privatization, frequent changes of government, the absence of a legislative base and of a strategic vision of the role of S&T in a small state led to the spontaneous dissolution of the former R&D system, and with it the loss of certain positive aspects of the old regime.

In 1990–1991, new decision-making bodies elected or nominated by scientists were established by the government; at the same time, the task of creating a new legislative basis for R&D got under way and a grant system based on peer review was introduced in order to ensure that science and research could be self-regulating. The reform of higher education and of the system of academic degrees was launched.

In 1990, the *Estonian Science Council* (ESC) was founded as an advisory body to the Estonian Government on science policy matters. At the same time, three state funding agencies, the *Estonian Science Foundation* (EstSF), the *Informatics Foundation* and the *Innovation Foundation*, were established to manage the financing of all stages of R&D and innovation.

During the dissolution phase, the government bodies played no part in analyzing the situation, developing S&T policy or initiating radical structural change. Unfortunately, the ESC, chaired by the President of the AoS, did not display sufficient initiative in matters of national S&T policy. The main lever of S&T policy was the distribution of funds. Until 1998, the EstSF made decisions on the distribution of the entire state budget for basic and applied research.

The first and most important political task facing the EstSF was to change the funding paradigm from one based on supporting entire research institutes to one designed to support project groups and individual researchers through the award of grants. In order to develop a criterion of excellence and to obtain an unbiased opinion on the quality of science in Estonia, a thorough evaluation of science and research results for 1986–1990 was carried out by Swedish experts in 1991 (SNSRC, 1992). This work was supported by the Swedish Government. It was followed by an evaluation of research institutions organized by the ESC (1994). The share of grant money in total research funding was increased from 5 per cent in 1991 to 32 per cent in 1995. Institutions engaged in basic research were guaranteed a certain level of core funding (Martinson, 1995). The Innovation Foundation provided financial assistance in the form of subsidies or low cost loans for technological research and development and the improvement of production technology and product quality.

During the years of transition to a market economy, the most distressing problem for R&D was a drastic decrease in financing. Although the money allocated to research in the state budget grew in absolute terms by more than 30 per cent per annum between 1992 and 1995, this growth was much slower than that of GDP. As a result, the expenditure on research from the state budget was only 0.37 per cent of GDP in 1995. The state budget allocations to development projects (via the Innovation Foundation) were even more limited and did not exceed 0.05 per cent of GDP. The restructuring of industry led to the disintegration of large enterprises and to an almost total absence of demand for R&D from the reconstituted productive sector, particularly since the new enterprises did not have any disposable resources. The result was that allocations to R&D from the business sector decreased to about 10 per cent against 30–40 per cent of total R&D expenditure in the 1980s (Table 7.1).

Because of insufficient financing, as well as an internal and external brain drain, the number of staff in research institutions decreased by 30 per cent between 1990 and 1993 and the number of researchers by more than 35 per cent (cf. Table 7.2). In 1985, the percentage of researchers in the total labor force was 0.91 in Estonia. By 1995, this share had dropped to 0.55. These data are not fully comparable as, from 1993 onwards, the statistical data on personnel engaged in R&D, collected according to the Frascati Manual, includes 'researchers and engineers'. At the same time, the dynamics of the age distribution of research personnel took an alarming turn. Thus, in 1991 the share of researchers over 50 years old was 33.7 per cent, in 1994 40.6 per cent and in the year 2000 42.8 per cent. The share of researchers under 30 years of age has constantly been about 10–11 per cent (SOE, 1993–2001).

During the first phase of reform, one vital undertaking was the *reform of the higher education system*. The following are some of the key concepts guiding this reform: intercommunication between academic and applied higher education, private higher education institutions, regional colleges, increasing share of students among the relevant age cohorts, changes in the structure of the student body, postgraduate studies, open university, continuing education and distance learning. In the course of university reform, the number and diversity of higher educational institutions changed considerably. One important step was to give the universities the status of public institutions. Public universities are guaranteed a great deal of autonomy. A number of special bodies dealing with higher education problems, including the University Rectors' Conference and the Higher Education Quality Assessment Council were established.

Higher education was divided into vocational and academic streams (bachelor's, master's and doctoral studies). Two types of higher educational institutions – public universities and state (applied) higher educational institutions – were established. A totally new category of institution – private universities

ESTONIA: TRANSFORMATION OF THE R&D SYSTEM 139

Table 7.1 Estonia: R&D financing by source of funds (thousand EEK/Estonian kroons)*

	Government funds	%	Business sector	%	Non-profit organizations	%	R&D institutes–own funds	%	Foreign funds	%	Total (100%)
1992	89 076	88.1	1 682	1.7			391	0.4	9 930	9.8	101 079
1993	100 032	74.9	19 077	14.3	5 839	4.4	4 257	3.2	4 333	3.2	133 538
1994	165 320	76.3	20 825	9.6	6 807	3.1	8 743	4.0	15 103	7.0	216 798
1995	180 608	71.4	32 706	12.9	6 086	2.4	9 428	3.7	24 217	9.6	253 045
1996	216 153	70.7	30 967	10.2	15 020	4.9	10 496	3.4	33 048	10.8	305 684
1997	260 243	67.2	33 142	8.5	19 422	5.0	19 199	4.9	55 890	14.4	387 896
1998	291 552	75.3	32 455	8.4	22 301	5.8	16 459	4.2	24 479	6.3	387 246
1999	344 597	79.1	41 259	9.5	9 298	2.1	3 858	0.9	36 783	8.4	435 795
2000	330 931	73.7	41 864	9.3	12 626	2.8	9 747	2.2	53 818	12.0	448 986

Source: SOE, 1993–2001.

* 1 EUR = 15.6466 EEK

Table 7.2 Estonia: number of researchers and engineers

	1990	1993	1995	1997	1999	2000
Higher Education Sector	3 174	2 942	2 734	3 294	3 134	3 347
– Universities	3 174	2 942	2 734	2 743	2 593	2 809
– Research institutes associated with universities				551	541	538
Government sector	4 729	2 054	1 769	893	758	675
– State research institutes	2 439*	1 624	1 422	741	627	
– Other R&D-related institutes	2 290	430	347	152	131	
Private non-profit sector				21	20	41
Total	7 903**	4 996	4 503	4 327	3 912	4 063

Source: SOE, 1993–2001.

* In institutes of the AoS: 1 342
** This figure includes all graduates working in R&D institutions; the number of researchers was 7 150.

and higher schools – appeared. As a result, the number and diversity of higher educational institutions changed considerably (cf. Table 7.3).

Table 7.3 Estonia: number of universities and R&D institutes

	1990	1993	1995	1999	2000
State/public universities	6	6	6	6	6
Private universities			1	9	9
State HE establishments		7	8	9	8
Private HE establishments	1	7	11	13	11
Public research institutions				1	1
Research institutes at public universities*				18	18
AoS research institutions	17	20			
State research institutes			32	16	17
State research institutions**		20			
Business enterprise sector and and others	18	18			
Private sector R&D institutions			6	19	24

* Due to the re-structuring of research enterprises in 1997–98 the legal status of institutions has diversified. "Research institutes at public universities" are mostly former AoS institutes integrated into universities. The data is from the Statistical Office and the list of R&D institutions registered at the Ministry of Education.
** Museums, wild-life reserves, etc. were not counted.
Source: SOE, 1993–2001.

The internal structure of the universities became much more flexible. Apart from faculties, chairs and institutes headed by professors, there are special (independent) research institutes and other units, such as competence centers and centers of excellence within these structures. At the same time, tens of former problem-oriented and branch laboratories established at the universities in the 1950s and 1960s for research in certain problem areas were closed.

In 1990, the re-organization of professional training for researchers in a new environment got under way and the system of *academic degrees* was reformed. Two academic degrees were established: the master's and the doctorate, only awarded by universities, not by research institutes as had been the case in the USSR. The degrees of 'Candidate of Science' and 'Doctor of Science' conferred on scientists in the years of Soviet power remained valid and did not require re-certification.

Between 1990 and 1994, there was a considerable amount of *spontaneous structural change* in the Estonian research system, mainly in governing bodies and former industrial R&D institutions. At the end of the 1980s, there were more than 50 branch institutes and planning and construction establishments.

In the course of industrial restructuring, the former branch institutes and other industrial research institutions established by all-union ministries were virtually eliminated. Since they had formerly served the interests of the Soviet industrial and military complex, they had no role in the new situation. Consequently, 'closed' laboratories and secret research institutes disappeared. By 1997, only the Oil Shale Research Institute, the Institute of Energy Research and six agricultural research institutes had survived as state research institutes carrying out applied research.

Industrial restructuring also led to the disintegration of the large enterprises that had formerly been part of the Soviet industrial complex. Construction and technology units at various enterprises were dissolved or switched to new areas of activity and given new organizational forms. Most of them were reorganized as joint-stock companies. The restructuring of these organizations represents the most radical change in the system of R&D institutions (Martinson et al., 1998).

In December 1994, the *Organization of Research Act* was passed by the Estonian Parliament (*Riigikogu*). This Act fixed the frameworks and organizational rules for carrying out research. It defined the R&D institutions, the rights and obligations of governing structures and the functions of the Research Councils, laid down the procedures for funding research and gave formal recognition to the old and new academic degrees.

In January 1995, this legislation was followed by the *Universities Act*. It later transpired that the *Organization of Research Act* was a failure, because it was inconsistent with other legislation. Nonetheless, it encapsulated the main trends and outcomes of the 'bottom-up' S&T policy phase, during which experience of the workings of an R&D system in a market economy and a multi-party democracy was acquired and a number of changes were made. However, by the end of this phase, the need for a consistent national S&T strategy and for structural changes had become evident. The most pressing problem for R&D in Estonia remained inadequate financing from the state budget and a lack of interest in and support for R&D from the business sector.

2. 1994 TO 2000: THE PERIOD OF STRUCTURAL CHANGE

By 1994, Estonia had halted its economic decline. The former centrally planned economy had become market oriented, with liberalized foreign trade and a privatized public sector. GDP grew by 4.3 per cent in 1995 and by 10.6 per cent in 1997. In mid-1996, the private sector's share in GDP was put at 70 per cent (MEARE, 1997). The cornerstones of Estonian economic policy became currency stability, a balanced budget, a 26 per cent income tax, and a liberal economic policy combining economic competitiveness and openness with a minimal degree of state intervention.

The development of international relations, the increasing influence of globalized culture and a decline in exaggeratedly nationalist sentiment began to influence the main strategic objectives of S&T policy. But the most significant catalyst in the development and approval of a state S&T policy was the *adoption of the unambiguous political objective of integration into Europe*.

In 1994, Estonia became an associate member of the European Union. Since then, the official policy of Estonia has been to seek membership of both the European Union and NATO. In accordance with this political objective, the main feature of S&T policy became internationalization. The strategy adopted was shaped by the overriding objective of integrating Estonian S&T into the EU S&T system and participating in EC S&T programs as an equal partner. In December 1997, the European Commission decided to make Estonia one of the first five post-communist countries selected to participate in the negotiations on EU accession.

In 1994, the ESC was reorganized into the *Research and Development Council* (RDC) and given responsibility for advising the government not only on science policy matters but on science and technology policy. The Council is chaired by the Prime Minister and composed of representatives of government, academia and business. The year 1995 can be regarded as the *start of the 'top-down' phase* of the construction of Estonia's reconfigured R&D system. The goal was to bring the R&D system into line not only with the general trend of political developments but also with the country's needs and opportunities.

One of the recommendations of the Swedish evaluation of Estonian science was the integration of research institutes into universities. In the first stage, a number of university professorships were established in AoS institutes. Between 1994 and 1995, a number of researchers were elected to the post of university professor. The political decision to transform the Estonian AoS into a classical type of Academy – an association of the academic elite – was made in 1994. Since 1995, the former AoS institutes have been state research institutes operating under the auspices of the Ministry of Education.

The next step – the integration of 17 research institutes into four universities – was accomplished between 1996 and 1998. It was the only way to maintain scientific potential, to slim down the oversized research system and to open up research institutions to an influx of young talent. As a result, only 7 of the 20 former AoS institutes have retained the status of state institutes (Martinson and Raim, 2001). Consequently, the higher education sector's share of R&D expenditure increased from 26 per cent in 1993 to 58 per cent in 1997, while the government sector's share fell from 74 to 37 per cent (Table 7.4).

The objective of higher education reform was to lay the foundations for a *knowledge-based society*. Scientific work is considered an essential part of university education. The scope of postgraduate studies was broadened to encompass the training not only of researchers but also of other high-level specialists. University and other HE study programs were accredited in accordance with the Universities Act. Degree programs were restructured in accordance with international criteria.

One important step was to prioritize *research areas* in accordance with EC criteria. The priority areas are biotechnology, innovation technology, materials science and environmental technology. Estonia has an advanced skills base and the necessary research infrastructure in these areas. Four new institutions – Centers of Strategic Competence – were established to serve as the bases for carrying out R&D work in priority areas.

At the end of 1993, the Baltic Technology and Business Development Center was established to support technological transfer and to draw up the Estonian innovation and investment policy. Furthermore, under the EU's Fourth Framework Program, the Innovation Relay Center and the Estonian Technology Center were established. A consulting program for small and medium-sized enterprises (SMEs) has been operating since the beginning of 1996.

Nevertheless, one of the most acute problems has been the weakness of the *innovation system* in Estonia. However, until 1998, the Statistical Office collected no data on R&D in the business sector. According to science statistics, the business sector's expenditure on R&D accounted for only about 8.5 per cent of the total between 1997 and 1998 (cf. Table 7.1). Expenditure on basic research was the largest single item and the share of expenditures on experimental development was far too low (Table 7.5).

It was not until 1999 that the Statistical Office conducted the first survey of enterprises carrying out R&D (this first survey refers to the previous year). The results for 1998 and 1999 have now been published (Vares, 2000 and 2001). The questionnaires were sent to all enterprises with more than 20 workers, as well as to enterprises whose main activities were R&D or who had reported R&D costs in annual surveys of financial statistics or reported intangibles in the annual investment statistics for 1997 – 1999. In 2000, 970 enterprises were surveyed. Since 2000, special attention has been paid to the collection of relevant, internationally comparable data on R&D and innovation. Since 2001, these data have been used to supplement the data on R&D statistics, thereby providing a more complete picture of the situation (SOE, 1993 – 2001). Thus in 1999 internal expenditure on R&D in the business enterprise sector totaled 137 041 thousand EEK and there were 651 researchers.

In 1997, a strategy for S&T policy was drafted and approved by the R&D Council and accepted by the government. *Short, medium and long-term or-*

Table 7.4 Estonia: R&D expenditure by type of institution (1 000 EEK)

	1993	1995	1997	1999	2000
Total	130 155	250 604	379 741	435 795	448 986
Higher education sector	33 942	70 707	220 196	293 511	303 717
– Universities	33 942	70 707	137 432	201 093	214 555
– Research institutes associated with Universities			82 764	92 418	89 162
Government sector	96 213	179 897	139 616	139 821	133 999
– State research institutes	70 354	145 535	131 354	129 003	
– Other institutions	25 859	34 362	8 262	10 818	
Private non-profit			1 948	2 463	11 270
Enterprises whose primary activity is R&D			17 981		

Source: SOE, 1993–2001.

Note: covered are only R&D institutions, not the business enterprise sector

Table 7.5 Estonia: R&D expenditure by kind of R&D activity (1 000 EEK)

	R&D expenditures	Basic research		Applied research		Experimental Development	
	total	total	%	total	%	total	%
1993	100 122	79 508	79.4	18 796	18.8	1 818	1.8
1995	250 604	132 014	52.7	89 042	35.5	29 548	11.8
1997	379 741	188 144	49.5	141 272	37.2	50 325	13.3
1999	435 795	216 918	49.8	150 515	34.5	68 362	15.7
2000	448 986	228 983	51.0	152 655	34.0	67 348	15.0

Source: SOE, 1993–2001.

Note: covered are only R&D institutions, not the business enterprise sector

ganizational preferences were defined. In March, 1997, the revised law on research organization – the *Organization of Research and Development Act* – was adopted. A month later, the Estonian Academy of Sciences Act passed the *Riigikogu* (Parliament).

The Organization of R&D Act is based on the following principles:

1. legislative diversity of R&D institutions;
2. diversity of sources and types of financing;
3. multiplicity of decision-making bodies;
4. regular international evaluation of R&D institutions by area of specialization (once every 7 years).

This Act defined clearly the roles of governmental institutions, funding bodies and research establishments.

Research funding is divided into

– targeted funding for research projects carried out by research institutions;
– research and development grants awarded on a competitive basis by the foundations;
– national programs;
– running costs (infrastructure).

Under the terms of the Organization of R&D Act, the Ministry of Education has the right to make decisions on targeted funding for research institutions by topic (about 47 per cent of total research money in 1999). *The Science Competence Council* was established as the advisory body to the Ministry of Education. The new Council makes proposals on targeted funding for long-term (5-year) research projects and on meeting the infrastructure costs of those research institutes operating under the auspices of the Ministry of Education. The EstSF and the Innovation Foundation were reorganized into independent, non-profit agencies distributing state budget allocations in the form of grants (Martinson, 2000). One very positive step was the establishment of special research grants for doctoral students as well as post-doctoral stipends. Since 1999, the EstSF has had the right to give special tax-free stipends to masters and doctoral students participating in grant-aided projects.

The task of building up an *innovation subsystem* in a rapidly changing (local and global) economic environment is an extremely complex one. Small countries like Estonia, with small domestic markets, limited human and financial resources, as well as little international market power and limited room for maneuver in their public innovation policy face a particular set of problems in seeking to come to terms with the globalization of production and the domination of large multinational firms in most sectors of the international economic arena. The relatively quick development of the Estonian economy was based mostly on trade, brokerage, stock exchange operations and other activities that do not directly create added value.

In the second half of the 1990s the amount of foreign direct investments (FDI) – an important input indicator of technology development for transforming economies – grew rapidly. Thus with FDI inflows totaling 10.6 per cent of GDP, Estonia headed the list of CEECs in 1998 (EBRD, 1999). Until the end of the 1990s, however, the main problems for Estonia were a foreign trade deficit, a weak technological base in its industrial sector and a lack of information on the real needs of firms' customers; these were further compounded by insufficient investment in R&D, a shortage of highly trained specialists and other workers in burgeoning high-tech areas, weak patenting activity and insufficient cooperation between university, industry and government (Hernesniemi, 2000).

Since 1998 – 1999, awareness of the potentially key role of innovation in furthering economic development and increasing international competitiveness has spread rapidly among various decision-makers in Estonia. All recent strategy documents identify the promotion of a knowledge-based economy, of innovation and of R&D activities as a key measure. In June 1998, the government adopted the *'Program on Innovation'*. This document proposed that state funding for development projects should rise to 0.3 per cent of GDP by 2000 and outlined a strategy for attracting support for R&D from the business sector and facilitating research and the implementation of promising initiatives. However, due to the economic slowdown, there was no increase in state funding for R&D in 1999 – 2000. The 'National Development Plan 2000 – 2002' (www.fin.ee/english), drawn up by a multiplicity of key institutions and actors and endorsed by the Government in 1999, has chapters devoted to industry and business development, including entrepreneurship, innovation and support for technology transfer.

One very positive achievement has been the rapid development of the *informational infrastructure* in Estonia. In 1996, a special program dubbed the 'Tiger Leap' was launched to support the computerization of Estonian schools. In May 1998 legislation on the fundamentals of information policy was approved by the *Riigikogu*. Development in this area has been impressive. In February 1998, there were 125 internet hosts per 10 000 inhabitants. By this index, Estonia was in 14th place among 50 European and Mediterranean region countries. By April 2001, the number of internet hosts per 10 000 inhabitants had grown to 306. There were 16.2 home PCs and 8.0 home internet connections per 10 000 inhabitants (Baltic Media Book, 2001).

3. THE THIRD PHASE OF TRANSFORMATION: THE FORMATION OF AN EFFECTIVE R&D AND INNOVATION SYSTEM SINCE THE YEAR 2000

The strategic goals of Estonian R&D policy were formulated in 1998, in the 'White Paper on R&D Strategy' approved by the government in January 1999. The strategy was aimed at developing a sustainable, balanced education, R&D and innovation system. From this starting point, a document entitled 'National R&D strategy 2002–2006: knowledge-based Estonia' was prepared by the Ministry of Economic Affairs and the Ministry of Education and presented by the Prime Minister to the *Riigikogu* for debate in December 2000. To coincide with the debate, the R&D Council published a review of R&D in Estonia for the period 1996 to 1999 (Kaarli and Laasberg, 2001). The strategy was adopted in December 2001.

The document submitted to Parliament for debate analyzed the state of the R&D and innovation system and the possibilities for its development in the light of social and economic requirements. The main strategic goals of R&D and innovation – revival of the knowledge base and increased corporate competitiveness – were formulated and the means to achieve these general goals defined. Three key areas of R&D were identified: user-friendly technologies for the information society, biomedicine, and materials technologies. The document set a target for increasing R&D funding from 0.75 per cent of GDP in 2001 to 1.5 per cent of GDP by 2006. It also stressed the need to promote international R&D cooperation and to stimulate partnerships between business and R&D institutions. One of the means of achieving this is to build up the network of bridging institutions, particularly centers of excellence in research and centers of competence specializing in developing know-how and technology transfer. In 2001 two centers of excellence set up as part of the EC's Fifth Framework Program and six national centers of excellence were established.

In April 2000, the Government took the decision to restructure the state system of foundations responsible for supporting business, innovation and R&D and promoting investments and exports. A total of nine foundations (incl. the Innovation Foundation) operating under the auspices of five different ministries were included in the restructuring exercise. The Estonian Enterprise Development Foundation (web site: *www.eas.ee*), with five agencies and two regional offices, was established. The Innovation Fund was re-organized into the Estonian Technology Agency (the counterpart of Tekes in Finland) (web site: www.estag.ee). It began to operate in 2000 (Government, 2000). At the same time, the Enterprise Guarantees Foundation was established. It comprises the Export Guarantees Foundation, the SME Loan Guarantees Foundation and the

Estonian Home Foundation. The re-organization was completed by 2001. The main difference over the former system is that now applications for funding for innovative projects will be dealt with separately from applications for funding for support structures (i.e. science parks and so on).

In 2001, there were about 60 spin-offs from public research institutions (including universities) in Estonia. The number of such ventures is in a state of constant flux. If all spin-offs involving public university staff (on a part-time basis) and using university equipment and library and other resources are taken into account, the total would be about 120. There are two science parks in Estonia. Tartu Science Park was founded in 1992 by Tartu City and County, Tartu University, the Agricultural University and the Institute of Physics. In 2000 it had 27 tenant companies, including 5 business incubators and 16 other R&D-based enterprises, plus 7 non-tenant associate members. Tallinn Technology Park, established by Tallinn Technical University (TTU), the Ministry of Economic Affairs and Tallinn City, functioned from 1989, but its activities died down in the middle of the 1990s. In 1998, the TTU established the Technical University Innovation Center Foundation. One of its functions was to manage the spin-off program and to develop incubation services for R&D-based start-up companies. On August 6, 2001, the Tallinn Technology Park resumed its activities.

In 2000–2001, the process of evaluating research by area of specialization and on a project-by-project basis got under way in accordance with the legislation on R&D organization. By December 2001, twenty-two areas had been evaluated. Foreign experts from many European countries are involved in this process. Each expert group consists of 3–5 persons. The results of this assessment are taken into account in making decisions on further restructuring of research enterprises and on research funding.

With strategic documents adopted and new organizations established, S&T policy in Estonia is about to enter its third phase, the objective of which will be the *formation of a sustainable, balanced R&D and innovation system* serving both the community and science in general. Interaction between the R&D sector and other sectors and the need to develop dialogue between science and the community, science and government and science and industry have become dominant concerns.

In recent years, participation in EU Framework programs has helped the Estonian R&D system become integrated into the European Union system. Estonia is participating in the Fifth Framework Program on a regular basis. The success of its applications to participate in EU programs is one indicator of the quality of the applicants. As of July 2001, the success rate was 24 per cent: of 425 proposals, 103 were adopted. Another positive movement was the admission of the EstSF and the AoS as members of the European Science

Foundation (ESF) in 1999 and the participation of Estonian researchers in ESF programs.

The complicated process of developing S&T policy in a small country such as Estonia during a period of dramatic changes in politics, economics and society as well as in the cultural, intellectual and moral spheres has advanced rapidly. In the ten or more years since the first attempts were made to plot the development of the R&D system in Estonia under the new conditions, these initiatives have grown from spontaneous 'bottom-up' actions into a strategic policy intended to safeguard Estonia's progress towards a knowledge and innovation-based learning society and to pave the way for its admission into the European Union. This primary goal of S&T policy is defined at government and parliament level.

The legislative base for the reform of university education and R&D has been created and new decision-making bodies have been established. The institutional restructuring of the R&D system has been launched, with research potential gradually being concentrated in the universities and a balanced system of education, research and development being built up. Key areas of R&D, considered vital for the future economic and social development of Estonia and corresponding to EC priorities, have been approved. New bridging structures at the interface between basic research and education, industry and other sectors of the economic and social infrastructure are being established. The development of support for innovation and high-tech SMEs has been approved by the government. The principles and methods of R&D funding have been transformed by the adoption of project financing in the form of peer-reviewed grants and targeted funding for certain projects based on scientific merit. Involvement in international research and participation in EC Framework Programs are especially strongly encouraged.

However, in spite of the definite progress that has been made in planning and implementing a strategic S&T policy, there is much to be done to implement this policy. Thus the dialogue between the scientific community and parliament, government and society must be intensified. State budget allocations for R&D must be increased in accordance with the proposals set out in the White Paper on R&D Strategy mentioned above, with primary attention and support being given to applied research and technical development. One basic problem that has to be resolved is how to initiate cooperation between research and industry and how to involve industry in carrying out R&D.

REFERENCES:

Baltic Media Book (2001) Tallinn/Riga/Vilnius: EMOR/BBF Gallup Media/SIC Gallup Media.
EBRD (1999) *Transition Report*. London: European Bank for Reconstruction and Development (EBRD).
Government (2000) "Ettevõtluse Arendamise Sihtasutuse asutamine ja põhikirja heakskiitmine" (On the establishment of enterprises in Estonia and the approval of their statutes – Decree no. 556-k). *Riigi Teataja* (State Gazette – Supplement) no. 76: 1166.
Hernesniemi, H. (2000) *Evaluation of Estonian Innovation System. Support to European Integration Process in Estonia*. Phare Project Report no. ES 9620.01.01. Helsinki: Etlatieto LTD.
Kaarli, R., and T. Laasberg (2001) *Research and Development in Estonia 1996 – 1999. Structure and Trends*. Tallinn: State Chancellery.
Martinson, H. (1995) *The Reform of R&D System in Estonia*. Tallinn: Estonian Science Foundation (EstSF).
Martinson, H. (2000) "Formation of research and development strategies for Estonia." Science and Government Series Vol. 5: *The Knowledge-Based Economy – The European Challenges of the 21st Century*. Ed. A. Kukliñski. Warsaw: Komitet Badan Naukowych (KBN), 115 – 121.
Martinson, H., and T. Raim (2001) "Strategic Approach versus Spontaneity in Restructuring the R&D System in a Small Country." *Government Laboratories – Transition and Transformation. NATO Science Series 4: Science and Technology Policy – Vol. 34*. Ed. D. Cox, et al. Amsterdam: IOS Press, 30 – 40.
Martinson, H., I. Dagyté, and J. Kristapsons (1998) "Transformation of R&D systems in the Baltic States." *Transforming Science and Technology Systems – the Endless Transition? NATO Science Series 4: Science and Technology Policy – Vol. 23*. Ed. W. Meske, et al. Amsterdam: IOS Press, 108 – 117.
MEARE(1997) *Estonian Economy 1996 – 1997*. Tallinn: Ministry of Economic Affairs of the Republic of Estonia (MEARE).
SNSRC (1992) *Evaluation of Estonian Research in Natural Science*. Stockholm: Swedish Natural Science Research Council (SNSRC).
SOE (1993 – 2001 *Science – Statistical Yearbooks*. Tallinn: Statistical Office of Estonia (SOE).
Vares, H. (2000) "Research and Development in Business Enterprise Sector." *Estonian Statistics* no. 1: 97 – 102.
Vares, H. (2001) "Research and Development in Business Enterprise Sector." *Estonian Statistics* no. 2: 121 – 125.

8. LATVIA: TRANSFORMATION OF THE S&T SYSTEM

Janis KRISTAPSONS

Latvia regained its independence in 1990–1991. The first parliamentary and presidential elections took place in 1993 and the last Russian (former USSR) troops left Latvia on August 31, 1994. In accordance with the renewed constitution, which had been originally adopted in 1922, elections were held in 1995 and 1998. Multi-party parliamentary democracy is functioning well in Latvia, giving the population a sense of political stability. The next political goal is integration into the European political, economic and security system (membership of EU, NATO, etc.).

1. ECONOMIC SITUATION AND THE TRANSFORMATION OF R&D

Latvia's GDP declined between 1990 and 1995. At comparative prices, Latvia's GDP in 1995 was 49.6 per cent of the 1990 level. Growth recommenced in 1996 (cf. Table 8.1). The share of industry and agriculture in GDP is declining, while the share of services is increasing year on year. The major factor in Latvia's successful development has been macroeconomic stability. According to estimates by European Commission experts, Latvia should not have any difficulty in implementing the Maastricht criteria in the medium term.

The present goal of economic development in Latvia is to create a Western-style system commensurate with the European Union's economic and market requirements (Gerhards et al., 2000). The national currency (Lat) is very stable – its value has remained constant in relation to the main world currencies since 1994. The inflation rate had fallen to 2.4 per cent by 1999. The state budget has almost always been in credit. However, according to expert estimations, it will be more than 20–25 years before per capita GDP in Latvia even approaches average European values, and even this goal is realistic only if annual GDP growth rates do not fall below 5 per cent (Bikis et al., 2000).

Privatization is one of the most important elements of the economic reforms that have taken place in Latvia. It has been the principal means of effecting the relatively rapid transition from a command economy based on public ownership of the means of production to a modern market economy based

Table 8.1 Latvia: basic indicators

	1990	1993	1994	1995	1996	1997	1998	1999	2000
Population, millions	2.67	2.60	2.57	2.53	2.50	2.48	2.46	2.44	2.38
Employment, millions	1.41	1.205	1.083	1.046	1.018	1.037	1.043	1.038	1.038
R&D personnel (head count), 1 000	17.7	8.7	6.9	6.8	6.1	6.0	6.1	6.2	8.2
R&D personnel (head count)/ 1 000 employment	12.6	7.25	6.4	6.5	5.9	5.8	5.8	6.0	7.9
Researchers (FTE), 1 000		4.0	3.0	3.1	2.8	2.6	2.5	2.6	3.8
GDP, millions USD				4 449.2	5 134.5	5 637.5	6 083.9	6 661.5	7 150.2
GERD as % of GDP	1.6	0.48	0.42	0.52	0.46	0.43	0.45	0.42	0.48
State budget for research as % of GDP		0.25		0.28	0.26	0.25	0.24	0.24	0.20
FDI, millions USD				615	936	1 272	1 557	1 813	2 105

Source: author's calculations based on CSBL, 1990–2001.

on private initiative and competition. The mass privatization program that was put into effect between 1994 and 2000 led to a continuous expansion of the private sector during that period. The privatization of small and medium-sized enterprises was largely completed by mid-1998. In the year 2000, the private sector's share in total value added was 68 per cent and the private sector accounted for more than 70 per cent of total employment. In the same year, the private sector's share of that part of GDP produced in manufacturing, construction, trade, hotels and restaurants exceeded 95 per cent (Gerhards et al., 2000).

When it came to building new S&T structures in the aftermath of the successful struggle for independence in 1990–1991, Latvia, like Estonia and Lithuania had to start from scratch. However, the approach to the transformation of the R&D system that was adopted in Latvia differed considerably from that taken in the two other Baltic States, and the results were different. The main instrument for the development of research and of the organizational structures within which it takes place is the level of funding and the mechanisms for the distribution of funds.

Total domestic expenditure (in absolute terms, per capita, etc.) and the state science budget in particular are substantially lower in Latvia than in Estonia and Lithuania and other CEECs. There is a close correlation between the number of researchers and their output (as measured by the number of published articles). In 1990, Latvia became the first and remains the only CEEC to allocate only a small proportion of government funding directly to research institutes. Most of the funds are allocated through the grant system to particular projects (basic, applied and market-oriented research), scientific programs, international cooperation (mainly for attendance at international conferences abroad) and doctoral students. The grants are allotted by special expert committees on which scientists themselves sit.

Table 8.2 summarizes the main events that occurred in the course of the transformation of the Latvian R&D system.

2. CHANGES IN THE S&T SYSTEM

2.1. THE LATVIAN FUNDING SYSTEM

In 1989–1990, a new system of science funding and management was established in Latvia. A working group set up by the Latvian Union of Scientists and the Latvian AoS worked out the basic principles of the new system in 1989. One of the first resolutions of the newly established democratic government of Latvia in 1990 was to accept the proposals drawn up by these scientists and to set up the Latvian Council of Science. As of January 1991, direct funding of

Table 8.2 Latvia: main events in the transformation of the R&D system

Date	Event
1988 November	founding of the Latvian Union of Scientists
1989	drafting of proposals on a science system reform
1990 May 4	declaration of independence of the Republic of Latvia
1990 July	government decision to change the science funding system and establish the Latvian Council of Science
1990 September	the Latvian Council of Science starts functioning, evaluation of the project applications (about 1 000)
1991 August 21	announcement of the full independence of the Republic of Latvia
1991 October	introduction of Latvian scientific degrees, large-scale procedure of receiving the new Latvian scientific degrees on the basis of the former USSR scientific degrees starts
1991 November	parliament adopts the Scientific Activity Act draft in the first hearing
1992 February	the Academy of Sciences adopts its charter and new statutes: classical type academy and independent institutes develop
1992	evaluation of research results by foreign experts
1992	Scientific Activity Act adopted by parliament (organizer: Danish Research Councils)
1992	Department of Science and Higher Education starts to function within the Ministry of Education
1994 July	research institutions of the Academy of Sciences transferred to the Ministry of Education and Science
1995 December	the Higher Educational Establishments Act adopted by parliament
1997 January	charter of the Latvian Academy of Sciences ratified by parliament
1996–1998	integration of 21 former academy and ministry research institutes with universities
1998 January	National Concept of the Republic of Latvia on Higher Education (developed by the Latvian Council of Higher Education)
1998 April	National Concept of the Republic of Latvia on Research Development (developed by the Latvian Council of Science)
1998 May	the revised Scientific Activity Act adopted by parliament
2000 November	the revised Higher Educational Establishments Act adopted by parliament
2000 December	the Latvian Industrial Strategy (developed by the Ministry of Economy)
2001 February	National Concept on Innovation accepted by government

scientific institutions was replaced by financing of selected projects, that is a grant-based system.

"Although we can criticize the shortcomings of a grant system and a certain inconsistency in its implementation, its undoubtedly positive role should be emphasized – scientific criteria became the main factor in determining the allocation of financial resources, eliminating fluctuations caused by changing economic circumstances and the unsubstantiated promises which were so popular in the age of 'developed socialism' science. Looking back over the years, it seems obvious that such radicalism was necessary to shatter the old administrative system of research management before it could recover and adapt to the new conditions" (Grens, 1995).

According to the guidelines for the new funding system, scientists in each branch elect a branch committee. Thus, 14 branch committees of the Latvian Council of Science have been formed to distribute money in the form of project grants on the basis of a thorough evaluation of the proposals submitted to the committees. About 20 to 30 per cent of project applications are rejected. The committees make their decisions on the basis of domestic peer review.

Scientific institutes receive no core funding from the state budget. The resources required for their general maintenance are obtained as percentages of each grant they receive. The institute researchers themselves determine the specific amount. As a result, some Latvian scientists believe that the role of the institute as an independent scientific unit has more or less disappeared and that a conflict of divergent topics and interests is starting to prevail. New research programs have in fact been introduced in order to counter this tendency. Plans have also been made to establish several research centers that will receive general funding. There are scientists who feel that each branch committee is currently acting as a kind of branch scientific council, setting up different branch policies for Latvia. Others are less worried, and note that the "main thing is to get the money. Our research is so significant that it will be funded regardless of the funding system, and we will determine the optimal organization within the institute ourselves" (Kristapsons et al., 1995, p. 53).

2.2. THE LATVIAN ACADEMY OF SCIENCES

Today, the Latvian AoS is a not-for-profit scientific institution, with elected members. It has no absolute power over research funding anymore and its role is defined by law (Scientific Activity Act, adopted by Parliament on November 10, 1992). Article 23 of this act states that "the Latvian Academy of Science is an autonomous state-subsidized non-profit-making scientific institution with elected members. The Academy of Sciences functions in accordance with its charter and statutes. The state may, by decision of the Parliament of the Republic of Latvia or the Cabinet of Ministers, delegate special functions or powers in the field of science." The charter of the Latvian AoS was ratified by Parliament on January 27, 1997 (Stradins et al., 1997).

Many of the most prominent so-called 'exiled' scientists of Latvian origin, now living and working abroad, have been elected as foreign members. The Latvian AoS has also established a wide circle of honorary members. In the year 2000, there were 55 such honorary members, some of them Latvians who have established themselves abroad. Thus the country's main scientific potential, including scholars of Latvian origin scattered all over the world, is united around the AoS. The honorary members include writers, poets, artists, religious figures and people of culture, as well as several scholars of the older generation. Their election serves two purposes: it unites the country's whole intellectual potential around the AoS and compensates for the small number of prominent scholars in the social sciences and humanities.

Until 1992, the scientific institutes of the AoS were fully subordinated to it. The Scientific Activity Act fundamentally changed the status of the research institutes. The act states that they are independent in terms of their scientific work and their administration. From 1992 to mid-1994, the institutes were totally independent, but since July 1994 they have been affiliated to the Ministry of Education and Science.

The Latvian AoS has found its new place in the scientific community (Stradins, 1998). In general, the AoS has a significant influence on the development of science. Most of the scientists holding both elected and appointed posts in the new R&D system represent the elite of their profession and are also active members of the AoS.

2.3. RESEARCH INSTITUTES

There are 32 research institutes in Latvia, including 20 university research institutes with independent legal status, 12 state research institutes and 15 other state research institutions. There are 17 state higher educational establishments, including 6 universities. One of the main targets of Latvia's research restructuring policy was the integration of individual research institutes into the universities in order to strengthen and modernize them. Those state research institutes not integrated into the university system will remain or become national research centers (centers of excellence).

This integration was initially a formal affair, with the universities simply replacing the AoS. The involvement of highly qualified scientists in teaching was, however, hampered by the fact that the vacant professorships and lectureships had already been filled by other university employees, most of whom were not actively involved in research. In 1993, it was suggested that all top-level university positions should be opened up to competition. However, such a radical step was not compatible with the interests of those who were already in the system, nor, for that matter, with the spirit of the law.

The government opted for another strategy. A special category of professors – the so-called state professors – was to be introduced. These professors would receive considerably higher salaries than those presently in post. Moreover, the number of state professors was to be strictly fixed at around 400 for the whole country. Many of the professors currently in post were expected to lose their positions by the end of the term for which they had been appointed; there were high hopes that these vacancies would subsequently be filled by top-level scientists. Things did not go entirely according to plan, however. Most of the vacancies have in fact been filled by scientists of less than the highest quality. High-level scientific skills have been a main criterion only for state professorships, although they are important for future development.

The first proposals for reform in Latvia raised the question of how to integrate and consolidate science and university education. A few years later, the scientific institutes were formally integrated into the universities. The term 'integration' has been retained, yet the very essence of the process has changed (Kristapsons and Millers, 1995; Millers, 1997). This may be attributed to two related factors: this is the first time in Latvia's history that such a process has taken place, and it is happening in a small country.

The democratization of the R&D systems has always been closely linked to funding policy. The latter has been and remains the most important issue when it comes to the restructuring of science in Latvia. In the mid-1990s, the role of the Ministry increased in Latvia. To summarize, the Latvian research system has gone through a cycle from state management before 1990 to complete internal democracy in 1990–1992 and then on to partial state management and partial democracy after 1992 (Kristapsons and Tjunina, 1995).

2.4. R&D STRATEGY IN LATVIA

The generally high level of research in Latvia was acknowledged in the evaluations of international experts (Sorensen, 1992). The large share of articles written today in collaboration with Western researchers is characteristic of Latvia and is obviously encouraging from the point of view of the globalization of science. The high level of R&D is still being maintained in many fields of science today. The UNESCO International Center of Biomedicine and Biotechnology began its activities in Riga in 1999. The national Genome Database of the Latvian Population project began in 2001. An EU Center of Excellence was opened in 2001 as part of the Institute of Solid State Physics in Riga. In addition, many Latvian scientists have been involved in EU Fifth Framework Program projects.

The overall R&D strategy was spelled out in a document approved by the Latvian Council of Science in March 1998 (Ekmanis et al., 1998). In this document, the three main tasks of science are formulated as follows:

1. to create an intellectual environment for the development of higher education and of society at large;
2. to lay the foundations for the development of innovative technologies and technology transfer and to encourage the wider application of scientific methods and practices in public administration and the national economy;
3. to promote dynamic and sustainable social development as well as economic growth and to preserve Latvia's national identity and cultural heritage.

The Cabinet of Ministers has determined which are the five most important research fields for the possible creation of national research centers:
- organic synthesis, biomedicine, pharmacy;
- material sciences;
- information technology;
- forestry and wood processing;
- *Lettonica* (studies of Latvian language, history, ethnography, art, etc.).

3. INDUSTRIAL R&D

Up to 1990, Latvia had a highly developed tradition of producing inventions, and this has been preserved to a large extent (Kristapsons et al., 1999). However, one change that did take place was the sudden disappearance of demand for invention-based innovations from Latvian industry, previously the basic consumer of inventions.

Latvia was one of the few regions in the USSR that sold licenses for research products, and in doing so brought foreign currency into the USSR. This reflected not only the high quality of some of the research work carried out in Latvia but also the good performance of the patent offices in several research institutes and companies. For example, the Institute of Organic Synthesis in Latvia was one of the leading organizations in the former USSR when it came to the invention of new medications. This institute was actually one of the major research institutes in the USSR's entire pharmacological complex. Sixteen original substances were synthesized at this institute, substances that have served as the basis for many medications. Actually, one fourth of all original medications that were eventually available from Soviet pharmacies were synthesized at this particular institute.

We have compared the statistics on patents before 1990 and afterwards. First, the total number of patents has decreased from approximately 1 000 per year in 1988 to approximately 100 per year in 1999. This trend is due to an overall decrease in industrial research. Secondly, the distribution of patent applications across the various fields of research has shifted. In some fields, virtually no applications are now being made, which means that the corresponding areas of research are no longer being kept alive in the country (Kristapsons and Tjunina, 1995a; Adamsone and Kristapsons, 2000).

On the one hand, cooperation between universities and industry is very weak at present. At the same time, the Latvian Council of Science has allocated funds (approximately 30 per cent of the total science budget) to various so-called cooperation projects (previously called programs), which are oriented towards practical applications and in which groups of scientists from different research institutions and some companies are involved.

4. INNOVATION PROCESSES

The Ministry of Economics noted in a report that patterns of economic development in different countries indicate that the growth of GDP is directly proportional to that of exports of innovative, new technology-based products. In the developed countries, the share of knowledge-based and research-intensive industries in exports is between 30 and 50 per cent. Yet in Latvia this figure does not exceed 6 per cent (Gerhards et al., 2000). Also the number of small and medium-sized enterprises per 1 000 residents in Latvia is only 15 whereas in the member states of the European Union there are between 30 and 50. This is due to the inadequacy of all elements of the innovation system – critically low funding for R&D, exclusion of manufacturing companies from research, poor technical equipment for research work in the universities, weak links between research laboratories and industry, insufficient and low-quality technological education in universities, legislation not conducive to innovative activity, problems in raising funds for research work with new products and services, etc. (Dimza, 1998).

Moreover, the problems that will dog innovation processes in Latvia in future are already beginning to emerge: low consideration of scientific recommendations by the decision-making institutions; non-allocation of financial resources to innovative development on the pretext of the budget possibly not being balanced; disinterest on the part of local enterprises in conducting R&D; lack of respective finance instruments and incentives (tax policy, etc.); difficulties with the realization of inventions developed in a small country due to the high costs of patenting and problems making contacts with potential purchasers in other countries (Adamsone and Kristapsons, 2000).

"The traditional components of the innovation process – knowledge creation (R&D, education) – are underestimated and underdeveloped in Latvia. Firms do not consider cooperation with research organizations important. At the same time, firms also state that the lack of sources of external know-how and difficulties in accessing consultants/specialists are significant barriers to innovation." (Karnite et al., 2001, p. 34)

Data on innovation activities in Latvia were gathered in a special survey carried out by the Central Statistical Bureau of Latvia in cooperation with Eurostat. The survey included enterprises in the extractive industries and opencast pit management, manufacturing, construction and the electricity, gas and water supply industries. From a total of 1 330 enterprises in the industries investigated, 562 were selected at random. Responses were received from 474 enterprises (86 per cent). The survey results show that on average only 31 per cent of enterprises are innovative. The most innovative are in manufacturing industry, where 37 per cent of firms have been engaged in innovation activities (Behmane et al., 1999).

There are some non-governmental agencies involved in the management of innovation processes, including the Latvian Association of Technological Parks, Centers and Business Incubators, the Latvian Academy of Sciences and the Academy of Intellectual Property & Innovation (founded in 2000).

There are some specific instruments in place for the promotion of innovation processes, including grants from the Ministry of Education and Science for market-oriented research projects (10 per cent of the state science budget) and the activities of the Innovation Relay Center, the Latvian Technology Center and the Latvian Technology Park.

In 2001 the *National Innovation Plan* was drawn up in order to draw together all the disparate elements of the state support system for innovation (Stabulnieks et al., 2001). The aim of the plan is to promote an economy open to innovation.

The next step after the approval of the National Innovation Plan by the Cabinet of Ministers is the development of the *National Innovation Program*. To ensure implementation of all the activities mentioned in the plans there are moves afoot to initiate a new budgetary program called Support for Innovative Activity.

Of particular importance is state support both for the establishment of small and medium-sized enterprises and improvements in the competitiveness of existing enterprises, with particular emphasis on their ability to compete internationally.

The government declaration on the planned activities of the Cabinet of Ministers refers to the priority status of external market studies, participation in international exhibitions and fairs, patent acquisition, the introduction of modern technologies and the introduction of energy-saving technologies.

The main problem hindering and restricting the efficient implementation of business support measures is insufficient government funding for the implementation of business support activities. A number of business support measures were developed and approved by the Cabinet of Ministers but not implemented due to the lack of government funding (Gerhards et al., 2000).

5. CONCLUSION: AN ASSESSMENT OF THE CURRENT SITUATION

Our analysis has demonstrated the vital importance of the qualitative and structural changes that have taken place in the Latvian science since 1990. The basic research potential has been preserved and provides a solid foundation for future development (e.g. within the framework of EU programs). However, there has been a decrease in the numbers of research staff and in funding and industrial research.

Inadequate state funding and a lack of investment have brought the infrastructure of Latvian science to a critical point, beyond which the irreversible degradation of many areas of scientific activity is to be expected, as well as an outflow of specialists to other countries (Ekmanis, 2001; Stradins, 2001). About 1 000 scientists from Latvia are already working abroad (the loss to the country is estimated at 100 million USD).

The loss of so-called industrial science is possibly the most significant. Before 1990, industrial science was a traditional part of scientific activity in Latvia and took place not only in academic institutes but also in design offices and separate applied research institutes. The international evaluation of 1992, the attitude of the government at that time and the initial policy of the Latvian Council of Science actually led to the whole field of industrial science being ignored. As a result, the research basis of some giant factories was damaged. Only a few former high-tech centers have survived. It is this field of practical science that has suffered the greatest decline.

Analysis of the current possibilities for the development of science in Latvia reveals contradictory prognoses, some pessimistic, others optimistic.

On the pessimistic side, a report compiled by the Ministry of Economics points to the shortage of funding for supporting and developing innovation processes and foresees no new additional sources of funding emerging in the near future.

On the other hand, there are some optimistic trends. The Ministry of Economics has prepared and the government has adopted a national innovation plan and industrial development program. An opinion increasingly propagated in all the mass media is that in order to be included in the group of developed nations and maintain a position there, the further development of Latvia must be based on education, science and new technologies. The program for scientific and academic staff development has been named a priority for the next state budgets.

Some scientists retain their optimism; it is based on the fact that despite meager financing, a high level of R&D is still being maintained in many fields of research.

REFERENCES

Adamsone, A., and J. Kristapsons (2000) "Inventing Activity and Transition: A Case of a Small Country." Unpublished paper, presented at the conference 'Worlds in Transition: Technoscience, Citizenship and Culture in the 21st Century', September 26–30, Vienna, Austria.
Behmane, M., et al. (1999) *Inovacijas apsekojums Latvija* (Survey on the innovation activities in Latvia). Riga: Central Statistical Bureau of Latvia (CSBL).
Bikis, J., et al. (2000) "Latvija: no viijas uz darbību. Ilgtspējīgas attīstības koncepcija" (Latvia: from vision to action. A concept for sustainable development). Riga. (Internet: www2.acadlib.lv/grey/valstsparvalde.htm#koncepcija).
CSBL (1990–2001) *Statistical Yearbooks of Latvia*. Riga: CSBL.
Dimza, V. (1998) "Par Latvijas ilglaicigas attistibas modeli" (Sustainable development model for Latvia). *Latvijas Vestnesis* (The Latvian Herald) no. 347/348: 8.
Ekmanis, J. (2001) "The present situation of science in Latvia." *National Strategies of Research in Smaller European Countries*. Tallinn: Estonian AoS, 47–52.
Ekmanis, J., et al. (1998) *National Concept of the Republic of Latvia on Research Development*. Riga: Latvian Council of Science (LCS).
Gerhards, K., et al. (2000)*Economic Development of Latvia*. Riga: Ministry of Economy of the Republic of Latvia (MERL).
Grens, E. (1995) "Zinatne musdienu Latvija: ar bazam un ceribam" (Science in contemporary Latvia: concerns and hopes). *Latvijas Vestnesis* (The Latvian Herald) no. 136: 12.
Karnite, R., S. Zilko, M. Klava, and M. Traubergs (2001) *Innovation Networks and Industrial Modernization – A Study on Armenia, Latvia and Russia (St. Petersburg Region). Country report: Latvia*. Riga: Institute of Economics.
Kristapsons, J., and T. Millers (1995) "Institutes of the Latvian Academy of Sciences and institutions of higher education: problems of integration." *Higher Education in Europe* 20, no. 4: 163–167.
Kristapsons, J., and E. Tjunina (1995) "Changes of the Latvia's science indicators in the transformation period." *Research Evaluation* 5: 151–160.
Kristapsons, J., and E. Tjunina (1995a) "Changes in the Latvian research system." *Science and Public Policy* 22, no. 5: 305–312.
Kristapsons, J., E. Tjunina, and K. Gedina (1995) *Transformation of Selected Institutes of the Latvian Academy of Sciences (1989–1995)*. Riga: Latvian AoS.
Kristapsons, J., A. Adamsone, A. Edzina, K. Kalviskis, and E. Tjunina (1999) *Inventions and inventors of Latvia*. (Internet: inventions.lza.lv). Riga: Latvian AoS/Patent Office of Latvia.

Millers, T. (1997) "Research at universities and participation of academies in the educational process." *Contributed papers of the 'International Workshop on Academies in Transition: Transformation of Science and Society', September 27 – 30, 1997.* Koshice: Slovakian AoS.

Sorensen, H. (Ed.) (1992) *Latvian Research. An International Evaluation.* Copenhagen: The Danish Research Councils.

Stabulnieks, J., et al. (2001) *National Concept on Innovation.* (Internet: www.lem.gov.lv/En/nat_conc.stm) Riga: Ministry of Economy of the Republic of Latvia.

Stradins, J. (1998)*Latvijas Zinatnu akademija: izcelsme, vesture, parvertibas* (Latvian Academy of Sciences: origin, history, transformation). Riga: Zinatne Publishers.

Stradins, J. (2001) "Guidelines of science in Latvia: historical development and contemporary real facts." *National Strategies of Research in Smaller European Countries.* Tallinn: Estonian AoS, 41 – 46.

Stradins, J., et al. (1997) "Charter of the Latvian Academy of Sciences." *Latvian Academy of Sciences Yearbook.* Riga: Zinatne Publishers, 10 – 14.

9. LITHUANIA: THE SCIENCE SYSTEM FROM 1989 – 2001

Ina DAGYTE

Lithuania is the largest of the three Baltic states in terms of both territory and population. It covers 65.3 thousand square kilometers and had a population of 3.7 million in 1999 (DSRL, 1989 – 1990; DSRL, 1999a, p. 39). Against the background of 'perestroika', the Lithuanian revival movement 'Sajudis' was formed in 1988. Among the founders were numerous representatives of the Lithuanian intelligentsia and members of the Lithuanian Academy of Sciences (AoS). The main political aim was the restoration of Lithuanian political independence, which was regained in 1990. Between 1988 and 1991, Lithuanians were very active politically and determined to achieve their goals. However, the subsequent development of the economic and social situation brought the Lithuanian economy to the brink of poverty, and some sections of society fell below even that level. This reduced the level of political involvement and undermined trust in political parties and governmental structures. Between 1990 and July 2001, the country had a total of 12 governments. In October 2000 elections to the Lithuanian parliament took place. At that time there were 37 political parties in Lithuania, encompassing only a small part of the population, about 2 per cent, as their members. The dynamics of political life in Lithuania are like the movement of a swinging pendulum, with left and right alternating in power. This has complicated the social and economic situation, which is now subject to strong external influences, such as the Russian economic crisis, as well as internal ones, such as a non-transparent privatization mechanism and uncertain foreign investments. In the absence of any deeper intellectual 'guiding spirit', the planned accession to the European Union and NATO tends to harden political attitudes because of the radical decisions that have to be taken, such as the closure of the main energy source, Ignalina Nuclear Power Plant.

However, there are also positive tendencies in the country's political and socio-economic development, for example the strengthening of Lithuanian contacts with European countries and non-governmental structures, the creation of independent defense structures and the dynamism of the privatization process (in 1999, of 25 000 operational entities designated for privatization 95 per cent had been privatized).

The transformation from central planning to a market economy caused serious economic decline and a structural crisis in Lithuania in the early 1990s, as it did in the other Baltic states. Radical changes in the economic environment that created the necessary preconditions for economic stabilization and revival were introduced mainly between 1992 and 1994 and were accompanied by economic decline. The introduction of national currencies and withdrawal from the 'ruble zone' were the most significant factors in curbing inflation and creating the basis for growth of GDP and production.

From 1995 to 1997, the trend in GDP growth was positive; from 1998 to 1999, this trend slackened and even became negative as a result of internal and external factors. Some growth in GDP was observed in 2000 and 2001 (cf. Table 9.1).

This is the general political and economic context in which science and technology (S&T) developed and continues to develop after the restoration of independence.

1. SCIENCE POLICY – LEGAL AND STRUCTURAL BASIS

Throughout the past decade, with all of its complications, science in Lithuania has been undergoing a process of reorganization and restructuring in order to bring it into line with the new economic and social conditions.

By the end of the Soviet period (1985–1989), Lithuanian science had essentially been subjected to the USSR's constitutional doctrine, laws and some of its institutions (e.g. Committee of Science and Technology of the USSR, Academy of Sciences of the USSR) and regulations, which were almost identically replicated by the Lithuanian by-laws governing the corresponding institutions. Thus, Lithuanian science policies were an organic part of Soviet science policies.

It must be granted that during the Soviet period R&D activities were diversified and that research and teaching professionals had strong positions. Science received extensive funding from the national budget and from contracts with industries and other institutions, while researchers were one of the highest paid groups in the country (the highest paid groups now include lawyers, finance specialists and energy economists).

At the beginning of R&D reform, when the Baltic states were still part of the USSR, issues related to science policy and organizational reform were debated in various forums. Thus the reforms were launched from the bottom up, through initiatives taken by the Unions of Scientists that were founded in the Baltic states in 1988–89. In October 1989, the Lithuanian Scientists Union had over 8 000 members (Mokslo Lietuva, 1990). The government actively supported the structural changes proposed by scientists in order to adapt the

Table 9.1 Lithuania: gross domestic product (GDP)

Year	1994	1995	1996	1997	1998	1999	2000	2001
Million litas (1 USD = 3,2 litas–01.2003; 1 EURO = 3.4527 litas after 01.02.2002)	16 904	24 103	31 569	38 340	42 990	42 655	45 148	47 958
Comparison with previous year (%); estimated at constant 1995 prices	-9.8	3.3	4.7	7.3	5.1	-3.9	3.8	5.9

Source: DSRL, 1999a; 2002.

R&D system as much as possible to the 'Scandinavian model'. A Science and Higher Education Act was passed by the Lithuanian parliament at an early date, in February 1991. This law regulated the structure and autonomy of the university systems and other research institutions, established science councils and determined their functions and established new principles for financing research and new academic degrees. Lithuanian law concentrated on organizations, stressing in particular their autonomy. At the end of 1991, the Science Council was created in Lithuania as an advisory body to the government on matters of science policy. Two thirds of the Council members are elected by scientists and one third appointed by parliament. The Council has no expert commissions and is not charged with financing research but rather with creating the prerequisites for this. In 1994 a Ministry of Education and Science was established.

The most important instrument of research policy in every country is *funding*. There has been a drastic reduction in state funding for research since the Soviet period; direct input from the business sector into R&D is very low. To prepare for the introduction of the grant system, Lithuania organized an evaluation in 1995/96, which was carried out by the Research Council of Norway. A program for the periodic evaluation of research institutions, conducted by local experts, was initiated.

In Lithuania, funds are still allocated directly to research organizations by the Ministry of Education and Science. Grants distributed by the State Research and Education Fund (established in 1994) account for about 1 per cent of all state funds allocated to research and higher education. These funding schemes and responsibilities are enshrined in legislation governing scientific activities. The Lithuanian government continues to rely on direct support to research organizations, distributing additional funds for research in new areas through competitive processes. The country's research institutes have not been fundamentally reorganized or reduced in number; research activities have not been consolidated and research funds are still not being distributed on the basis of scientific merit.

It must be stated that Lithuania still needs a stable science policy based on scientific research and on multifunctional analysis of various parameters as well as the interrelation between them. Available documents only partially meet this need.

On April 14, 1997, the Science Council of Lithuania, which is an independent scientific and research body that acts as scientific arbiter in the competition for government funding, approved a document entitled General Policies of Lithuanian Science and Higher Education Development. It deals with such topics as the goals of development, the system of science and higher education institutions, the integration of science and university studies, the autonomy of science, university studies and national regulatory processes, the financing of science and university studies, standards of science and quality of university studies and the need for legal provisions pertaining to R&D. This 9-page document attempts to look in a more systematic way into Lithuanian science. The need to shape society on the basis of scientific, informational and cultural advancements, which is one condition for equal partnership on a global level, clearly defines the goals of contemporary Lithuanian science and higher education:

1. to aim for international levels of competence in the main fields of science, especially in those of vital significance to Lithuania;
2. to give all able candidates an opportunity to get such higher education and analogous professional qualifications as are recognized throughout the world;
3. to secure an effective scientific potential and to develop, in education and research, the necessary prerequisites for Lithuania's economic, cultural, and social development;
4. to develop a society that is open for education, culture and science.

It is interesting that experts from Lithuania and abroad, notably those of the Research Council of Norway who analyzed the level attained by Lithuanian science in the period from 1994–1996, stated that, despite the inadequate funding of Lithuanian science and universities and the country's weak and outdated technological base, the level of education in Lithuanian society is sufficient to provide a basis for the improvement of living conditions. In many areas, the research that is carried out meets international standards. This situation makes it necessary to devote a greater share of national resources to science and higher education. The aforementioned document also recommends:

– the development of a binary system of higher education;
– the completion of national screening of the higher education program;
– the training of specialists in different areas;
– the forecasting of demand for university graduates until 2005;
– development of the infrastructure;
– the raising of qualification levels among the personnel involved in these studies, and
– the promotion of inter-institutional competition.

The system of scientific institutes and of science organizations is to be further reorganized, with the aim of concentrating funding in the areas of greatest significance to the country's needs, promoting sophisticated competitive science and enhancing the status of university-level institutes.

Certain aspects of Lithuania's declared science policy are very difficult to implement under present-day conditions. These relate not only to the considerable financial problems, but also to the methods of allocating funding to higher education and research, which have to be improved.

A crucial part of policy on the national development of science and higher education is the establishment of a national information system on scientific and innovation activities, the strengthening of international cooperation and the integration of research and teaching. It must be noted that in the last decade of the 20th century Lithuania suffered (and continues to suffer) from a lack of funding for research institutes as well as from an inadequate legal basis for its institutions of higher education. The Institutions of Higher Education Act came into force in July 2000. It defines the national goals of higher education, describes the varieties of higher education institutions, lays down in detail the terms of their autonomy, defines their place in the Lithuanian education and science system, specifies qualification requirements for study programs and teachers, regulates internal and external financing and tackles the crucial issue of free and fee-paying higher education in Lithuania.

This act, together with other legal measures, shows that Lithuania is moving away from abstract autonomy and 'freedom' in the system of science and higher education towards concrete regulation. These measures, which in essence create an autonomous higher education system, limit somewhat the constitutionally guaranteed right to education for a vast majority of Lithuanian citizens (Constitution, 1993).

The Civil Service Act has been in force in Lithuania since July 1, 1999. It pertains to employees of higher education and scientific institutions. Civil servants are subject to rather strict professional and social requirements, e.g. loyalty to nationally approved policies, etc. However, the status of the researcher as a representative of a creative profession is not defined. The law should be amended because of this and other clauses. The problem is being worked on.

An analogous situation can be seen in the forthcoming change of science management and national regulatory structures. After 1992–1993, political activity on the part of scientists abated. The Lithuanian Scientists' Union began to function in a more formal way and membership numbers fell. The subordination of scientific and technical associations and of other scientific organizations was becoming less pronounced. In 1999, a Congress of Lithuanian Scientists was organized to mark the tenth anniversary of the Lithuanian Scientists' Union. On the agenda were such topics as the relationship between science and the state, science funding etc.. In the opinion of the representatives of higher education institutions, Lithuanian science is being transformed. Change is being driven not only by the needs of the state, but also by factors such as qualification, erudition, intellectual orientation and values and often by concrete pressure brought to bear on the science system.

The Soviet mentality has dominated the regulatory processes of Lithuanian science, and has possibly grown even stronger over the last few years. Legislation is determined by the Culture Committee and the Education and Science Committee of the Seimas (parliament) of the Republic of Lithuania, executive power is exercised by the Ministry of Education and Science; in addition, the parliament, the government and the Science Council of Lithuania influence science and higher education in certain areas. The function of national experts falls to the Lithuanian Academy of Sciences (now converted into a very independent institution), the Board of Rectors, the Board of Directors of the National Education Institutes and around twenty other institutions. Thus, by the end of this period in the transformation of Lithuanian science, the country's S&T system is still struggling to become integrated into Western research and economic structures, to meet the needs of the national culture and economy, to secure the financial means necessary to support research and higher education and secure a skilled science base, to be able to promote international cooperation, restrict unwanted emigration etc. At the same time, a line of action of sorts has emerged and is reflected in the laws currently in force, legal judgments and proposed legislation on the regulation of science and higher education. Moreover, the country's research and regulatory system strikes a balance between academic freedom, autonomy and the quest for maximum cost-effectiveness and financial control.

2. STRUCTURAL CHANGE IN THE ORGANIZATION OF RESEARCH AND HIGHER EDUCATION

In the Soviet period there was an established system with three types of organizations: institutes of the Academy of Sciences (AoS), higher education (HE) institutes and (branch) institutes of various types. At the end of the Soviet period in Lithuania, there were no fewer than 12 AoS institutes, 12 HE institutes and 32 branch institutes.

Over the past decade, essential changes have been made to the structure and size of the R&D system (cf. Tables 9.2–9.4). The statistics paint a diverse picture. Around 1990, none of the requirements of the Frascati Manual (OECD, 1994) was being met. In 1996 the statistical service started to implement OECD methods. The organizations included in national statistics were broken down by sector (higher education, public sector, business and industry, non-profit organizations), a full-time equivalence workforce was defined, etc. However, data from the transitional period should be interpreted very cautiously and conclusions should be drawn only when additional studies are available.

Table 9.2 Lithuania: number and structure of R&D institutions

	1990	1995	1999	2000	2001
Number of organizations of which:	62	86	104	147	175
– State universities	3	6	8	9	10
– State HE establishments	10	9	8	6	7
– Private HE establishments	.	.	3	3	3
– AoS research institutions	17	.	.	1	1
– State research institutes	.	29	29	29	29
– Scientific (departmental) research institutions	.	14	.	.	.
– Enterprise sector	32	28	26	60	88
– Private non-profit sector	.	.	4	.	1

Source: DSRL, 1998–2002.

Table 9.2 and some supporting data suggest that the number of research institutions is growing because many big research institutes are being split up into smaller units. Certain sectors, i.e. the Academy (later referred to as the public sector), HE institutions and branch institutions (later referred to as the business and industry sector) have been transformed in many different ways.

At the beginning of the transition process in 1990, the number of HE institutions shrank somewhat to 8. Subsequently, however, the sector gained strength and by 1992 the number of HE institutions had risen to 14, compared with 12 in 1989. In the 1996–1999 period the figure was 19. The sector consisted and still consists mainly of state educational institutes. Private schools of higher education face many difficulties in establishing themselves. The first private HE establishment to obtain recognition was the Business Management School. It was founded with Norwegian investment funds and developed out of the Business Management Center at Kaunas Technological Institute.

In the last few years there has been a slight decline in the number of organizational units in the public sector (1996 66 institutions, 1997 65, 1998 62, 1999 59). The attempt to integrate higher education institutions and research institutes produced virtually no results. However, in the 2000–2001 period more resolute action was taken to merge the higher education sector with the research institutions. On June 22, 2000, the Lithuanian government approved the Proposal on Structural Alterations put forward by the Department of Science and Education, part of the Ministry of Education and Science

Table 9.3 Lithuania: R&D personnel and researchers

	1990	1992	1993	1994	1995	1996	1997	1998	1999	2000	2001
R&D personnel of which:	31 000	11 786	11 420	12 645	13 632	16 067	15 436	*15 561	*15 296	*14 592	*14 980
– Researchers (only 1990)/ scientists (with PhD)	15 400	5 652	6 389	5 872	6 133	5 769	5 495	5 588	5 663	*5 377	*5 130
– in %	50	48	45	46	45	36	36	36	37	37	34
– R&D personnel per 1 000 labor force		.	7.7	7.3	7.8	9.0	8.7	8.8	8.5	8.1	8.6

*Number of employees in the main working place only
Source: DSRL, 1998–2002.

(MES, 1999). In accordance with this Proposal, 13 out of a total of 29 state research institutions remain, some having been merged, while 6 have become science and research institutes within the universities. The remainder continue as State Research Organizations without core state funding. There was a similar downward trend in the number of institutions in the enterprise sector: 1996 30, 1997 33, 1998 29, 1999 26 institutes. There was virtually no shrinkage in the non-profit sector: 1996 6, 1997 3, 1998 5, 1999 4. Table 9.2 presents an overall figure for the enterprise and non-profit sector.

The largest number of researchers is employed in the HE sector. However, there are more technical and assistant personnel in the government (public) sector (with technical personnel in the Academy institutes in 1996 accounting for 54.6 per cent, in 1997 56.2 per cent, in 1998 65.4 per cent and in 1999 61.5 per cent of the overall number of technical personnel). The greatest number of other supporting staff was also employed in the government sector: 1996 49.9 per cent, 1997 55.0 per cent, 1998 53.0 per cent, 1999 57.7 per cent and 2000 55 per cent.

This whole period from 1991 to 2001 was characterized by an attempt to bring some stability to personnel numbers (researchers/scientists). As has already been indicated, the researchers were concentrated mostly in the HE sector (1996 3 911, 1999 3 983). Personnel levels fluctuated in the various sectors. In 1996, there were 1 769 researchers in the government (public) sector, with the number falling to 1 634 in 1999; in the enterprise sector the numbers dropped from 77 in 1996 to 39 in 1999. These fluctuations were influenced by science policy and overall economic developments.

The difficulty of analyzing and comparing personnel figures from the 1990s and later years is further compounded by the use of different statistical methods. For example, in the 1990s the number of researchers includes scientists with PhD and people working in science without a science degree; after the 1991 Science and Higher Education Act came into force, only researchers with PhD were taken into account. Incidentally, the number of scientists hit its first low in 1997, falling to 5 495. In 1997–1999 it went up slightly, without ever exceeding 5 700; in the years 2000 and 2001 a new decline to 5 060 ocurred (cf. Tables 9.5 and 9.6).

In Lithuania – as in Latvia – a politically motivated attempt was made to review the former Soviet academic degrees. This was a complicated procedure, involving much paperwork and producing little significant gain. Lithuanian science degrees and academic titles were subject to validation (nostrification). The renaming of degrees was part of Lithuania's attempt to shake off its Soviet past: in place of the former Soviet degree of Doctor of Sciences, that of Doctor habil. (PhD with post-doctoral university teaching qualification) was introduced, while the degree of Candidate of Science was replaced by the Doctor

(PhD). Table 9.5 shows the shifts in these qualifications over the past years and the gender proportions within the groups.

Table 9.4 Lithuania: distribution of researchers by type of organization (%)

	HE	Public Sector*	Other Institutions
1995	69.4	25.2	5.4
1999	71.8	26.5	1.7
2000	71.2	25.4	3.4
2001	71.0	28.0	1.0

*State research institutes
Source: DSRL, 1996–2002.

There have been changes in the age structure of scientists/researchers that might indicate a less than certain future for Lithuanian science, since the country's scientific community, quite predictably, has aged (cf. Table 9.7).

Specialists in the natural sciences still account for the largest share of scientists (although the trend here was significantly downward from 1992 to 1996, for example), while the second largest group consists of those engaged in technology research, who have maintained this position despite a 4.6 per cent drop between 1992 and 1996. The share of humanities specialists is growing (up 5.6 per cent between 1992 and 1996) and the share of the social sciences has grown by 2 per cent. Agronomists have retained a fairly stable share. Researchers in the medical sciences have grown in number by 0.7 per cent [1]. There are several reasons for the growth tendency in the humanities. The two most important ones seem to be 1) the 'humanization of science', which is one of the hallmarks of Lithuanian scientific institutions and 2) that fact that, as the complex task of economic restructuring continues, areas of research that require expensive technological support are less likely to be able to develop quickly and may even slow down.

Immediately after the restoration of independence in Lithuania, there were difficulties in finding students willing to undertake doctoral studies in the country (this being the period when the total number of students in the higher education institutions of Lithuania decreased too). Later, in 1997–1998, the numbers tended to grow (cf. Table 9.8). The cost-effectiveness of doctoral studies

---

[1] The Lithuanian Department of Statistics has started restructuring data on science, so that technological sciences now appear in place of technical sciences, physics alongside natural sciences, etc. This makes analysis of the data more difficult. However, the above mentioned tendencies can still be said to be discernible. Besides, the natural science community is mainly concentrated in the AoS institutes, and technology in the sector of higher education institutions.

Table 9.5 Lithuania: scientists, scientific degrees and academic titles

	1993		1994		1995		1996		1997		1998		1999		2001	
	Total	In which women	Total	in which women	Total	in which women	Total	in which women	Total	in which women	Total	in which women	Total	in which women	Total	in which women
Total number of scientists	6 389	1 950	5 872	1 828	6 133	1 903	5 769	1 844	5 495	1 789	5 588	1 828	5 633	1 866	5 060	1 802
Habilitated doctors	812	88	856	94	948	115	843	114	768	110	794	113	833	116	741	102
of which																
– Professor	550	50	568	44	650	54	612	55	542	54	583	55	612	62	574	58
– Lecturer	102	7	122	9	138	24	90	17								
Doctors/ PhD	5 289	1 791	4 777	1 688	4 869	1 704	4 619	1 643	4 423	1 601	4 494	1 639	4 572	1 670	4 219	1 673
of which																
– Professor	80	8	60	5	59	7	56	6	55	4	55	5	52	4	39	6
– Lecturer	2 563	751	2 298	674	2 399	704	2 305	693	2 186	679	2 293	729	2 356	781	2 076	711

Source: DSRL, 1998–2002.

Table 9.6 Lithuania: scientists by field of science (without business sector)

Areas of science	Number of scientists	
	2000	2001
Humanities	895	787
Social sciences	825	824
Technological sciences	1 108	1 069
Physical sciences	1 011	997
Agricultural sciences	359	308
Natural sciences	546	515
Medical sciences	589	560

Source: DSRL, 2001, 2002.

Table 9.7 Lithuania: age structure of scientists (in %)

Age in years	< 30	30–39	40–49	50–59	> 60	Total
2001	2.0	16.0	29.0	29.0	24.0	100
2000	2.0	15.0	28.0	31.0	24.0	100
1999	0.9	12.8	27.6	30.9	27.8	100
1998	0.5	13.7	27.2	34.2	24.4	100
1993	0.6	23.7	27.8	37.9	9.0	100

Source: DSRL, 1998–2002.

is not optimal: in 1997 only 157 individuals completed a PhD program and in 1998 there were only 171 (DSRL, 1999b, p. 38). Preferences for fields of study have also changed. There has been a rapid growth in the number of doctoral students in the social sciences, although these figures are not matched by any similar increase in the number of completed dissertations. The same trend is evident to a lesser extent in the humanities, while technical sciences and engineering are struggling to remain in the lead, with natural sciences lagging behind in this regard.

Thus between 1991 and 2001, the structure of research and education changed markedly. HE institutions expanded and grew stronger, while the universities also gained in importance. After the restructuring of the institutes of the Lithuanian Academy of Sciences, the Academy became a public-sector entity, retaining its status, albeit with some losses. It was in the industrial branch sector that developments were most negative. Restructuring led to a decline in R&D and it was this sector that suffered the greatest loss of scientific personnel. In the other sectors, there is a marked trend towards stability in the number of research personnel.

Table 9.8 Lithuania: PhD students by field of science

Areas of study	Number of doctoral students		Women doctoral students		Completed doctoral studies	
	1997	1998	1997	1998	1997	1998
Humanities	251	344	169	245	41	58
Social sciences	464	600	275	359	47	46
Natural sciences	133	195	41	70	25	20
Biology and medicine	249	307	128	177	21	24
Technology	379	429	104	121	23	23
Total	1 476	1 875	717	972	157	171

Source: DSRL, 1999b.

3. THE FUNDING OF SCIENCE

Among the most important problems in post-socialist Lithuania, as well as other post-socialist countries, is the lack of adequate funding. The directors of public-sector scientific institutions view reducing the number of employees as a partial solution to this problem. Though the practice of institutional funding does, in fact, continue in Lithuania, individual researchers and teams are becoming more active in searching for funds on both a competitive and contract basis inside and outside Lithuania.

However, in reality, the underlying trends in the financing of Lithuanian science remain the same and they are related to economic development. Both Lithuanian politicians and scientists declare that, ever since the country regained its independence, science has been a top economic priority in Lithuania either in conjunction with university studies or as a means of serving the needs of the national economy.

There are several reasons why it is so difficult to analyze the real funding situation in Lithuanian science. First, in the period under investigation here, different currencies were in circulation (the ruble, coupons, litas...) and the inflation rate varied dramatically. Second, the data presented in statistical reports were collected by different methods. Thus we are dealing in essence only with fragmentary or relative data. It should also be acknowledged that R&D funding often used to be combined with higher education funding in national accounts. Since the number of higher education institutions in Lithuania has been growing, the general tendency has been for the expenditure figure to increase. In 1998, the combined figure had reached 1.28 per cent of GDP.

However, it is necessary to analyze the actual financing of R&D separately. Let us establish our starting position in 1989, when the portion of the national budget that was allocated to science was 0.4 per cent (DSRL, 1989–1990). However, if we add here the money obtained from self-financing contracts, the sum was much greater, almost one third higher in fact (estimated 0.5–0.6 per cent of GDP). In 1989, state funding was still increasing (e.g. funds for the Academy of Sciences rose by 11.8 per cent in that year). In 1990, though the national budget was shrinking, there was no major impact on the financing of R&D (DSRL, 1991). Serious financial problems began to emerge in the 1992– 1993 period.

Since our aim here is statistical reliability, Table 9.9 includes figures from 1993 onwards. The intensive efforts to stabilize research funding are evident here. Nevertheless, financing from the national budget should increase to 0.8– 1.0 per cent of GDP and the figure for science and higher education expenditure should reach 2 per cent of GDP. These figures appear to be of vital importance in the 21st century (Kaulakys, 1999).

Considerable attention has been paid to the ratio of R&D expenditure to the GDP. However, it is important to analyze that share of the national budget covering science and university studies. In 1992 this figure was 7 per cent, in 1993 5.7 per cent and in 1994 6.4 per cent (DSRL, 1997, 1998). The reference point subsequently shifted to 7 per cent. Though the share of the national budget allocated to scientific research was 7.5 per cent in 1999, this still appears to be insufficient.

The increased funding of science and higher education became evident only in one instance, when scientists' salaries fell to a disproportionately low level and the Prime Minister came under pressure from various groups (Lithuanian Science Union, Conference of Rectors of the Higher Education Institutions, or the Board of Directors of the Science Institutions) to raise them. The last time such a raise was implemented was in 1997. In 1998, higher education institutions received 31.4 per cent more funding from the state budget than in 1997 (DSRL, 1998, 1999). Though these measures placated the workforce to some extent, they have said little about the material basis provided by the fund for higher education institutions equipment and libraries. All state research institutes, 29 in total, received funding at a level of 0.18 per cent of GDP. Table 9.10 shows the changes in the sources for R&D financing since 1996. The Russian crisis had a major impact on many potential customers of Lithuanian science, both within the country and abroad. In consequence, core state funding became even more important and accounts for three quarters of all funding for both the higher education institutions and the governmental sector. It has also become more important in the business sector.

It must be noted that Lithuanian participation in international projects is expanding, but much more slowly than was expected (MES, 1999; 2000). It appears difficult to put in place a grant-based system of funding in Lithuania

Table 9.9 Lithuania: expenditure on R&D

	1989	1993	1994	1995	1996	1997	1998	1999	2000
Total in USD, thousands (1 USD = 4 litas, after 01.02.2002; 1 EURO = 3.4527 litas)		12 770.7	21 820.2	28 729.7	41 216.4	54 305.9	61 123.8	55 075.0	67 450.0
Share (%) of:									
– Basic research	.	.	53.6	52.6	39.5	41.1	46.6	55.7	41.7
– Applied research	.	.	37.0	39.5	41.6	44.1	43.3	34.5	36.3
– Experimental development	.	.	9.4	7.8	18.9	14.8	10.1	9.8	22.0
Share of GDP (%)	(0.5–0.6)	0.35	0.52	0.48	0.52	0.57	0.57	0.52	0.60 estimated

Source: DSRL, 1998–2001.

LITHUANIA: THE SCIENCE SYSTEM FROM 1989–2001 181

Table 9.10 Lithuania: sources of finance by sector (in %)

	Total				Sector of Higher Education		Public sector		Business sector	
	1996	1998	1999	2001	1996	1998	1996	1998	1996	1998
In Total	100	100	100	100	100	100	100	100	100	100
– From national budget	70.4	74.4	72.4	57.9	77.0	74.9	72.3	76.6	0.7	3.6
– From contracts	22.4	17.2	14.7	12.1	19.7	17.0	22.9	16.4	21.4	30.0
with:										
+ Lithuanian institutions	89.5	84.6	83.3	.	88.5	85.3	89.7	85.4	86.7	55.5
+ CIS countries	0.8	.	0.7	.	.	0.2	0.4	1.0	13.3	1.4
+ other countries	9.7	14.7	16.7	.	11.5	14.5	9.9	13.6	.	43.1
– Through international projects	1.8	2.5	3.4	3.7	2.0	3.6	1.2	1.9	.	.
– Other sources	5.6	5.9	9.5		1.3	4.5	3.6	4.8	77.9	66.4
– Own funds				23.1						

Source: DSRL, 1997–2002.

because the tradition of institutional funding is very strong. The ratio between these two types of financing is 1:5.

Analysis of the expenditure on research in different areas of science reveals the preeminence, in absolute terms, of technical and natural sciences. Social sciences are financed more generously than the humanities. For example, in 2000 expenditure on R&D by field of science was: technology 25.4 per cent, biomedical science 30.5 per cent, natural sciences 20.9 per cent, humanities 12.5 per cent, social sciences 10.7 per cent.

In the HE sector medical and technical sciences each receive slightly more than one fifth of all funding, while in the governmental sector about one third goes to the technical and to the natural sciences. The amount of financing is correlated with the size of the scientific community within the field. A more dynamic change in the situation is seen in the enterprise sector. In 1996, funds allocated to natural and technical sciences were almost equal (49.5 per cent and 44.8 per cent respectively), but in 1998 technological sciences (including technical sciences) received 79.7 per cent of all financing. Clearly, account has to be taken of the new structure of scientific disciplines with their new orientation towards the creation and application of technologies.

Thus any evaluation of the science funding in Lithuania must be differentiated according to discipline. The optimal balance between the different areas of science and in relation to GDP has not been achieved. This is due to both interior and exterior factors and also to government policies, which have failed to take full account of future development needs. Over the last few years, budgetary appropriations have mainly covered the researchers' salaries and research institute maintenance costs. In 1995, these two items accounted for 95.1 per cent of expenditure, which left very little for the renewal of equipment. Equipment is being renewed to a limited extent by other means, e.g. through competitive financing. Also, the influence of the Science Fund of Lithuania is growing in different areas, though not in the funding of new equipment.

There are plans to allocate about 116 million US dollars from the National Privatization Fund mainly to support the fund for small and medium-sized enterprises (SMEs) (about 75 million dollars). Lithuanian science will receive about 5 – 7 million US dollars from this fund. This is a relatively small sum but is, nevertheless, absolutely indispensable for the purchase of new equipment, textbooks and scientific journals. Thus, the funding of Lithuanian science, like that of science in other post-socialist countries, is dependent not just on economic factors, but also on the value orientation of society and scientists themselves, i.e. the ability to see the role of science in shaping the further growth of the country. Lithuanian science has to produce results if funding is to be increased. In 1994 – 1995, the Science and Technology Union produced about 0.5 per cent of GDP and, according to the frequency of citation index, Lithuania stood in third place among CEECs, being surpassed only by Hungary and Estonia and followed by Slovenia, Poland, Latvia and the Czech Republic.

4. SUMMARY

A general survey of the transformation of Lithuanian science from 1989 to 2001 reveals that early revolutionary changes in Lithuanian politics and in its economy were followed by evolutionary shifts within the Lithuanian science system. The pivotal point in the process of change was and remains a scientific community which, forced to operate under deteriorating economic conditions and to deal with dramatic alterations in the funding of science, struggled to survive, both physically and intellectually, and to safeguard academia. This is reflected in the restructuring of the governmental science sector, during which the AoS institutes almost turned into independent institutes within the Academy. Individual institutes sought to protect their autonomous status by rebuffing the attempts of state functionaries to merge them with higher education institutions. The integration of science and higher education in Lithuania has proceeded on the basis of subject rather than structure, though in the year 2000 some steps were taken towards greater structural integration (cf. EC, 2001).

The most complicated matter appears to be the fate of industry-oriented science. Here we encounter a vicious circle: the business-oriented institutes await support in order to be able to function, while industry in Lithuania is expecting solutions to emerge during the stagnation crisis. Here we should emphasize how inadequate the results of participation in international programs and projects are.

The academic freedom of scientific institutes is being increasingly restricted by new legislation and new management structures. On the other hand, it is being used positively in the area of international cooperation, which is likely to expand in future.

New science policy priorities will appear along with the economic priorities in the country. At this point the vitality of Lithuanian science depends on its ability to perform two functions: to develop a national scientific culture and to apply the results of research for the good of the national economy.

REFERENCES:

Constitution (1993) *Constitution of the Republic of Lithuania*. Vilnius: Seimas of the Republic of Lithuania Publishers. (in Lithuanian)
DSRL (1989–1990) *Statistical Yearbooks of Lithuania 1988 and 1989*. Vilnius: Department of Statistics of the Republic of Lithuania (DSRL). (in Lithuanian)

DSRL (1991 – 2002) *Research activities – Collection of Statistics*. Vilnius: DSRL. (in Lithuanian)
DSRL (1999a) *Economic and Social Growth in Lithuania*. Vilnius: DSRL. (in Lithuanian)
DSRL (1999b) *Education – Collection of Statistics*. Vilnius: DSRL. (in Lithuanian)
EC (2001) *Regular Report on Progress towards Accession – Lithuania*. European Commission (EC). Internet: europa.eu.int/comm/enlargement.
Kaulakys, B. (1999) "Financing of Science and the Development of Society." *Mokslo Lietuva* (Lithuania Scientific), October 21 and November 4. (in Lithuanian)
MES (1999) *White Book on Lithuanian Higher Education*. ES PHARE Program. Vilnius: Ministry of Education and Science (MES).
MES (2000) *Financing of Lithuanian Higher Education in the New Millenium*. ES PHARE Program. Vilnius: MES.
Mokslo Lietuva (1990) *Mokslo Lietuva* (Lithuania Scientific) nos. 5 – 60. (in Lithuanian)
OECD (1994) *Proposed standard practice for surveys of research and experimental development – Frascati Manual 1993*. Paris: OECD.

10. POLAND: RESTRUCTURING S&T WITHOUT RADICAL TRANSFORMATION

Jan KOZLOWSKI

After Poland regained its independence in 1918, Polish science experienced a period of strong institutional development, especially from the mid-1930s onwards. A significant number of higher education institutions and (mostly publicly funded) research institutes were founded. During the occupation by Nazi Germany and Stalin's Soviet Union (at the beginning of World War II), Polish science lost its base. Universities and scientific societies were disbanded and scientists were subject to a special form of persecution and not permitted to do research. After World War II, a reconstruction that can be characterized as the 'nationalization' of science took place. This reconstruction was based broadly on the general Soviet model. However, there were also some significant deviations from this model. Firstly, Poland's S&T system was less isolated from the West than that of most other Eastern European countries; provided they were not engaged in opposition movements, Polish scientists were free to travel to Western Europe and the USA. Secondly, there was less of a gap between scientific research and training and teaching than in other socialist countries. The higher education sector was quite heavily involved in both basic and applied research and scientists from the Polish Academy of Sciences (AoS), higher education institutions and R&D institutions were active in so-called 'branch research' programs. Thirdly, although most industrial and/or applied R&D was carried out in industrial (branch) R&D units, there were also 'development units' undertaking R&D at enterprise level. Finally, in contrast to other CEECs, Poland experienced a severe economic recession in the early 1980s and had already had to reorganize its S&T system at that time under conditions of decreasing financial resources and personnel.

1. A NEW INSTITUTIONAL FRAMEWORK FOR S&T

Right from the beginning of the transformation period, there seems to have been a kind of consensus within the new post-communist government and among large parts of the scientific community in Poland as to the general goals of the reform in this field. These were to democratize the S&T system and to introduce greater autonomy; radical transformation was not considered in the hope of avoiding a breakdown of the whole S&T system as had occurred during the early 1980s. Following debates in the Polish Parliament in 1990 and 1991, significant legislative changes were implemented (Beek, 1995). All previous legislation in the field of S&T (except for that governing the Polish AoS) was abolished, the law governing R&D institutes was amended and new legislation on higher education and academic titles and degrees and on the establishment of a state committee on scientific research (the so-called KBN) was approved.

The KBN Act created a new body that acts simultaneously as a government agency (its chairperson being a minister and member of the government) and as a national council representing the scientific community (with most of its committees elected by every Polish citizen possessing a PhD degree). The KBN is responsible for the formulation of national science policy and for the funding of the entire R&D/S&T sector. In reality, the Act regulates the basic relationship between science, the state and the central administration and determines the basic mechanisms for the budgetary financing of S&T activities. Thus the Act has radically changed the entire system of funding basic and applied research in all S&T sectors, introducing a structured procedure for the competitive and differentiated distribution of public funds for S&T projects. Except for some military R&D, which is funded by direct transfers from the Treasury to the Ministry of Defense, all government support for separately budgeted research is channeled through the KBN. Finally, the process of allocating public R&D funds has been made subject to the twin principles of evaluation and competitiveness. An evaluation procedure for R&D institutions has been introduced, the outcome of which determines the amount of funding each institution receives. This procedure is based mainly on academic criteria, such as the volume of publications and so on. Secondly, a competitive grant system has been introduced, in which project proposals are accepted or rejected on the basis of peer review.

Since as early as 1991, these laws, the establishment of the KBN as the key state S&T institution and binding regulations governing the state funding of institutions and projects on the basis of evaluation have formed the fundamental political framework within which S&T has been adapted to meet the demands of the market economy and of democratic government.

KBN policy has consisted mainly of altering the general conditions for state support of (state-owned) R&D institutions and using the evaluation process and the grant system to identify those institutions with the highest levels of competence and performance. Indirect means of stimulating innovation through tax breaks, customs policy, protection of consumer rights and so on remain relatively underdeveloped. The Polish Government and the KBN, as the body mainly responsible for formulating S&T policy, went one step further in 1997 by publishing a document outlining the preferred direction of R&D with a view to increasing the innovativeness of the Polish economy. This document identifies those areas to be prioritized in the allocation of state financial and administrative assistance. However, it also contains some critical comments on the new S&T system, and especially on the KBN. One criticism relates to the representation of the various areas of S&T activity on the KBN committees. Since higher education institutes comprise the largest share of the scientific community (in terms of personnel at least), their elected representatives now dominate the various decision-making bodies of the KBN. Critics argue that its committees are therefore not competent to make informed judgments on applied or industrial research.

2. SHIFTS IN THE STRUCTURE OF THE S&T SYSTEM

The changes in the political framework of S&T occurred relatively quickly and on the basis of a broad consensus agreement between the political and scientific spheres. They resulted in changes in the institutions engaged in scientific work that were evolutionary rather than revolutionary in character, with facilities being largely preserved but with reductions in funds and personnel (cf. Tables 10.1 and 10.2)

The total number of industry or branch S&T units remained roughly constant between 1990 and 1995. The number of higher education units (HEUs) increased to 286 in 1999 (103 state universities and 183 non-governmental institutions). The number of AoS units has remained constant at 81. The number of branch R&D units fell to 240 in 1999, including 115 industrial laboratories managed by the Ministry of Economy. The number of in-house R&D units in enterprises employing more than 5 persons has increased in recent years and totaled 402 in 2000.

As regards the scientists, the reductions appear to have been less drastic than in other countries. However, there have been severe reductions in industrial branch R&D units. Since 1994 there has been a tendency toward stabilization. In 2000 there was a total of 55 174 researchers in Poland (FTEs) (cf. Table 10.3).

The Higher Education Act stipulates that all higher education institutions should operate in accordance with the principle of freedom of research and teaching and with respect for human rights, patriotic values and democracy. The HE system has become highly differentiated. New organizational units have been set up in established higher education institutions and levels of education have become more differentiated. There has been a radical change in the relations between university authorities and government as well as within universities. Many new institutions have been founded (most of them not controlled by the state) with the right to teach at different levels (bachelors' de-

Table 10.1 Poland: number of S&T units

Category	1988	1989	1990	1991	1992	1993	1994	1995	1996	1997	1998	1999	2000
HE units (conducting R&D activity)	*	*	80	85	76	88	100	104	104	104	114	115	114
AoS Units	78	81	79	79	81	82	81	81	81	81	82	81	81
Branch R&D Units	297	297	260	296	252	310	274	252	255	256	246	240	240
In-house R&D	679	*	*	*	*	*	366	295	344	373	438	498	402
S&T Service Units	58	33	51	66	56	44	7	4	10	6	25	25	23
Total Number of S&T Units	(1 112)	(411)	(470)	(526)	(465)	(524)	828	733	794	820	905	955	860

* data not available

Source: GUS Scoreboards; data since 1995 according to OECD methodology, therefore not comparable to previous years.

Table 10.2 Poland: total employment in S&T units

Categ.	1988	1989	1990	1991	1992	1993	1994	1995	1996	1997	1998	1999	2000
HE Units	47 742	50 524	50 048	51 385	50 688	32 745	32 745	34 883	39 046	40 977	42 478	42 948	41 499
AoS Units	11 593	11 339	10 690	9 905	9 202	8 782	7 777	8 089	7 705	7 262	7 600	7 486	7 233
Branch R&D Units	90 698	80 610	72 060	60 556	55 935	49 586	29 720	30 900	27 836	26 158	25 160	23 918	23 044
In-house R&D	47 783	*	*	*	*	*	8 929	8 857	8 621	9 220	9 083	7 879	6 906
S&T Service Units	2 970	4 785	4 606	3 253	3 490	2 351	77	72	140	187	189	137	43
Total	200 786	(147 258)	(137 404)	(125 099)	(119 315)	(93 464)	79 248	82 801	83 348	83 804	84 510	82 368	78 925

* data not available

Source: GUS Scoreboards; data since 1995 according to OECD methodology, therefore not comparable to previous years.

Table 10.3 Poland: total number of researchers

Category	1988	1989	1990	1991	1992	1993	1994	1995	1996	1997	1998	1999	2000
HE Units	47 742	50 524	50 048	51 385	50 688	50 861	52 106	27 388	31 134	32 846	36 472	35 284	34 242
AoS Units	4 538	4 587	4 388	4 385	4 020	3 972	3 992	4 812	4 767	4 861	4 745	4 776	4 710
Branch R&D Units	12 718	10 912	10 552	9 314	8 396	7 820	7 874	14 491	13 394	13 893	13 107	12 376	12 386
In-house R&D	995	*	*	*	*	*	*	3 041	3 112	3 900	3 605	3 911	3 691
S&T Service Units	175	160	148	117	98	60	*	7	68	102	113	86	141
Total	66 168	(66 183)	(65 136)	(65 201)	(63 202)	(62 713)	(63 972)	49 739	52 475	55 602	58 042	56 433	55 174

Note: Data on 'development units' (in-house R&D) is not available between 1989 and 1992; these units also include laboratories, construction bureaus and experimental farms.
Source: GUS Scoreboards; data since 1995 according to OECD methodology, therefore not comparable to previous years.

grees, specialized diplomas and so on). The humanities, economics, management/business and legal studies have increased in popularity, while student numbers have dropped in technical, medical and agricultural sciences. Research activity, including that undertaken in research centers in the higher education sector, has been less affected by these changes; there has been a reorientation towards criteria introduced by the KBN and research is becoming increasingly commercialized. The Ministry of National Education funds institutions' statutory teaching activities, while their research activities are funded by the KBN. The economic transformation is expected to open up new sources of funding, but effective interaction with industry appears very unlikely in the near future.

In the Polish AoS, there have been no significant changes in relations between management and institutes since 1989, although the power of the President of the Academy was strengthened in 1997 and the institutes have acquired the status of 'legal entities' in their own right. Most AoS institutes have adapted, more or less successfully, to the new 'rules of the game' imposed by both the KBN, which involve competing for the highest category of evaluated institutes in statutory funding and for research, and the market, with work being commissioned by government and industry. In particular, they have become involved in teaching, with five university-level institutions being established under the auspices of the AoS, and have extended the scope of their applied research. Applied research predominates in over 30 AoS institutes (10 of which specialize in technical sciences, 8 in agricultural sciences, 8 in earth sciences and 5 in medical sciences), whereas most AoS institutes are engaged mainly in basic research. However, attempts seem to have been made recently to introduce a commercial element into the institutes' activities and to establish closer links with enterprises.

There was a dynamic growth of non-profit organizations after 1989, and 454 of them declared themselves to be undertaking "research activities" in the year 1994 (KLON/JAWOR, 1995). It seems that these activities relate primarily to expert opinions, consulting and advisory services and far less frequently to original research. After 1989 most of the important institutions liquidated during the communist era were re-established, for instance the Polish Academy of Sciences and Arts. There are also a growing number of foundations. Some of them have become an important source of funding for S&T, while others have acquired a strong reputation for excellence in research.

Industrial R&D activities are mainly carried out in two categories of organization: industry or branch R&D units and in-house development units. As of the year 2001, the Ministry of Economy supervised 115 R&D units. The institutes have been forced to adapt, scaling back research activities for which there was no prospect of demand from industry, and/or extended their S&T activities to meet the market demand for standardization, testing, measurement, etc. They have also substantially reduced the number of people they employ. The basic pattern of restructuring during the period of transition was to increase non-R&D activities at the expense of R&D activities. As a consequence of the poor financial situation and the low salaries paid to researchers in comparison with other professional categories, the ageing of the workforce is a serious problem. Privatization of institutions is voluntary, and the process has to be initiated by the research institution itself ('the unit's managers and employees'). Certain areas are excluded from privatization, namely defense, health and environmental protection, agriculture, education, mining safety, epidemiology and meteorology. So far none of the independent research institutes in Poland has been privatized. This is mostly due to the institutes' unwillingness to lose the privileges they are guaranteed in law and to the absence of a relevant legal framework. It is, however, common for units to become partially privatized by setting up new or joining existing commercial companies and transferring a specific group of employees from the R&D unit in question to the new organization. As a rule, it is production and service activities that are spun off, with the newly established companies cooperating with the original R&D unit and drawing on its research output. It is estimated that approximately 50 per cent of all R&D units have undergone such changes. A general reorientation of activity profiles has also taken place, with R&D activities being reduced in most cases and production and/or service activities increased.

The industry or branch R&D institutes have seen a decline in the share of their income coming from public funds. The extent of this decrease depended largely on the outcome of the evaluation that was carried out by the KBN. Total employment fell by 45 per cent between 1988 and 1993 and then stabilized. The number of scientists working in these units, however, fell by a smaller percentage (38 per cent) and within the various subcategories of scientists, the number of professors (as opposed to lecturers and assistants) almost doubled. This may be interpreted as a reaction by the independent research institutes to the new modalities of state funding, which are based mainly on scientific achievements. The relatively significant R&D potential contained in the industry or branch R&D units is still rather disconnected from the wider economy. Continued state funding enables these institutes to carry out research without much regard for the needs of end users, whether consumers or firms (Jablecka, 1995). This is indicative of the continuing fragmentation of the R&D and innovation system. Furthermore, neither the research institutes themselves nor the government have any plans for privatization. On the one hand, most of the old branch institutes have been able to retain an important part of their scientific and technical potential from the socialist period. On the other hand, they have had to introduce a commercial component into their activities, mainly in the form of direct contracts with private companies. As a result, they are now half research institutes and half commercial enterprises. They are effectively caught between two stools: their R&D activities are too extensive for them to survive as a private firm in a market economy without public subsidies and their commercial activities are now too significant for them to be considered solely as research centers (at least by Western standards). The R&D units operating within companies are almost exclusively financed by the company to which the unit belongs (cf. Tables 10.4 and 10.5).

Table 10.4 Poland: current expenditure on R&D activities
(in mill. National Currency)

Category	1995	1996	1997	1998	1999	2000
Higher Education	561.0	768.5	961.9	1 106.7	1 274.3	1 201.4
AoS	265.6	318.5	382.6	429.2	497.0	477.1
Branch R&D	1 010.5	1 178.8	1 390.8	1 576.5	1 815.3	1 666.4
In-house R&D	292.9	489.2	619.0	881.5	987.7	604.8
S&T Service Units	2.8	6.4	6.7	11.2	16.2	31.8
Total	2 132.8	2 761.4	3 361.0	4 005.1	4 590.5	3 981.5

Source: GUS Scoreboards.

Table 10.5 Poland: sectoral structure of current expenditure on R&D activities (%)

Category	1995	1996	1997	1998	1999	2000
Higher Education	25.6	27.8	28.6	27.6	27.8	30.2
AoS	12.7	11.5	11.4	10.7	10.8	12.0
Branch R&D	47.8	42.7	41.4	39.4	39.5	41.9
In-house R&D	13.8	17.8	18.4	22.0	21.5	15.1
S&T Service Units	0.1	0.2	0.2	0.3	0.4	0.8
Total	100	100	100	100	100	100

Source: GUS Scoreboards.

The transformation of the industrial R&D sector took place against a general economic and industrial background characterized initially by deep recession, which hit manufacturing particularly hard. Between 1989 and 1991, Poland's GDP decreased by about 15 per cent; it has been increasing continuously since then, however, finally returning to its 1989 level at the end of

1995. Poland was the first of the CEECs to experience positive growth (cf. Figures 10.1 and 18.2). GDP grew at rates of 6 to 7 per cent per annum between

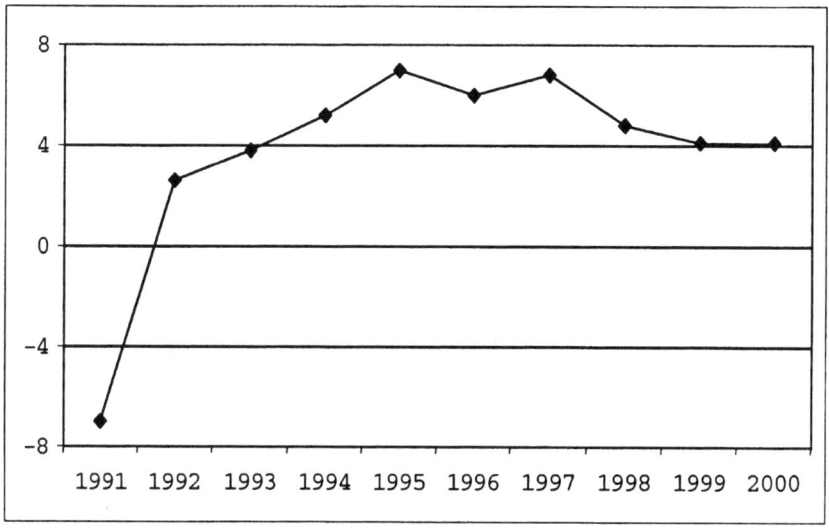

Fig. 10.1 Poland: annual growth rate in GDP (%)
Source: GUS Scoreboards.

1995 and 1997, inflation declined gradually and living standards improved. This growth largely reflected the dynamism of the private sector, which far outstripped the mediocre results of the bloated state-owned enterprises. The entire business environment needed to be compatible with the requirements of a market economy, and all institutions had to be created almost from scratch. Poland's lengthy period of economic expansion has been underpinned by several waves of structural reforms. The price and trade liberalization of the early 1990s was followed from 1998 onwards by large-scale privatization and product market deregulation and the launch in 1999 of four important structural reforms related to public finances (public administration, pensions, education, and health care). The tax system has undergone significant changes in the last ten years and performed well in that time. It has produced strong tax revenue on a continuous basis. Deregulation and small-scale privatization have led to the emergence of a vibrant small business sector. The share of SMEs (companies with up to 249 employees, almost entirely privately owned) in GDP was 48.2 per cent in 1999 (PAED, 2000). More than two million entrepreneurs are now operating in such sectors as retail and trade, construction and light manufacturing industry. They contribute substantially to economic growth, job creation and the formation of a new class of consumers. Poland has the highest level of foreign direct investment (FDI) of all the CEECs in absolute dollar terms, which is another important factor in the country's economic recovery. There was a shift in FDI in Poland during the 1990s, connected with privatization policy, from food industry, through labor and capital-intensive sectors to services. The structure of exports has also been changing, with a gradual increase in technology intensity (FTI, 2000). According to WTO data, Poland is ranked 35th on the list of world exporters (0.48 per cent of world exports) and 26th on the list of world importers (0.76 per cent of world imports) (FTI, 2000a).

The strongest point in the innovation system is the growing innovation effort in the manufacturing sector. During the last years of the 1990s, expenditure on innovation accounted for more than 4 per cent of turnover in this sector. In 1996, innovation intensity in the Polish manufacturing sector was on a par with the EU average. In 1999 one quarter of all industrial enterprises (manufacturing, mining, energy, gas and water supply) surveyed by the Main Statistical Office spent money on innovation activities. The largest share of innovation expenditures was devoted to the purchase of embodied technology. However, the increase in the share of enterprises spending money on R&D was considerably greater, reaching 18.5 per cent of all enterprises in 1999, compared with 11.8 per cent in 1998 and 12.9 per cent in 1997 (GUS, 2001). The 1990s saw the development of a basic business and technology transfer infrastructure in Poland.

3. S&T AND THE INNOVATION SYSTEM

While Poland has a significant research potential, its scientists make only a limited contribution to improving the competitiveness of its industry. Despite numerous reforms, the Polish R&D infrastructure continues to be divided into the three main sectors inherited from the socialist time, and scientific cooperation among these three sectors is one of the weakest links in the Polish system. This problem is aggravated by the further reduction in in-house capacities and the continuing organizational separation of enterprises and branch R&D institutes.

The mode of technology transfer that dominated in the past (research commissioned by enterprises and performed by Polish research institutes) has been declining since the start of the transition. For financial and other reasons, the number of research orders placed with research institutions in all three sectors, but especially in the independent branch units, has decreased dramatically. Technology transfer mediated by new firms seeking to commercialize technologies developed in Polish laboratories is emerging, but remains very weak. A certain number of such firms have emerged during the last eight years, although their output is small in volume. However, the emergence and further development of such firms are hampered by a structural deficiency – the lack of an infrastructure for commercializing new technologies. As a result, rather minor forms of technology transfer have emerged over the last years, in the form of innovation centers, so-called technology incubators, foundations and technology agencies, the organizational and cultural conditions for which are currently just developing (Jasinski, 1997).

The Polish economy is facing the challenge of sustainability in closing the technology gap. In view of the severe capital constraints, economic preference is being given to sectors that promise profitability over the relatively short term, such as services, trade, food production and the manufacturing of end products made from imported components. Sectors requiring high capital inputs and a high level of R&D are facing an exceptionally difficult situation. Consequently, a further major objective of S&T policy is to strengthen its strategic component and to reinforce the links with industrial policy. The present innovation policy documents formulated in 1994 and 1999 by the KBN and in 2000 by the Ministry of Economy seek to establish the optimal conditions for the development and functioning of industry by creating an organizational and legal framework for proprietary, organizational and technical restructuring. Within this industrial policy framework, technology policy is the element of state policy that is most directly related to science and research activities. The main goal of technology policy is to make the products of Polish industry competitive and more innovative by raising technical standards, decreasing material and energy consumption in the production process and, finally, by making Polish products technically and legally compatible with EU standards.

The KBN and the Ministry of Economy will in future be mainly responsible for formulating and coordinating this policy, which will be implemented by the relevant government authorities (such as the Technology Agency and the Polish Agency for Enterprise Development).

REFERENCES:

Beek, U.J. van (1995) "The Case of Poland." *Transformation mittel- und ost-europäischer Wissenschaftssysteme.* Ed. R. Mayntz, et al. Opladen: Leske+Budrich, 256–299.
FTI (2000) *Foreign Investment in Poland.* Warsaw: Foreign Trade Institute (FTI).
FTI (2000a) *Zagraniczna polityka gospodarcza i handel zagraniczny Polski 1999– 2000* (Foreign trade policy and foreign trade of Poland, 1999–2000). Warsaw: FTI.
GUS (2001) *Nauka i Technika w 1999 roku* (Science and Technology in Poland 1999). Warsaw: Glowny Urzad Statystyczny (GUS).
GUS Scoreboards (1988–2001). Warsaw: GUS.

Jablecka, J.(1995) "Changes in the Management and Finance of the Research System in Poland: A Survey of the Opinions of Grant Applications." *Social Studies of Science* 25, no.4: 727–753.

Jasinski, A. (1997) "New developments in science – industry linkages in Poland." *Science and Public Policy* 24, no. 2: 93–100.

KLON/JAWOR (1995) *Informator o Organizacjach Pozarządowych w Polsce – JAWOR 1994/95* (Reference book on the non-governmental organizations in Poland – JAWOR 1994/95). Vol. 2. Warsaw: KLON/JAWOR.

PAED (2000) *Raport o stanie sektora małych i średnich przedsiębiorstw w Polsce w latach 1999–2000* (Report on the state of the small and medium sized enterprises in Poland in 1999–2000). Warsaw: Polish Agency for Enterprise Development (PAED).

11. CZECH REPUBLIC: TRANSFORMATION OF R&D – FROM RESEARCH POLICY TO A NATIONAL S&T POLICY

Karel MÜLLER

Czechoslovakia was founded in 1918, in the aftermath of World War I. During the 1920s and 1930s, the country, particularly the Czech territories, was among the leading industrialized countries of the time. During this period the country had a well-developed science system (of a German type), including independent research institutes supported by the state. The outbreak of World War II and the German occupation brought about the closure of all universities; some university scientists moved into industrial laboratories. After the war, in 1948, Czechoslovakia became a member of the Soviet bloc and introduced the 'Soviet model' of S&T. This was accompanied by extensive development of the R&D base. Most academic research was transferred from the universities into the Academy of Sciences (AoS). After 1948, branch institutes were established on the basis of the former industrial laboratories, with only a small share of R&D remaining in enterprises. In-house R&D began to develop in domestic industrial companies in the 1960s. In the 1980s, and particularly in the second half of the decade, the industrial R&D system became more and more decentralized. Branch R&D institutes became integrated into combines; central state funding of these institutes via state technology programs decreased, while financing through R&D contracts with manufacturing enterprises increased. On the other hand, there were 'non-independent' R&D units, which had no legal status independent of the production and economic units to which they belonged, being organized as sub-units, departments, divisions, workshops, laboratories, etc. In contrast to most of the socialist countries, the in-house industrial R&D potential in Czechoslovakia was twice as great as the external R&D potential. Contract research with industries was also carried out by the AoS institutes and the higher education sectors, sometimes in joint research units with enterprises. Even before the beginning of the transition period, the Czechoslovak industrial R&D system was relatively close to the Western model, and quite different from that in the former Soviet Union.

The aim of this chapter is to describe the evolution of S&T policy and of the regulatory framework during the transition period in the 1990s and to identify the factors shaping that evolution. This will be done in three stages. In the first section, the initial political and regulatory efforts will be described. The second section is given over to an analysis of the structural factors influencing developments. In the third section, finally, the regulatory efforts will be assessed with reference to both the existing domestic conditions and the challenges of accession to the EU.

1. DEMOCRATIZATION, ECONOMIC REFORM AND THE SEARCH FOR A NEW R&D POLICY

The effect of systemic transformation on Czechoslovakia was twofold. First, it lost its links with the CMEA countries; second, on January 1, 1993, the country separated into the Czech and Slovak Republics. After the 'velvet revolution' of November 1989, a pluralistic and democratic political system was introduced. Shortly after separation from the Slovak Republic, the newly elected government of the Czech Republic embarked on a radical transformation of the economy, including privatization. Similarly, a new S&T system (STS) was to be introduced, based on the principles of scientific freedom, institutional autonomy, pluralism of funding sources and institutions and competition. A series of measures was introduced throughout Czechoslovakia in 1991/92 and came into full effect in the Czech Republic after its separation from Slovakia in January 1993.

From the very outset of the transformation, a lack of consensus as to the course transformation should take was apparent among both the research community and political decision-makers. Two options were debated: an 'organized' and 'politically governed' transformation of the entire STS or a reorganization of single facilities by way of a primarily spontaneous and 'natural selection process.' The latter scenario won the support of the state authorities as well as of the majority of the scientific community in the AoS and universities. The course of transformation was not centrally directed and differed according to the conditions prevailing in the individual R&D institutions (Provaznik et al., 1998, p. 26). Many industrial R&D institutes wished to adopt a model similar to that of the German Fraunhofer-Gesellschaft, but failed. Between 1992 and 1994, the legal framework for crucial R&D sectors was established through the Academy of Sciences Act, the Governmental Support for Scientific Activities and Technology Development Act and the Higher Education Act. These acts abolished central planning and management of science and research as well as direct state intervention in the activities of research institutions.

In 1993, all federal institutions were abolished and legislation was introduced that delegated the main government responsibilities in the field of S&T

to the Ministry of Education, Youth and Sport. There were also various ministries with departmental competences for S&T in their field of activity (particularly industry, health, agriculture and the environment). The AoS retained its independence, which was laid down in law and supported by 'ring-fenced' funding in the state budget. Nearly all industrial research institutes, about 110 in total, were 'transferred' to the private sector in two waves of mass privatization. In 1994, the Czech Republic Grants Agency was set up to fund research on a competitive basis. At the same time, the government's advisory council for scientific research and technologies was established. One of its tasks was to improve the coordination of the national R&D institutions. All these measures related primarily to 'academic' science, although they have not been sufficient to bring about fundamental change. No progress has been made, for example, in transforming the institutional structure of academic science, whose resources are divided between the AoS research organizations and higher education (HE) research units located within faculties. It had been expected that reducing funding would solve the problem; this has not, however, proved to be the case.

Underlying the policy measures is an unwritten assumption that academic and industrial science should be strictly differentiated: the government took responsibility for funding public science (AoS and HE research sectors) while no provision was made for state support for industrial research and technology transfer. In terms of the evolution of S&T policies in Western countries, this can be described as a science policy of a type that was implemented in the West in the 1950s and 1960s. The reasons why the initial search for a new approach led only to an antiquated form of R&D policy and proved incapable of establishing one more appropriate to the conditions found in modern societies are structural in nature, relating as they do to the distribution of R&D resources and the institutions in which they are embedded. These two factors have hampered the development of a more radical policy and regulatory framework for R&D. These issues will be dealt with in more detail in the following section.

2. CHANGES IN THE R&D SYSTEM BY SECTOR

One initial consequence of the political and economic reforms in the S&T system was a very sharp reduction in R&D funding and employment. By 1994, employment was less than 30 per cent of what it had been in 1989. The largest decreases occurred in 1990 and 1991; from 1993 onwards, employment levels stabilized somewhat, albeit at a very low level relative to the pre-transition era. The greatest downturn took place in industrial R&D. The HE institutions and AoS institutes also experienced a downturn, but to a lesser extent.

At the beginning of the 1990s there were 72 faculties in the *higher education* sector of the Czech Republic (then still part of the Czechoslovak Federal Republic); between 1990 and 1992, a further 27 faculties were established and by 1997 another nine. Today, 23 HE institutions, with 108 faculties, are operating in the Czech Republic. The number of newly enrolled students is also growing while the number of HE teaching staff has been stagnating. New HE institutions have mostly been established in the *regional universities*. An important goal in the transformation of Czech university education was substantial growth in university research and an improvement in its quality. Progress toward this goal has, however, been constrained by the fact that university personnel have been overloaded with teaching commitments. Moreover, the level of laboratory and computing equipment still lags behind that of AoS research institutes.

The transformation of the *Academy of Sciences* began with a distinct and clear-cut concept immediately after November 1989, on the initiative of its scientific community. The Academy's privileged position in the STS was abolished. Its research units underwent an independent evaluation of their scientific level and performance, with input from foreign scientists, and all scientists were evaluated on the basis of their individual performance. The total number of research centers has fallen from 85 to 59, while the overall number of staff dropped from 13 896 in 1989 to 6 972 in October 1993 (a reduction of about 50 per cent – see Table 11.1). Staff reductions were more drastic in the humanities and social sciences than in natural sciences. AoS researchers were generally very well prepared professionally and linguistically for the new challenges and many of them found attractive and well-paid jobs, primarily outside the research sector, in business, banking and policy-making.

In changing its management system, the AoS has adopted a combination of institutional funding with a competitive project-funding model. The share of project funding in total research funding has become so crucial that teams that fail in their grant applications face almost immediate demise. One of the key areas determining the success of the transformation of the AoS has been cooperation with foreign academic institutions. Such contacts have reached considerable proportions, and AoS scientists are generally integrated into the international communities in natural and social sciences. To strengthen cooperation among academic institutions, the Grants Agency began in 1995 to support joint projects proposed by scientists from the AoS and universities. On the other hand, AoS involvement in solving the problems faced by industry is very poor. The AoS has many patents and is seeking contracts; it also provides cheap services and carries out research with little scientific value. The increasing shortage of funds has left the AoS with a choice: it can either en-

Table 11.1 Czech Republic: selected R&D indicators, 1989–1994

Indicator	Research and development, total						
	1989	1990	1991	1992	1993	1994	1994 % of 1989/(1990)
Independent R&D organizations (number)	207	233	266	224	155	149	72
Total R&D staff (persons) of whom:	137 927	105 916	76 487	57 227	40 214	38 752	28
– engaged in R&D activities	77 850	62 268	41 668	31 543	23 336	23 741	30
University teachers and research personnel of whom:	.	11 532	12 762	12 907	13 463	12 625	(109)
– researchers	.	3 578	3 112	2 322		1 317	(37)
AoS personnel	13 896	12 360	9 729	8 881	7 127	7 275	52
R&D expenditure, total (million CZK, current prices)	21 420	12 415	15 211	14 499	9 750	12 983	61

Source: CSO, 1996.

force another personnel cutback or become more involved in contract research in order to acquire new resources.

The *industrial research sector* has been going through a complex transformation. The privatization of the enterprise-based research sector in the Czech Republic was completed in the mid-1990s. Initially, the institutes' desire for independence concurred with the political aim of privatization. In 1991 industrial research institutes were transformed into state limited companies and later privatized in two waves of voucher privatization (109 institutes with an R&D workforce of about 30 000). The privatized research institutes have become joint stock companies. Their shareholder structure is mostly quite diffuse, with each research organization being owned on average by 5 – 10 investment funds and numerous private shareholders. In 1996, more than 90 per cent of funding came from privatized firms; this represents a crucial change in the financing of industrial research, which before 1990 came from state R&D programs and funds. In addition, the number of employees in this sector slumped in 1991 to some 65 per cent of the 1990 total, and fell further in 1992 to 46 per cent. In 1995, about 20 000 people were employed in the enterprise-based research sector. By 1996, the overall number was about one third of the 1990 figure. In subsequent years, the number of employees in industrial science stagnated (see Tables 11.2 and 11.3).

Table 11.2 Czech Republic: R&D personnel by sector (in FTE)

	1995	1996	1997	1998	1999	2000
Business enterprise sector (BES), total	11 346	11 048	11 464	11 287	12 283	11 527
– researchers	4 936	4 822	5 120	5 067	5 811	5 533
Government, total	7 643	7 891	7 751	7 390	6 963	7 148
– researchers	4 315	4 560	4 597	4 596	4 281	4 424
HE, total	3 689	4 403	3 981	4 026	4 736	5 331
– researchers	2 685	3 504	2 848	2 884	3 380	3 768
Private non-profit sector (PNP), total	.	32	34	37	124	192
– researchers	.	30	15	19	63	127
Total	22 678	23 374	23 230	22 740	24 106	24 198
– researchers	11 936	12 916	12 580	12 566	13 535	13 535

Source: CSO, 2000, p. 27; 2001, p. 16.

Table 11.3 Czech Republic: R&D workforce by selected manufacturing industries (physical persons, % of total industrial R&D workforce)

Branch	1990 Personnel	1990 %	1995 Personnel	1995 %	% to 1990	1999 Personnel*	1999 %
Fuel	1 509	2.35	211	2.7	14	11	0.1
Ferrous metals	2 039	4.58	276	3.4	14	251	3.2
Non-ferrous metals	1 384	2.16	22	0.2	2	31	0.5
Chemicals	6 031	9.40	1 311	16.4	22	1 019	13.0
Machinery	32 348	50.46	4 133	52.2	13	3 811	48.6
Electro-technology	9 451	14.75	1 178	15.7	12	1 559	19.9
Building materials	1 365	2.12	92	1.1	7	116	1.5
Metalwork	1 152	1.8	340	4.2	30	437	5.6
Pulp/paper/wood	1 069	1.66	30	0.2	3	177	2.2
Textiles	1 666	2.6	59	0.6	4	98	1.3
Clothing	237	0.36	8	0.0	3	75	0.9
Leather	1 255	1.96	82	1.0	7	20	0.2
Food/refrigeration	1 126	1.75	75	0.8	7	66	0.8
Other industries	2 115	1.8	129	1.5	6	166	2.2
Total workforce	64 197	100.0	7 946	100.0	12	7 837	100

* personnel in FTE
Source: CSO, 1991, 1996, 2000.

As far as the impact of privatization on industrial research is concerned, experts claim that the privatization process did not take sufficient account of the specific features of research institutions. Privatization led to a considerable decrease in the scope of research activities or even to the closure of whole research branches. The direct cause of the reduction in research was primarily the absence of demand for R&D from industry. In many cases, the changes in the activity profile of research institutes were such that the former branch institutes could no longer be considered research centers, since research no longer constituted their main activity, having been replaced by testing, measuring, training, production and so on.

The changes that took place following the restructuring of the privatized research institutes, and in particular the very significant decline in R&D activities, gave rise to much criticism. The critics pointed out that the transition process had proceeded too quickly, with rapid and irreversible results, namely the total disappearance of industrial R&D potential in some fields and branches. On the other hand, those who defend the process and its outcomes argue that ultimately the research institutes' activities are linked to the needs of industry and that the state had to abandon its role in industrial R&D. They see the cause of the massive decline in research activities as lying not in rapid privatization but in the lower level of activity in the national economy as a whole (see Table 11.4), as well as in the ineffective links between R&D organizations and industrial firms.

Changes were made to the internal structure of research with the aim of keeping pace with international developments in S&T and coping with the research problems posed by domestic innovation processes. In many instances, however, the evaluation criteria applied in R&D sectors have been diverging. In *academic research*, different approaches were followed. AoS research organizations introduced a system of independent peer reviews (including citation by other scientists in the case of individual performance). These reviews have become the key criterion for the assessment of organizations' research performance. Consequently, research excellence and orientation to the research problems of world science have become important factors in the restructuring of academic science. The implementation of such an evaluation framework in the universities has been constrained by the increased teaching commitments that have resulted from the expansion in student numbers, which has not been matched by a corresponding increase in lecturing staff. University funding has reflected this situation and student numbers have been accepted as a decisive criterion of evaluation. Only recently have research performance and the number of PhD students been adopted as additional evaluation criteria. In 1998–99, a scheme was introduced to establish research centers in faculties, with the aim of facilitating cooperative research in HE. Industrial research insti-

Table 11.4 Czech Republic: general economic and social indicators

Indicator	1989	1990	1993	1995	1997	1999	2000
Population (in 1 000)	.	10 360	10 330	10 331	10 304	10 280	10 273
Active population (average number in 1 000)	.	5 351	.	5 012	4 946	4 760	4 664
GDP*	.	579	1 002	1 381	1 680	1 870	1959
GDP**	.	1 461	1 257	1 381	1 439	1 391	1 431
(%; 1990 = 100)		100	86	95	98	95	102.9
Unemployment rate	0.0	0.7	3.5	2.9	5.2	9.4	8.78
Average real wages in %	100	94.8	79.9	93.5	103.5	108.1	.
Average real pensions in %	100	94.0	70.0	83.6	96.2	98.4	.
Inflation rate in %	.	9.60	20.8	9.1	8.4	2.1	3.9

* current prices
** constant prices 1995
Source: CSO, 2001.

tutes and AoS research organizations have been invited to participate in this scheme. Following an evaluation process, more than 30 research centers have been founded.

Since all research organizations have been privatized, the volume of contract research and profitability have become the crucial points of reference in *industrial research*. However, in a environment of economic transition and recession, these criteria merely reflected short-term customer requirements (Kubík et al, 1997). The government support schemes that began to be put in place in the second half of the 1990s should have eased this imperfect market situation to some extent, but they were implemented on a wholly inadequate scale. Only in the last three to four years, following an improvement in their economic situation, has firms' demand for R&D begun to grow and the funding of industrial research begun to improve. The government support schemes have also become more generous.

The restructuring of the S&T system was accompanied by the development of new links between academic research and education and the practical application of research results. However, there was substantially less cooperation between academic research, the enterprise-based applied research sector and the domestic manufacturing industry than in the past (during the 1980s, for example). Instead, organizations had to compete for limited R&D funds.

Limited sources of R&D funds, the prevailing competitive environment and the low level of coordination among governmental R&D agencies were all crucial factors limiting the S&T policy framework. This framework was very narrow, with its scope limited to the regulation of research in public institutions through the allocation of R&D funds from the state budget. The distribution of funds was largely in the hand of the scientific community and was based on internal evaluation criteria. Such a policy framework can be regarded only as a research policy. Due to the absence of an interactive relationship between the R&D system and its users, the pace of R&D restructuring was rather slow. Traditional disciplinary patterns (with engineering and technical sciences playing a key role) still prevailed (see Table 11.5).

Table 11.5 Czech Republic: R&D personnel by scientific discipline and type of activity (1999; FTE)

Discipline	Total	Researchers	Technicians	Others
Natural science	6 074	3 784	1 536	754
Engineering, technology	13 033	6 775	4 396	1 862
Medical science	1 325	781	426	118
Agricultural science	1 828	1 000	573	255
Social science	1 539	1 010	385	144
Humanities	307	184	88	35
Total	24 106	13 534	7 404	3 168

Source: CSO, 2000.

In the second half of the 1990s, the factors influencing S&T were improving. Economic recovery increased demand for industrial R&D output; political decisions were taken to increase government funding of R&D and positive changes in the institutional framework were facilitated by the prospect of accession to the EU. The following section will assess the changes in the S&T policy framework with regard to structural dependencies on the inherited centralist regulatory practices, the strategy of radical economic transition and the regulatory challenges that are emerging as the Czech Republic prepares for accession to the EU regulatory framework.

3. EVALUATION OF S&T POLICY SHIFTS FROM THE PERSPECTIVE OF EU ACCESSION

There are two fundamental developments in R&D regulation during the transition from a centralist to a democratic regime that should be emphasized. First, the institutions have gained their traditional autonomy in law and have been able to restore the traditional instrument of meritocratic evaluation. Second, the democratization of the public and political arenas has forced all institutions, including S&T institutions, to justify their missions and claims within a pluralistic public environment; they now have to engage with other political actors, compete for public respect and resources and form public platforms and alliances with other actors pursuing similar missions.

The impact of these aspects of institutional change has, of course, been shaped by various factors, such as the evolution of the political environment, shifts in political expectations, the formation of a pluralistic political system and the strategies adopted by the most powerful political actors (Müller, 1997). Three periods can be identified, each one characterized by a specific approach to key issues in the development of a new regulatory framework (such as the formation of a private sector, the transformation of the public sector and the role of the private non-profit sector) and hence to the restructuring of S&T institutions.

The first period lasted from 1990 to 1992. The leading political actors (from the Civic Forum) were keen to ensure a gradual transition from state ownership to private and public ownership. First, industrial enterprises, including S&T organizations, gained legal autonomy, following which the process of privatization was to be started. It was expected that the formation of a public sphere would play an important role in social transformations. The challenges and opportunities of such a political situation were taken up by the academic agencies. These agencies were, first, the faculties, which gained crucial competences (in relation to university rectors and the Ministry), and, second, those AoS research organizations that gained crucial competences in relation to AoS headquarters. As already mentioned, the autonomy of academic organizations was legitimated in different ways. The AoS introduced a system of evaluating organizations, teams and scientists based on their scientific excellence (as indicated by publications, citations, etc.), while the HE institutions followed the principle of teaching capabilities. The industrial science organizations took advantage of the opportunity to become so-called state public limited companies, which became autonomous legal entities managed independently of their former supervising authorities (ministries and industrial 'combines').

The second period began in 1993, after right-wing parties won the elections. A new strategy for managing the transformation was adopted, with preference being given to mass privatization, accompanied by monetary regulation and a restrictive fiscal policy. The assumption was that private actors would be better able to restructure enterprises. Such political decisions had crucial implications for the transformation of the S&T system. First, the industrial research institutes became subject to privatization. Second, public science institutions became less of a priority for politicians; they were not considered a strategic issue in the new wave of reforms and it was expected that their restructuring would require only limited funding. As a result, there were radical reductions in research staff, direct cuts in AoS institutes and indirect cuts in HE faculties because of the increased pressure of teaching. The negative implications of the radical strategy began to manifest themselves as early as 1995, when it became clear that privatization was not leading to industrial restructuring. The right-wing government attempted to identify these negative impacts and to take countermeasures. A sort of *research policy* was formulated in 1994, in which attempts were made to indicate how public R&D funding should increase and what proportion of that money should be used to cover research institutions' operational costs and to fund agencies, universities, international cooperation projects and so on. The Czech Republic Grants Agency, together with the government's advisory R&D Council, were given the task of supporting and overseeing the implementation of this policy. However, the measures introduced were aimed only at academic and public science, to the exclusion of industrial R&D, support for which was shifted lock, stock and barrel to private business. Even with these adjustments, however reluctantly made, the right-wing approach was unable either to halt the deterioration of the economic situation, including that of S&T organizations, or to prevent the emerging recession. The deteriorating economic and social situation was one of the factors that brought the Social Democratic Party to power after the 1998 election.

The third period began with the election of the Social Democrats, who were faced with the unexpected consequences of radical economic reform. The Social Democrats had been critical of the concept of mass privatization and its negative impact on industry as well as on the public sector, including education and S&T institutions. In government, they have sought to introduce a more active industrial policy. Constraints on capital formation and availability were identified as the main reasons for a slowdown in the transformation of the economic system. From this it was concluded that the key to economic recovery lay in the privatization of the large banks and support for foreign direct investment (FDI).

The Social Democrats' S&T policy priorities were as follows:
- identification of research priorities geared to the needs of the most effective export-based manufacturing industry and to long-term and environmentally friendly economic growth;
- promotion of basic research in scientific disciplines that have high international standards and are deemed necessary for national education, health, the security of citizens and other social needs;
- promotion of the practical application of scientific and technological results;
- the introduction of legislation harmonizing national regulatory and legal frameworks with EU standards.

From 1996 onwards, the Social Democratic Party had been advocating a more specific mode of economic strategy with the twofold aim of eliminating the negative implications of radical economic reform and laying down a foundation for sustainable economic growth. This broad strategy was to be supported by two sets of measures: support for education, including science and research, and, secondly, support for the restructuring of domestic industries (backing for investment and exports and for the development of SMEs and regional economies) (MIT, 2000). The first strand of government efforts was formulated in the following documents:
- 1997 – Main Government Aims (Principles) in the Field of R&D;
- 1998 – The Decision of the Government of the Czech Republic (on the preparation of a national R&D policy); the framework of 'Science policy for the 21st century';
- 2000 – 'National Research and Development Policy of the Czech Republic', government resolution (issued on January 5, 2000).

The 1997 government resolution resulted in increased state funding for R&D, including industrial research and innovation-related activities (see Table 11.6). Indeed, the government's intention to increase public R&D funding from 0.4 to 0.7 per cent of GDP has almost been fulfilled, although not at the expected rate (0.65 per cent of GDP in 2001).

Table 11.6 Czech Republic: gross expenditure on R&D (millions of CZK)

	1995	1996	1997	1998	1999	2000
Total	13 982	16 526	19 477	22 864	23 646	26 487
– from state budget	4 513	6 117	.	8 423	10 077	11 788
Non-investment expenditure	12 431	14 030	16 870	20 135	20 815	23 066
– from state budget	3 951	5 077	6 359	7 219	8 736	10 165
Investments	1 551	2 226	2 607	2 729	2 830	5 420
– from state budget	561	1 039	.	1 204	1 340	1 623

Source: CSO, 2001.

A 1998 government resolution laid down the guidelines for debate on long-term science policy (the government paper under discussion was called 'Science policy for the 21st century'). Concurrently, a detailed evaluation of domestic S&T capacities relative to trends in EU countries was prepared (MEYS and CoG, 1999; MEYS, 1999). The outcome of public debate on this issue was reflected in a government resolution issued early in the year 2000. This resolution focused on the improvement of the regulatory framework for R&D. Apart from an increase in state R&D funding, several new provisions were proposed:

(i) concentration of state R&D funding on the most relevant areas (priorities suggested in the government resolution); to this end, national strategic research and development plans to be devised; (ii) better coordination of ministries and other executive agencies through the establishment of a central executive body for R&D; (iii) increased support for HE research; (iv) implementation of indirect regulatory provisions in the field of industrial research (tax and customs duty relief).

The second strand of state support for R&D has had a rather indirect impact on domestic S&T capacities. The key role has been played by the Ministry of Industry and Trade's set of support schemes, which are intended to boost FDI, exports, SMEs, regional development and quality control measures (see Table 11.7).

Table 11.7 Czech Republic: government funded programs supporting industrial R&D and innovation

Program	Objective	Funding sources
EXPORT	support of competitive capacities of domestic industries	50% public (subsidy/ 5 year loan) 50% Business enterprise sector (BES)
STRATECH	support of development and diffusion of key technologies in the field of defense and state security technologies	50% public (subsidy/ 5 year loan) 50% BES
CENTERS	support of centers of competitive products and technologies	50% public (subsidy/ 5 year loan) 50% BES
TECHNOS	support of SME and their technological capacities	50% public (subsidy/ loan, max. 9 mill. CZK for 5 years) 50% BES
PARK	foundation, operation and development of scientific and technological parks	25–50% public (subsidy/loan, max. 9 mill. CZK)

Source: MIT, 2000a.

The increased public R&D funding of recent years has been focused primarily on the promotion of R&D activities in HE and in the industrial research organizations (see Table 11.8). The distribution of funding has also been improved. Previously, public R&D funds had been channeled mainly through a competitive applications system. Even though this gave research teams opportunities to approach various funding agencies, it was a fundamentally short-term approach. This shortcoming impacted mainly on academic science, in which a substantial amount of research is necessarily long-term. In order to improve R&D funding in HE, a system of research centers was established in the year 2000. In addition to high research quality, the cooperative nature of the proposed centers (participation of research teams from HE, AoS and industrial research) was an important criterion in the evaluation of proposals for the establishment of such centers. Twenty-one basic research centers and 12 applied research centers are now in operation in HE faculties. The Ministry of Industry and Trade, whose R&D budget grew rapidly between 1997 and 2000, has taken the lead role in supporting industrial research.

Table 11.8 Czech Republic: gross expenditure on R&D by source of funds and sector of performance (1999; millions of CZK)

Source of R&D funds		Sector of R&D performance				
Sector	amount	BES	Government	HE	PNP	Foreign
BES	14 759	11 943	2 101	95	3	718
Gov.	5 736	394	5 210	7	0.6	123
HE	2 917	36	2 709	65	6.9	99.9
PNP	129.9	64.7	55.8	0.5	5.6	3.1
Total	23 646	12 439	10 077	168	16	944

Source: CSO, 2000.

Data for the second half of the 1990s indicate that R&D capacities have been slowly growing. This is reflected, firstly, in an increase in state R&D funding (see Table 11.6). The growth in R&D expenditures has been accompanied by a slight growth in R&D personnel, particularly in 1999; there has been a slight decline in the government sector and modest growth in HE and the private sector (see Table 11.2). Closer examination of changes in the distribution of R&D personnel by industry (private manufacturing sector) reveals that industrial R&D has recovered from the massive reductions that occurred in the first half of the 1990s and in some industries has even been expanding. Available data on the distribution of R&D by sector indicate, firstly, that it differs from the pattern in advanced countries and, secondly, that the growth of R&D capacities has not so far been accompanied by institutional changes. The first point is reflected in the comparatively low volume of R&D carried out in the HE sector, the second in the inadequacy of the coordination and cooperative links among S&T institutions. The data on the flow of R&D funds among the research sectors (see Tables 11.8 and 11.9) provide evidence of this situation.

Most R&D funds are spent within the funding sector, with only a small share going to the other R&D sectors.

Table 11.9 Czech Republic: distribution of R&D sources by sector of performance, 1999 (millions of CZK, %)

Source of R&D funding		Sector of R&D performance			
Sector	Total	BES %	Government %	HE %	PNP %
Business enterprise sector	12 439 100%	96.0	3.2	0.3	0.5
Government sector	10 077 100%	20.8	51.7	26.9	0.6
Foreign sources	944 100%	76.0	13.0	10.5	0.5

Source: CSO, 2001, p. 38.

Regulatory measures taken in the second part of the 1990s have been successful in restoring the resources devoted to R&D to earlier levels and have even led to increased spending on academic and industrial science. However, the regulatory framework is limited to support for R&D, in line with the objectives of an R&D policy. The challenges of EU regulatory measures are of a more advanced nature, since they go beyond R&D to embrace innovation policy as well. Our analysis of the evolution of S&T policy, as well as the EU evaluation reports, indicate that further regulatory and legal measures must be developed and applied in order to create an environment in which the various S&T actors can cooperate fully with each other. Closer cooperation between research and industrial communities, as well as more effective coordination among government agencies, seem to be crucial areas in need of improvement if the Czech Republic is to advance from its current R&D policy towards an innovation policy.

REFERENCES

CSO (1991–2001) *Ukazatele výzkumu a vývoje 1990–2000* (R&D Indicators 1990–2000). Prague: Czech Statistical Office (CSO).
Kubík, J., V. Neumajer, K. Müller, and S. Obst (1997) *Problems of Transformation of the Industrial Research Institutions*. Zlin: Faculty of Economics and Management (FEM).
MEYS (1999) *Technology Profile of Czech Republic*. Prague: Ministry of Education, Youth and Sports (MEYS).

MEYS and CoG (1999) *Analysis of Previous Trends and Existing State of R&D in the CR and Comparison with Situation Abroad*. Prague: MEYS/R&D Council of Government (CoG).

MIT (2000) *Podpora podnikání v CR* (Enterprising Support in the CR). Prague: Ministry of Industry and Trade (MIT).

MIT (2000a) *Politika podpory malého a středního podnikání v letech 1999–2002* (Policy of fostering SME in the years 1999–2002). Information bulletin. Prague: MIT.

Müller, K. (1997) "The Institutional Transformation of the S&T-System in the Czech Republic – Country Report." Unpublished Paper. Berlin: Wissenschaftszentrum Berlin für Sozialforschung (WZB).

Provaznik, S., A. Filacek, E. Krizova-Frydova, J. Loudin, and P. Machleidt (1998) "Transformation of science and research in the Czech Republic: the emerging research system and its role in the country's economic and cultural life." *Science and Public Policy* 25, no. 1: 23–35.

12. SLOVAKIA: S&T TRANSFORMATION WITHOUT A STRATEGY [1]

Stefan ZAJAC

As part of Czechoslovakia, Slovakia had a well-developed potential for research and development (R&D) during the socialist era. The conditions at the outset of the transformation were, therefore, similar to those in the Czech part of the country. The first changes within Slovak research institutes (elections of new management, new democratic bodies, etc.) occurred in 1990/91, as they did throughout the Czech region. New legislation on state support for S&T and the Slovak Academy of Sciences (SAS) was still in preparation, however. Unlike in the Czech Republic, and despite various efforts, no overall strategy or guiding principles for effecting R&D transformation had met with general acceptance. Some of the measures relating to the organization and funding of the system that were put in place were only partial measures, inadequately linked to broader science and technology (S&T) policies. For instance, the abolition of the 'State Plan for the Development of S&T', the main R&D policy tool in Czechoslovakia, disrupted long-term links between the national republics of the Czechoslovak federation. These links were further weakened by the division of Czechoslovakia into two nations in January 1993.

There are four salient facts to note about this split. First, the links between individual organizations within the national republics began to weaken after the abolition of S&T planning systems in 1990. As a result, neither republic could be said to have a functioning S&T system. Second, the same was true, and possible even more so, of relations between the republics, with the possible exception of military research. Third, since 1990, there had been increasing differences in the orientation of science and technology policy between the national republics, especially from the economic and legal perspectives. Fourth, the splitting of the federation only served to legitimize tendencies that were already established. In other words, the adjustment of S&T in Slovakia to the republic's new-found status as an independent nation was taking place in tandem with the overall social and economic transformation. The only exceptions were those areas that had previously been federal competences.

---

[1] Paper was supported by VEGA (Scientific Grant Agency), Contract No. 2/7096/20 "Science and technology policy in the Slovak Republic in the context of the EU enlargement".

Continuing political, economic and institutional instabilities during the transformation process complicate any interpretation of the development of science and technology in Slovakia. These instabilities first became apparent in 1988 and they are still making themselves felt, albeit on a different scale, as are feelings of existential uncertainty throughout the entire S&T system.

1. CHANGES IN S&T POLICY IN THE CONTEXT OF RADICAL SYSTEMIC CHANGE

Economic reform became the point of departure for R&D transformation in September 1990. It was widely accepted that the state should start to put in place a new science and research strategy. The main aim was to abolish obstacles to the development of new knowledge and to innovation. New development strategies became elements of the transformation process, but only after needless delay. However, the transition to a market-oriented science and technology policy model was under continuous time pressure.

As far as the general legal environment of S&T activities is concerned, the situation has not changed since 1990. The legal framework for S&T now needs to be consolidated in order to stabilize the sector and provide a clear perspective for future development. New legislation on research and development, on the Slovak Academy of Sciences (SAS) and the Agency for Support of Science and Technology has been in preparation for a long time. The bills were approved by the National Council in February 2002, together with a bill on higher education.

The leading authority in the sphere of S&T is the Ministry of Education and Science (MES). The Council for Science and Technology was set up in early 1991 to advise the government in this field. The Council consisted of representatives of the Slovak government as well as representatives of the SAS, universities and applied research institutes. There were no representatives from the business or banking sectors. The activities of the Council included:
- proposing and defining strategies for research and experimental development;
- reporting to the government on various documents and initiatives in the field of science and technology.
- The Council was reformed several times and was finally abolished, for various reasons, in November 1995. It was re-established in September 1999.

The creation in 1991 of the Grant Agency for Science, dealing with basic research, and the establishment in the following year of the Grant Agency for Technology, dealing mainly with applied research, signified a further important change in the Slovakian S&T system. The grant agencies enter into a tripartite contract with the main applicant and with the director of the relevant institute. The contract spells out the conditions for realizing the project for which the grant is awarded. Grants are awarded on the basis of peer review (two independent national and, on rare occasions, foreign experts). The initial purpose of these agencies was not only to introduce new forms of funding but also to promote a competitive environment within research institutes. Due to the absence of an effective market, the low priority given to research by government and inadequate funding for grants, they were unable to make a substantial contribution to S&T restructuring. It is important to note that from the very beginning, this grant system faced certain limitations because no special fund was created for its use. Since resources are scarce, the grant system may have only a limited impact on the effectiveness of research activities, although it has introduced greater rigor into the selection of research projects. The weakness of the grant system as currently constituted is exemplified by the short periods for which funding is allocated, namely from one to three years. This is rarely enough time to complete basic research. In 1995, the SAS and MES abolished the Grant Agency for Science and created the Scientific Grant Agency (Vedecká grantová agentúra – VEGA).

After Czechoslovakia split into its constituent parts in October 1993, the Agency for International Cooperation in Science and Technology Act was approved by the National Council of the Slovak Republic (Slovak Parliament). It is of great importance to note that the Slovak parliament approved an amendment to the so-called 'competence act' in March 1995 that established the Office for Developmental Strategy for Society, Science and Technology. This office is the central state body responsible for 'planning development strategies for society, science and technology and regional development'. Consequently, the Ministry of Education and Science became a central body with responsibility 'for elementary and secondary schools, colleges, other school establishments, life-long education, science, youth and sport'. In late 1999, the Office for Developmental Strategy for Society, Science and Technology was abolished and the MES again became the central body for S&T in Slovakia.

2. STRUCTURAL CHANGE IN THE S&T SYSTEM

The transformation of the S&T system was characterized by the spontaneous and uncontrolled degradation of S&T institutions. A restrictive budget policy and rapid economic decline also had a major impact.

R&D personnel and expenditure

Domestic analyses in the late 1980s and analyses by foreign experts in the early years of the next decade concluded that in the late 1980s the R&D base in Slovakia, with a 1.05 per cent share in the total labor force, had been overstaffed (Table 12.1). Thus at the beginning of the transformation, it seemed natural and necessary to reduce the number of personnel (Table 12.2). The reduction process, however, lacked guiding principles and a degree of randomness prevailed.

It can probably be assumed that this 'brain drain' tended to flow towards other sectors of the Slovak economy rather than abroad. The private sector tended to recruit scientists working in those areas in which there continued to be an interest in cooperation with S&T (e.g. information technologies, telecommunication, robotics, testing and consulting). In R&D the share of personnel with scientific degrees increased because their average age was higher and workers tended to be less mobile.

As already noted, the reduction in S&T personnel was considered desirable; however, this was not the case for expenditure on S&T (Table 12.3). An unfavorable trend in the evolution of funding during the transformation process is reflected in two indicators: a decline in the nominal volume of funding and the impact of inflation. These figures confirm the overall impression that conditions for S&T are worsening.

In the 1980s, state S&T expenditure rose from 1.5 to 2.0 per cent of GDP, peaking in 1989. However, in the first phase of the transformation between 1990 and 1992 a restrictive fiscal policy led to a decline in state expenditure to less than 1 per cent of GDP. This decline continued between 1993 and 1999, bottoming out in 1994 at only 0.4 per cent of GDP. Such a striking decline, which was once again 0.4 per cent of GDP in 1999, could be explained by the low priority given to S&T by the government. Nevertheless, it is true to say that state expenditure helped to moderate fluctuations in overall S&T expenditure. It is impossible to estimate this influence because there are significant differences between Statistical Office figures on expenditure and state budget data contained in the so-called state final account. It is extremely important to note in this regard that the share of gross domestic expenditure on research and development (GERD) was 3.88 per cent of GDP in 1989 but declined dramatically to 1.53 per cent in 1993 and to 0.68 per cent in 1999 (see Table 12.1).

Slovakia's gradual opening-up to the global economy led, among other things, to a sharp devaluation of the Czechoslovakian currency in 1990 and of the Slovak currency in 1993. At the same time, it also led to a decline in the resources invested in S&T and in opportunities for international cooperation. The devaluation made it more difficult to access not only information but also, and most importantly, foreign scientific instruments and materials for S&T (cf. Table 12.3). Despite the removal of political obstacles to the free flow of information, these economic factors have, ironically, created a situation that is actually much worse than the one that prevailed prior to 1989. Slovakia's participation in international projects might go some way to improving this unfavorable situation. Substantial changes are under way in the area of tech-

Table 12.1 Slovakia: basic macro-economic indicators

	1989	1990	1991	1992	1993	1994	1995	1996	1997	1998	1999	2000
Population (1 000)	5 276	5 298	5 283	5 307	5 325	5 347	5 364	5 374	5 383	5 391	5 395	5 401
Labor force (1 000)	3 188	3 208	3 155	3 148	2 534	2 481	2 471	2 509	2 522	2 545	2 573	2 624
Total R&D personnel per thousand labor force	10.5	9.0	6.8	5.2	4.3	6.9	6.8	6.6	6.5	6.5	5.8	5.8
GDP (SKK billion)	267.3	278.0	319.7	332.3	369.1	466.2	546.0	606.1	686.1	750.8	815.3	887.2
R&D expenditure as a percentage of GDP	3.88	1.75	2.25	1.88	1.53	0.96	0.98	0.97	1.13	0.82	0.68	0.69
FDI (USD million, end of year)	.	.	.	.	369	778	773	1.405	1.616	2.037	2.149	2.896

Source: Own compilation based on data from the Statistical Office of the Slovak Republic.

Table 12.2 Slovakia: R&D personnel (annual average*)

	1989	1990	1991	1992	1993	1994**	1995	1996	1997	1998	1999	2000
R&D personnel total (persons) in which:	33 535	28 745	21 404	16 286	13 459	17 255	16 182	16 613	16 365	16 461	14 849	15 221
– Researchers	21 301	15 550	12 576	10 681	8 927	10 642	10 247	10 613	10 549	10 801	9 204	9 955
– R&D personnel with scientific degree (PhD and higher)	3 797	3 652	3 224	2 803	2 769	4 142	4 081	4 348	4 451	4 791	4 242	4 458

* Data for 1989 – 1993: average recalculated number of full time personnel (so called R&D Base), this average is not identical with FTE (full-time equivalent); data for 1994 – 2000 in FTE (OECD, 1994).

** The increase in personnel in 1994 requires several explanatory notes. Up to 1993, in the higher education sector, personnel only engaged in research and development were reported, but in statistical terms the research activity of the higher education teaching staff was not reported. Such a methodology lowered, although not by much, the real magnitude of personnel. In the nineties, this meant 2 000 research scientists and engineers annually (FTE–full time equivalent) in Slovakia. However, this figure does not include technicians and an auxiliary personnel, as well. Therefore, when Frascati Manual methodology was consequently applied in 1994, the number of personnel simply increased due to methodological reasons.

Source: Statistical Office of the Slovak Republic.

Table 12.3 Slovakia: expenditure on R&D (SKK million, current prices)

R&D Expenditures	1989	1990	1991	1992	1993	1994	1995	1996	1997	1998	1999	2000
Total	10 370	4 859	7 185	6 241	5 662	4 473	5 374	5 905	7 744	6 154	5 552	6 086
in %	100	46.9	69.3	60.2	54.6	43.1	51.8	56.9	74.7	59.3	53.5	58.7
– from the State budget		1 593	2 275	2 254	2 274	1 620	2 011	2 334	2 657	2 790	2 658	2 592
Structure												
– Capital expenditures	784	571	732	684	645	385	681	825	918	700	619	514
in %	7.6	11.8	10.2	11.0	11.4	8.6	12.7	14.0	11.9	11.4	11.1	8.4
– Current expenditures	9 586	4 288	6 453	5 557	5 017	4 088	4 693	5 080	6 826	5 454	4 933	5 572
in which:												
+ Labor costs	2 465	2 233	1 881	1 735	1 626	1 495	2 006	2 162	2 651	2 488	2 379	2 517
in %	25.7	52.1	29.1	31.2	32.4	36.6	42.7	42.6	38.8	45.6	48.2	41.4
+ Other costs	7 121	2 055	4 572	3 822	3 391	2 593	2 687	2 918	4 175	2 966	2 554	3 055

Source: Statistical Office of the Slovak Republic.

nical standardization and testing. Technical norms are being harmonized with international standards, starting with European Union standards. Industrial and intellectual property rights are an important part of the S&T system for which the Office for the Industrial Property of the Slovak Republic is responsible.

S&T activities

The structure of S&T activities seems, at first glance, to be quite clear (Table 12.4). The declining volume but growing share of basic research could be explained by the decline in financial support for governmental and academic institutions due to budget constraints. At the same time, there is a lack of alternative, non-state sources of funding, for example from industry for experimental development. Developments in applied research are influenced by government support for those research units that are necessary for the functioning of the state. Experimental development was very much concentrated in heavy industry and military production. Clearly, it is the loss of former Soviet Union markets and the conversion of military production to civil production that caused the decline in business ventures of this type.

Table 12.4 Slovakia: structure of activities in R&D organizations (1989: by performance; by current expenditure for the period 1994–2000) (%)

	1989	1994	1995	1996	1997	1998	1999	2000
Basic research	12.1	24.7	24.3	22.9	19.5	24.8	28.8	24.9
Applied research	30.8	54.0	55.7	50.6	59.4	51.3	47.9	52.3
Experimental development	57.1	21.3	20.0	26.5	21.1	23.9	23.3	22.8

Source: Statistical Office of the Slovak Republic.

SAS and higher education institutions have been foremost in carrying out basic research. In the SAS, research is concentrated in medical and pharmaceutical sciences, biological and ecological sciences, molecular and cellular biology, electrical and mechanic engineering sciences and mathematical and physical sciences. University research is dominated by medical and pharmaceutical sciences, agriculture and forestry, veterinary sciences, electrical and mechanical engineering sciences and chemical and chemical engineering sciences. However, the social sciences are insufficiently supported, in spite of their importance in the production of new knowledge during a period of systemic transformation.

Branch institutes are more involved than the business enterprise sector in applied research and experimental development. Research is oriented towards advanced chemistry and the discovery of new materials, agriculture and foodstuff production, mechanical and electrical engineering and geological research.

S&T structure by sector

The institutional structure of the S&T system seems to have changed little since the previous period, in spite of formal changes at the top level of the system. Essentially, the whole S&T system was and continues to be divided in four subsystems: the SAS, higher education institutions, branch institutes and commercial research organizations, which have the greatest potential for innovation.

These four sectors remain in place, with some changes in relative size and organizational structures, especially in industrial R&D.

The higher education (HE) system is relatively clearly structured and has changed significantly. In 1999, according to educational statistics, there were 22 institutions of higher education (one of which is private) with 95 faculties in Slovakia. There are two multi-disciplinary universities, four technical universities and various specialized universities. Within the universities, the traditional separation between teaching and research, conducted in special laboratories, has been abolished. The establishment of the grant system to fund research encouraged teaching staff to conduct research.

Despite very strong criticism of the SAS, initial attempts to dissolve it were not successful. SAS scientists resisted efforts to transfer them to much weaker organizational structures in the university system, while the universities felt that they were not able to assimilate specialized Academy research institutes, even if the corresponding financial funds were to be transferred. There is no doubt that the universities have a relatively large and hitherto insufficiently exploited research potential in their disproportionately large number of teaching personnel. One of the ideas that seems to have emerged from the lengthy debate on this issue is that most of the inefficiency in Slovak science is due not to the existence of the SAS and its bloated administration or the weak potential for R&D in higher education, but rather to the absence of flexible links and mutual cooperation between universities and the Academy. The management structure of the SAS has been democratized. It is funded through the state budget, with an increasing proportion of funds, between 15–20 per cent, being raised through grants and contracts in order to ensure the survival of the various institutes (cf. Table 12.5). In 1992 the presidium of the SAS began to evaluate research facilities on the basis of key research criteria and to differentiate institutional funding accordingly. The eight weakest of the 61 institutes were closed. The marked reduction in the number of SAS personnel, which had fallen to about half of the 1989 level by 1999, is due primarily to the departure of many scientists from research activity although and only in part to migration to foreign research facilities (see Table 12.6).

The restructuring of industrial R&D in Slovakia is part of a complex process of economic transformation (Zajac, 1998). The success of this transformation depends to a large extent on developments in the industrial sector. However, industrial R&D has suffered drastically during the transformation. The decline of production in individual industries between 1990 and 1993 did not result in restructuring, nor did it make the economy leaner and more efficient. The course of the transformation process between 1994 and 1996 created favorable conditions for traditional heavy industries that consume large quantities of raw materials and energy but have an adverse environmental impact. The nascent private enterprise sector has lacked the resources to invest money in R&D activities which do not bring an immediate return. In general, industrial enterprises lack adequate R&D backing. Those able to export have proven to be successful in adopting advanced domestic and foreign technologies and gaining the relevant know-how. Nevertheless, the vast majority rely on inadequate in-house R&D. This severely limits their capacity to adapt to new production requirements or to meet the demands of structural readjustment.

This disadvantageous development is also reflected in the fact that in 1991 the number of R&D personnel in the business enterprises was outstripped by those in independent industrial R&D institutes. The causes of this decline were a lack of interest in the entrepreneurial sphere due to structural changes in production, the consequences of enterprise insolvency and the focus on social problems within enterprises to the detriment of support for innovation. The adverse financial situation in research institutes was caused mainly by the inability of customers to pay, as well as by an already high debt burden. This was seen mainly in institutes linked to industries struggling as a result of the military conversion. Further causes included the rising cost of energy and materials and the limited funds available for applied R&D either from the state budget or from business. The decline in the monetary value of such funding was compounded by inflation. Should these trends continue, they may not only lead to a re-orientation of research within the industrial research institutes but also undermine the prospects for international cooperation with EU member states.

The privatization of R&D organizations was another important element of the Slovak economic transformation, with investment vouchers being issued during the first wave of privatization in 1992. A shortage of resources and a restrictive fiscal policy limited the demand for R&D and led to a dramatic re-orientation of individual institutes between 1990 and 1992. The activities in the branch institutes evolved differently according to whether or not they had changed their internal structures. The most frequent change was to replace R&D activities with production and other non-research activities (mainly ser-

Table 12.5 Slovakia: gross expenditure on R&D in the SAS (SKK million, current prices)

Expenditures	1989	1990	1991	1992	1993	93:89 in %	1994	1995	1996	1997	1998	1999	2000	% 1994
Total	806	701	714	524	437	(54)	539	597	677	704	772	791	848	(157)
of which:														
– from State budget		511	655	378	350		388	520	574	597	657	652	701	
share (%)		73	92	72	80		72	87	85	85	85	82	83	
of which:														
– Capital expenditures	168	184	161	86	32	(19)	42	26	35	41	41	37	61	(158)
in %	21	26	23	16	7		8	4	5	6	5	5	7	
+ Current expenditures	638	517	553	438	405	(63)	497	571	642	663	731	754	787	(158)
+ of which: labor costs	275	274	278	263	248	(90)	274	387	436	467	523	541	548	(200)
in %	43	53	50	59	61		55	68	68	70	72	72	70	

Source: Statistical Office of the Slovak Republic.

Table 12.6 Slovakia: R&D personnel in the SAS (annual average* 1989 – 1999)

R&D personnel	1989	1990	1991	1992	1993	(% to 1989)	1994	1995	1996	1997	1998	1999	2000	(% to 1994)
Total	4 573	4 349	3 749	2 979	2 489	(54)	2 829	2 668	2 744	2 703	2 807	2 744	2 858	(101)
in which:														
– Researchers	3 178	2 409	2 573	2 210	1 972	(62)	1 867	1 774	1 811	1 886	1 919	1 758	1 864	(100)
– Personnel with (PhD)	1 524	1 459	1 390	1 198	1 221	(80)	1 112	1 054	1 072	1 210	1 244	1 205	1 199	(108)

* Data for 1989 – 1993: average recalculated number of full time personnel (so called R&D Base), this average is not identical with FTE (full-time equivalent); data for 1994 – 2000 in FTE (OECD, 1994).

Source: Statistical Office of the Slovak Republic.

vice and commercial activities). One very important way in which the transformation of industrial R&D manifests itself is in support for emerging small and medium-sized enterprises (SMEs), particularly through the development of technological centers and science parks. All these measures constitute the beginnings of a much more highly developed system of innovation promotion.

R&D institutes are now working under new judicial regulations, have adapted to the conditions of the market economy and more or less stabilized their research programs. However, their position and responsibilities within the Slovak R&D system as a whole need to specified in greater detail. Legislation on support for science and technology has been in preparation since 1990 (!); the consequent absence of a new legal framework means that conditions for the future development of industrial R&D activities in Slovakia are still far from clear.

Furthermore, the development of new S&T structures in the various sectors was hampered by inertia in the public sector (mainly state-funded research), by the failure rigorously to implement science policy principles that had already been approved (higher education institutions) and a general decline in the socio-economic conditions for S&T development, not only in manufacturing but in other areas as well, with the exception of agricultural research. Higher education institutions account for only a small proportion of the country's S&T potential, despite the proclaimed support for this sector (see Table 12.7). However, this situation reflects not only the neglect of this sector in the past but also the strong influence of the incomplete reporting of total R&D capacities and expenditures (see methodological note in Table 12.7). Compared to its share of the overall R&D personnel, the higher education sector has reached a disproportionately low share of the total R&D expenditures (Table 12.8). The most unfavorable developments were reported in industrial R&D (Tables 12.9 and 12.10). There was a particular reduction in the share of R&D performed in-house by industrial firms.

The situation in primarily state-funded departments seems at first sight to be more favorable. In 1991, there was a qualitative change in manufacturing industry when the share of independent industrial R&D institutes in R&D personnel outstripped that of the business enterprises. This is a considerable difference from advanced market economies, where the situation is just the opposite. Radosevic and Auriol found that "independent R&D institutes are difficult to classify, as they do not fit with any of the institutional sectors. In Western countries, independent R&D institutions also exist but on a much smaller scale and many of them belong to the PNP sector. Only in CEECs is this type of institution found as the dominant form of organization" (Radosevic and Auriol, 1998, p. 5).

Table 12.7 Slovakia: R&D personnel in the higher education sector (annual average* 1989 – 2000)

R&D personnel	1989	1990	1991	1992	1993	(% to 1989)	1994	1995	1996	1997	1998	1999	2000	(% to 1994)
Total	2 148	2 113	1 754	1 717	1 685	(78)	4 284	4 543	4 521	5 041	5 514	5 063	5 860	(137)
in which														
– Researchers	1 474	1 022	967	1 268	1 229	(84)	3 773	4 097	4 048	4 325	4 821	4 255	5 009	(133)
– Personnel with PhD	485	476	432	392	566	(117)	2 200	2 279	2 322	2 430	2 726	2 302	2 542	(116)

* Data for 1989 – 1993: average recalculated number of full time personnel (so called R&D Base), this average is not identical with FTE (full-time equivalent); data for 1994 – 2000 in FTE (OECD, 1994).
Source: Statistical Office of the Slovak Republic.

Table 12.8 Slovakia: gross expenditure on R&D in the higher education sector (SKK million, current prices)

Expenditures	1989	1990	1991	1992	1993	1994	1995	1996	1997	1998	1999	2000
Total	244	229	299	286	286	219	316	302	520	580	551	579
– of which: from State budget		218	282	282.4	277	214	307	289	504	566	529	530
+ Capital expenditures	52	39	80	58	65	24	29	34	58	66	51	54
+ Current expenditures	192	190	219	228	221	195	287	268	461	514	500	525
– of which: labor costs	85	92	104	172	127	144	198	192	302	332	351	371

Source: Statistical Office of the Slovak Republic.

Table 12.9 Slovakia: R&D personnel in the business enterprise sector (BES) (annual average*) 1989–2000

R&D personnel	1989	1990	1991	1992	1993	1994	1995	1996	1997	1998	1999	2000
Total (manufacturing)	18252	14814	10026	6094	5023	5695.0	4859.4	5295.3	7407.8	6708.1 (3264)	5692.4 (2478)	5171.8 (2176)
– Researchers	11159	8123	6059	3744	1972	2648.0	2102.8	2258.4	3386.6	2902.9	2521.7	2420.3
– Personnel with scientific degree	711	636	528	354	221	233.3	207.1	221.9	565.7	535.7	491.3	447.7

* Data for 1989–1993: only industry; average recalculated number of full time personnel (so called R&D base), this average is not identical with FTE (full-time equivalent); data for 1994–2000 in FTE (OECD, 1994).
Source: Statistical Office of the Slovak Republic.

Table 12.10 Slovakia: gross expenditure on R&D in the business enterprise sector (BES) (SKK million, current prices)

Expenditures	1989	1990	1991	1992	1993	1994	1995	1996	1997	1998	1999	2000
Total (manufacturing)	7235	2483	4145	3316	2839	2359	2897	3296	5854	4049 (2477)	3473 (2016)	4005 (2308)
of which: from State budget		204	330	464	226	324	312	406	967	909	848	823
– Capital expenditures	369	173	260	224	193	139	397	463	556	358	267	265
– Current expenditures	6866	2310	3885	3092	2646	2220	2500	2833	5298	3691	3206	3740
– of which: labor costs	1494	1244	1055	900	748	650	850	909	1678	1348	1233	1323

Source: Statistical Office of the Slovak Republic. Data for 1989–1993 only 'industry'.

3. QUALITATIVE CHANGES IN THE S&T SYSTEM

There are several salient points that need to be addressed. First, it should be stressed that the renewal of academic autonomy has not led to any significant change in the institutional structure of S&T. It is true to say that under the past regime the scientific intelligentsia was gradually marginalized. However, this marginalization process did not stop after November 1989. Indeed, it could even be said to have accelerated due to the economic recession and 'brain drain' from the S&T system.

Second, the current situation is influenced by differences of opinion. One school of thought takes the view that political change will lead to a gradual renewal of interest in R&D and to the political conviction that S&T is a necessity and must be supported. However, the opposing school of thought, supported by some scientists who have rejected the practical role of science, insists on the autonomy of science and the freedom of research from political influence. Discussions of S&T policy in Slovakia between 1991 and 1992 focused on the division of research between the SAS and higher education institutions. Two misconceptions frequently arose in the course of the debate. The first was that teaching and research activity need always to take place alongside each other, the second that most research in advanced countries is carried out in universities.

The assumption that teaching and research necessarily have to take place alongside each other is based on the notion that research is a precondition for good teaching. There is little concrete evidence to substantiate this statement. There are numerous examples to suggest that a diversity of possible arrangements for the location of research and development activities is possible in advanced market economies; in some, basic research is predominantly university based, whereas in others it is not. As far as the second assumption is concerned, research facilities are, for economic and historical reasons, located in different locations in different countries. In some countries in which universities are long established, there is a tendency for research and teaching to be combined. In countries where universities are more recently established, there has been a tendency to separate research and teaching, mainly because such a division is simpler to manage.

In late 1992, the SAS institutes were accredited on the basis of international evaluations of their scientific research. The institutes were divided into four groups defined as follows:

A a very good institute whose work is absolutely necessary for the advancement of basic research. The best research teams working there should be given opportunities to develop;

B a good institute. Its work is necessary. The institute should improve its standard. Good working conditions should be established for its best research teams. A positive selection of research topics and scientific teams is needed;

C an institute whose work may be necessary. Its standards must be improved and there are chances that they will be. Restructuring is needed in order to reduce costs, ensure that research is more selective and sharply focused, primarily on strategic applied research, and to prepare for possible future merger with an SAS or branch institute working in a similar field;

D an institute that should not be funded by the SAS and should, in time, be shut down.

Groups A and B comprised institutes that could report very good results in basic research activities. Group C included the institutes that were successful in applied research, Group D the institutes with weaker results in either basic or applied research. Seven institutes fell into this last category and were closed between 1993 and 1994. In 1992 through 1994, a similar process of accreditation, with similar criteria, was carried out in the institutions of higher education, although without such drastic consequences. The aim of these selective measures was to improve the performance of the SAS and universities and to adapt their functions to the conditions of the emerging market economy. Other departmental research institutes, with the exception of those in the agricultural sector, were not accredited at all.

Finally, an important part of Slovakia's economic transformation was the privatization of research and development organizations. The assets of state research organizations in the various industrial branches were privatized through the issue of so-called investment vouchers. The first wave of privatization occurred in 1992 and had a significant effect on research institutes' internal structures. It also forced institutes in industrial branches into numerous adjustment procedures. The most common adjustment was the substitution of production activities and other non-research activities (mainly service and commercial) for research and development activities. This obviously meant that some research institutes became organizations with very different core activities. In some industrial branches, change also took the form of supporting the development of small and medium-sized firms by harnessing Western experience and know-how to establish technological centers or science parks. Such parks already exist in Bratislava, Piešt'any and Košice and plans are afoot to construct others in Prievidza and Spišská Nová Ves.

4. FUTURE TRENDS IN THE DEVELOPMENT OF S&T

In early 1997, the former Office for Strategy established a draft committee comprising experts from the SAS, universities and government bodies to draw up proposals for the future development of S&T policy. The committee used existing analyses as a starting point for its deliberations. The draft report was discussed by various government bodies and was approved by the Slovak government in July 1998 (Government, 1998). It was updated by the new government in September 2000 (Government, 2000). For the period 2001 to 2005, there are to be two sets of priorities in the field of S&T, which were approved by the Slovak Parliament in December 2000:

I. Cross-sectoral R&D programs:

1. Building an information society
2. Quality of life – health, nutrition, education
3. Advanced technologies for an efficient economy
4. Utilization of domestic raw materials and resources
5. Use of progressive principles for energy production and the transformation thereof
6. Social sciences for social development

II. Cross-cutting directions in R&D

1. Economic competition
2. Human resources
3. External and internal national security
4. Integration of research and development practices into the European Research Area (Government, 2000).

A further draft proposal relies heavily on a study of the development of S&T systems in Slovakia up to the year 2015 (Zajac, 1999, p. 852–858). According to this proposal, a number of steps must be taken to improve S&T performance and management during this period:

– A sustained S&T effort is necessary for the creation of an intensive knowledge-based economy and to support Slovak efforts to increase its contribution to world-class scientific research. This clearly implies that the Slovak government's target of allocating 1.8 per cent of GDP to R&D by 2005 should be maintained, and budget allocations are needed to assure that this goal is met;

– One of the persistent problems for which a satisfactory solution should be found quickly is the need to establish an efficient and effective framework within which institutions can put in place innovative policy structures. This problem has its roots in the very foundation of the post-socialist society in Slovakia, which is concerned with the egalitarian distribution of resources

among all its members. This inevitably has repercussions, chief among them being the lack of long-term focus and the scattering of resources over a large number of S&T activities, including those of sub-critical size and importance;
- The government's support of S&T should be more clearly focused and co-ordinated. To this end, a national S&T strategy should be established and implemented, so as to provide a clear, medium-term framework for all concerned. This strategy, based on a clear view of Slovakia's economic and technological development, should define national goals and key research areas that require priority support;
- The Slovak Government should support public and academic research more selectively, with support going to strategic research that can be exploited by local industry and to teams and individuals considered outstanding by international standards;
- Continued efforts need to be made to improve the efficiency of public-sector S&T institutes and their ability to respond to the needs of the Slovak economy. It would also be inadvisable to remove responsibility for public-sector S&T institutes from sponsoring ministries. It is important, therefore, that institutions responsible for S&T (ministries, SAS, etc.) have been given an explicit role in determining their budgets, research programs and method of evaluation;
- As far as institutions of higher education are concerned, the primary issue is financial support. Thus it would be advisable for all departments to seek to maintain international standards and for the budget allocation to be divided equally between education and research. Industry funding of these institutions seems excessively low and should be stimulated with appropriate tax incentives;
- The national S&T effort will bear fruit all the more readily if the right framework exists in the economy as a whole – in terms of a fiscal and tax system, trade and competition regulations, information infrastructure, development of human resources and so on. Such a strategic effort, however, should not lead to support for innovation activities in all industrial branches.

5. CONCLUSIONS

Despite the changes to the S&T system outlined above, the transformation has been very slow and controversial. The situation at the beginning of the new millennium is characterized by an absence of new legislation, which could be an important vehicle for institutional changes. In addition, there is a lack of resources to support economic development in S&T. This situation influences not only the orientation of research in the S&T system but also impacts adversely on future prospects for international cooperation due to the effects of the dramatic decline in capital and current expenditure (materials for laboratories, books, journals, post and telecommunications, electricity and fuel etc.). Slovakia's gradual integration into the world economy led to a substantial devaluation of the Slovak currency in 1993 and hence to a reduction in the resources available for investment in S&T and a decline in the possibilities for technology transfer and international cooperation in S&T.

The nature of the changes that have taken place in the S&T system mean that the period 1989 to 2000 cannot be regarded as very successful. Because of the sharp decline of industrial R&D (cf. Table 12.11), there has been a structural shift between sectors that has not always been effective or well planned. Nevertheless, the European Commission stated in its opinion on Slovakia's application for membership of the European Union that "in the perspective of accession, no major problems should be expected in this field. Accession would be of mutual benefit" (EC, 1997, p. 66).

In conclusion, Slovakia can be a member of the EU and a respected partner in international cooperation only if it has its own strong national R&D base and if its S&T system is attractive to partners beyond its borders.

REFERENCES:

EC (1997) *AGENDA 2000 – Commission Opinion on Slovakia's Application for Membership of the European Union*. Brussels: European Commission (EC).
Government (1998) *Koncepcia štátnej vednej a technickej politiky, uzn. vlády SR è. 494/1998* (Design of the state science and technology policy's principles – Decree No. 494/1998). Bratislava: Government.
Government (2000) *Koncepcia štátnej vednej a technickej politiky, uzn. vlády SR è. 724/2000* (Design of the state science and technology policy's principles – Decree No. 724/2000). Bratislava: Government.
OECD (1994) *Proposed standard practice for surveys of research and experimental development – Frascati Manual 1993*. Paris: OECD.
Radosevic, S., and L. Auriol. (1998) "Patterns of Restructuring in Research, Development and Innovation Activities in Central and Eastern European Countries: Analyses based on S&T Indicators." SPRU Electronic Working Papers Series no. 16. Sussex: SPRU.
Zajac, S. (1998) "Industrial R&D in Slovakia." *Transforming Science and Technology Systems – the Endless Transition? NATO Science Series 4: Science and Technology Policy – Vol. 23*. Ed. W. Meske, et al. Amsterdam: IOS Press, 235–243.
Zajac, S. (1999) "Sú casný stav a perspektívy výskumu a vývoja na Slovensku" (Research and Development in Slovakia: current state and prospects). *Ekonomický casopis* 47, no. 6: 835–862.

Table 12.11 Slovakia: the structure of R&D personnel and expenditure by sector (in %)

	1989	1990	1991	1992	1993	1994	1995	1996	1997	1998	1999	2000
Personnel:												
– HE	6.4	7.3	8.2	10.5	12.5	24.8	28.1	27.2	30.8	33.4	34.1	38.5
– SAS	13.6	15.1	17.5	18.3	18.5	16.4	16.5	16.5	16.5	17.1	18.5	18.8
– Industry/BES	54.4	51.5	46.8	37.4	37.2	33.0	29.8	31.9	45.3	40.8	38.3	34.0
– others	25.6	26.1	27.5	33.8	31.8	25.8	25.6	24.4	7.4	8.7	9.1	8.7
Expenditure:												
– HE	2.5	4.4	4.2	6.1	5.1	4.9	5.9	5.1	6.7	9.4	9.9	9.5
– SAS	7.8	14.4	9.9	10.7	7.7	9.0	11.1	11.5	9.1	12.5	14.3	13.9
– Industry/ BES	69.8	51.1	57.7	53.1	50.1	52.7	53.9	55.8	75.6	65.8	62.6	65.8
– others	19.9	30.1	28.2	30.1	37.1	33.4	29.1	27.6	8.6	12.3	13.2	10.8

Source: Author's compilation based on data from the Statistical Office of the Slovak Republic.

13. HUNGARY: FROM TRANSFORMATION TO EUROPEAN INTEGRATION

Judith MOSONI-FRIED

Hungary has a long tradition in science. The first university was established in Pécs in 1367, at much the same time as other universities, such as Vienna, Prague and Cracow, were being founded in Central Europe (Tamás, 1985). The widespread use of Latin in Europe during the Middle Ages facilitated the integration of Hungarian science. From the early 19th century onwards, science became the main tool for eliminating backwardness in the Austro-Hungarian Empire. Indeed, one of the aims of the Hungarian Academy of Sciences (HAS) when it was founded in 1826 was to bring about social change in the country. The scientific community in Hungary has traditionally enjoyed at least relative autonomy and it is no coincidence that Hungary never accepted the orthodox Soviet model of S&T and always tried to move away from its very strict management system. S&T in Hungary deviated significantly from the Soviet model in several respects. Scientific research was, for example, never neglected by the universities. Although it was not directly funded by the state and preference was given to the HAS institutes, where the young, 'revolutionary' generation was less evident than in the universities, research in the higher education (HE) sector was always held in high esteem both by society and the scientific administration. Hungary had a well-established system of patent regulation long before transformation. The decentralization of S&T management began in the mid-1980s, when a multi-channel funding system was put in place. The Hungarian National Scientific Research Fund (OTKA), established in 1986, was the first transparent system in a planned economy providing subsidies for basic research.

Our purpose in what follows is first to introduce the main actors in S&T policy over the last ten years and then to outline their actions during that time, which initially promoted and then hampered the transformation of the R&D sector in the transition period. In the final section, the decisive factors for the future development of S&T in Hungary are summarized.

1. A GRADUALIST APPROACH TO SYSTEMIC TRANSFORMATION

The stability of the Hungarian R&D management system during the transition period was attributable mainly to the fact that reform of the system had got under way in the 1980s. Consequently, there was some degree of continuity in the restructuring process, mainly in the financing system. As a result, academic institutions remained relatively undisturbed during the early years of systemic transformation and researchers were able to carry on with their work, even when it was impossible to do so elsewhere. The early years of transformation were characterized by a laissez-faire S&T policy. The political parties paid little attention to research institutions, which had very limited political power and, with a few exceptions, little monetary value as potential targets for privatization. It was the scientific community itself that initiated the development of new *science policy guidelines* in 1993. At the same time, all the principal actors in R&D management drew up their own proposals for restructuring, or merely reforming, their own particular areas of responsibility. Each actor strove to have at least one bill passed by parliament. Three bills were in fact passed: the Hungarian Scientific Research Fund Act of 1993, the Higher Education Act of the same year and the Hungarian Academy of Sciences Act of 1994. Many other acts of parliament and decisions influenced R&D in the transition period, including acts on business associations, foreign capital investments, the founding of joint ventures (1988–1989), on bank reform (1988), tax reform (1989), the liberalization of imports (1990), the transformation of state enterprises (1992), state property (1992), venture capital (March 1998), etc. Many of these pieces of legislation facilitated the transformation of business enterprises and strengthened academic institutions but did little to assist business sector R&D. Neither branch institutes nor company R&D laboratories were protected by legal or policy measures. In many cases, such measures would certainly have been unsuccessful anyway, since applied research and experimental development cannot be built up when the demand side is in ruin. However, there were also many R&D facilities whose long-term future could and should have been secured by appropriate support schemes. Such schemes failed to materialize, however, and many researchers were forced to leave R&D altogether. It was no thanks to the government that some of them were able to return reasonably quickly, this time in the employment of foreign investors, and make a very effective contribution to the process of technology transfer. Engineering staff in many green-field investment projects were recruited from liquidated or downsized R&D branch institutes and company laboratories.

The current growth in R&D personnel (cf. Figure 13.1) is due mainly to an

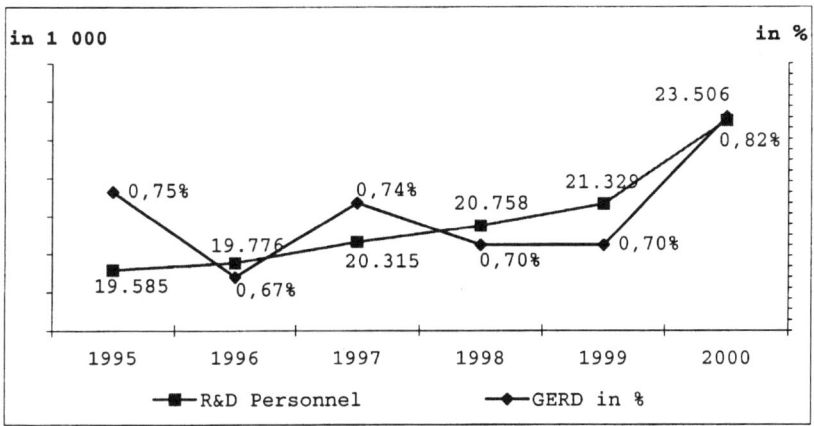

*Fig. 13.1 Hungary: R&D personnel (in 1 000 FTE), and GERD (as percentage of GDP)
Source: KSH, 2001.*

increase in the higher education sector. The number of scientists and engineers is growing, while that of support staff is decreasing.

2. CONCENTRATION OF POLITICAL POWER AND FINANCIAL RESOURCES

In the centrally planned economic system, academic research was directly controlled by the Council of Ministers and, even more immediately, by the science policy committee of the Communist Party. Parliament was excluded from this field of activity. Following the transformation, the Hungarian Academy of Sciences Act divided responsibility for academic research between government and parliament. Every two years, parliament is required to debate in plenary session a report presented by the HAS president on the state of Hungarian science. The debate results in a parliamentary decision in which representatives confirm acceptance of the report and request the government to take measures to promote the future development of science in Hungary. In addition, several permanent parliamentary committees (primarily the Committee on Education and Science) are now entitled to deal with special S&T policy issues. It was on a government initiative that the highest advisory and coordinating body, the government's Science Policy Committee, was dissolved in 1998. The *Science and Technology Policy Collegium*, headed by the Prime Minister, was set up to replace it. According to the government decree that established the body, it is an advisory, decision-making and coordinating body. Its members include the minister of education, the presidents of the HAS and OTKA, the minister of economic affairs, the minister of national cultural heritage, the minister heading the Prime Minister's Office, and the chair of the Council of the National Committee for Technological Development (OMFB). It is primarily a forum where the heads of the represented organizations are able to express their opinions on various issues. Recommendations and decisions made by the Collegium are submitted to the government by the minister of education. (The prime minister may delegate his or her seat in the Collegium to other distinguished personalities. Recently the ex-minister for cultural affairs has been in charge of the Collegium.) The new organization appears to be very similar to the dissolved Science Policy Committee, but in fact is quite different, since it seems to be merely a forum for irregular discussions on S&T policy issues. Policy itself is made by another body, the so-called *Scientific Advisory Body*. It is attached to the Collegium by government decree, its purpose being to assist the Collegium's work. Its members are appointed on the invitation of the prime minister, the head of the Collegium. They include distinguished scientists and leading figures in the business sector. Moreover, a dedicated secretariat was set up within the Ministry of Education; this marked a significant departure from the conditions under which the Science Policy Committee operated during the early years of transformation. Its responsibilities include preparatory work and monitoring, not only for the Collegium (and the Scientific Advisory Body) but also for the Minister of Education, who is a member of the Collegium and president of the Advisory Body. The secretariat is headed by a vice-minister, who is appointed by the Minister of Education. In this way, S&T policy making has been largely taken out of the hands of a coordinating body not directly under government control and transferred to the Ministry of Education. Although S&T policymaking is still formally a cooperative effort involving the main actors in R&D management (as it used to be for a few years in the 1990s), in practice it is now a central government function. Some actors, in particular all the elected leaders of the HAS, are much less involved in the decision-making processes than they used to be. S&T policy is drawn up and presented by the Ministry of Education as if it were explicitly a ministry of science and technology.

Public funding is divided into two streams: core institutional funding from the state budget and competitive funding, also derived largely from the state budget but channeled into different funds. These funds are administered by two large institutions. The *OTKA* allocates money for basic research on a competitive basis. Its budget is about 5 per cent of government expenditure on R&D (Figure 13.2). Although OTKA's budget is formally part of the HAS budget, it operates autonomously. Its president and vice-president are appointed by the prime minister, on the basis of a proposal drawn up jointly by the president of HAS and the minister of national cultural heritage. In recent decades, the

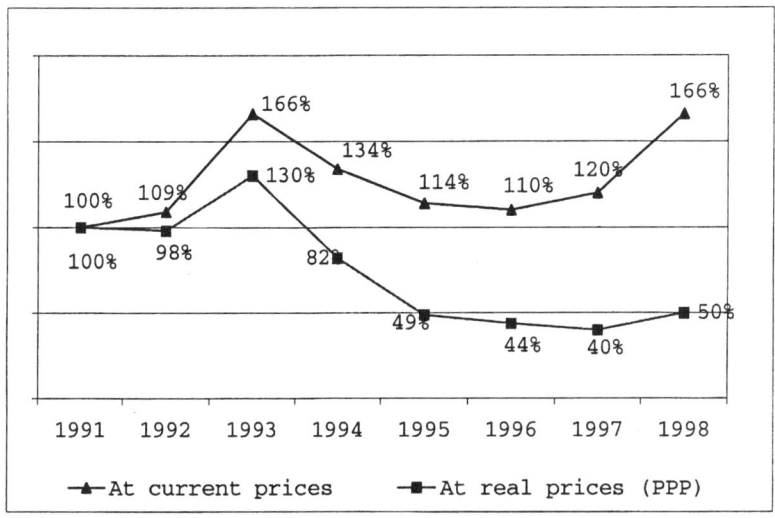

Fig. 13.2 Hungary: National Scientific Research Fund (OTKA) spending at current and real prices (in %; 1991 = 100)
Source: OMFB, 1999

OMFB was the biggest government organization responsible for funding applied research and technological development. The situation started to change in 1994, when money from the main central body for technological development (KMÜFA), which was for several decades allocated by OMFB, was partially transferred to other funds. At the same time, the whole system of fund-building was restructured. Government decision no. 143/1994 (XI.10) stipulated that companies were no longer required to contribute to the Central Technology Development Fund (KMÜFA). Instead, money was to flow into the fund directly from the state budget (Figure 13.3). This seemed to be a rational measure, since the firms' ownership status had changed completely during the transition period. In KMÜFA's 'golden age', nearly 100 per cent of firms were state-owned in Hungary. By the middle of the 1990s, the share of state-owned enterprises had fallen to about 20–30 per cent. Companies were now free to take independent investment decisions, including those related to technology. They also pay corporate taxes at relatively high levels, which should be enough to finance national priorities in technology development as well. This was the thinking behind this structural change.

OMFB's position became rather difficult toward the end of the 1990s. First, government decree no. 147/1998 (IX.16.) directly changed the supervision of the OMFB and indirectly changed its status (cf. HAS, 1998). This central body engaged in R&D administration lost its relative independence within the pub-

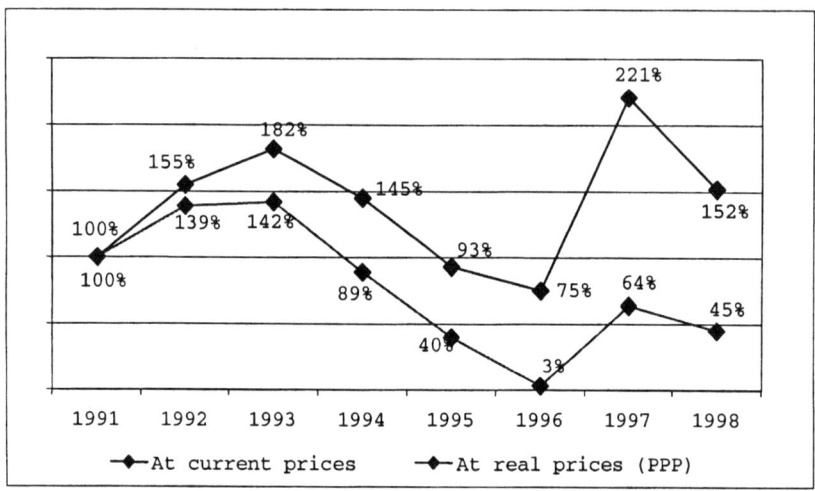

Fig. 13.3 Hungary: Central Technology Development Fund (KMÜFA) spending at current and real prices (1991 = 100%)
Source: OMFB, 1999.

lic administration and was subordinated to the minister of economic affairs. Its authority for independent decision-making was thereby curtailed, with regard both to operational and other matters. Traditionally, the OMFB had been answerable directly to the government. This decree was not the end of the changes. A new government decree, no. 226/1999 (XII.30), passed responsibility for the OMFB to the Ministry of Education from January 1, 2000. It is now operating as the Ministry's R&D division. The aim of this decision was to centralize S&T policy and concentrate the various sources of the R&D budget, an aim that was successfully realized.

There have also been changes to the *branch ministries'* responsibilities for S&T policy. Only one of them, namely the Ministry of Education (formerly the Ministry of Education and Culture), saw its responsibilities in this area strengthened during the transition period. Others were deprived of a great deal of their responsibilities, firstly through the transfer of branch R&D institutions to the State Privatization and Property Agency in 1992/93 and, secondly, when the 1993 Higher Education Act transferred responsibility for all universities to the Ministry of Education and Culture. Prior to 1993, the universities of agricultural sciences had been under the authority of the Ministry of Agriculture, while the Ministry of Health had been responsible for universities of medical sciences. The only exception now is the University of National Defense, which operates under the control of the Ministry of Defense. (Responsibility for the Police Academy lies with the Ministry of the Interior.) Since 1995, the Property

Agency has been involved only in the privatization of branch R&D institutes. Since most of them are still state-owned, they remain under the control of the various ministries.

To sum up, the transition period started with rather insignificant institutional changes in the R&D management system at government level, while the actual restructuring processes, which had begun in the mid-1980s, proceeded quietly, gradually and largely untroubled. In the early years of transformation, little political attention was paid to the S&T sector and limited government resources were allocated to it. On the one hand, this was not so bad for the scientific community. Their scientific work was harmed by financial restrictions, but since the institutions were preserved, scientists in the academic sphere could continue to do their jobs. On the other hand, laissez-faire policies proved to be very harmful. No concerted actions were taken to preserve S&T resources (human resources, accumulated knowledge, networks and facilities) in, for example, branch institutes or privatized companies. R&D policy was not sufficiently 'rational' to defend and develop capacities of great importance for economic competitiveness (Balázs, 1994). Adjustment to the new conditions took place primarily at the level of individual (i.e. personal) strategy. The end of the transition period saw a shift away from laissez-faire policy and the decentralization of decision- and policy-making towards an active S&T policy based on the concentration of political power and financial resources in public-sector R&D. This new policy does not include any special assistance for branch institutes; at government level they are not treated as a single group of R&D institutions. Company laboratories, whether new ones or those that were not liquidated, are not given any preferential treatment either; only their profile, economic situation and capability count when they compete for state grants, subsidies, etc.

3. INTEGRATION, CONSOLIDATION, RESTRUCTURING

Institutions in the *higher education* sector have become autonomous organizations with the right of self-government. Moreover, responsibility for training researchers and awarding scientific qualifications has been given to accredited universities. The (State) Committee for Scientific Qualifications has been abolished. In its place, universities have established doctoral councils responsible for postgraduate training and awarding PhD degrees. (The HAS is entitled to award the degree of 'Doctor of the Academy' in all branches of sciences.) As already noted above, new legislation placed all universities and colleges under the supervision of the Ministry of Education. The Council of Higher Education and Research (FTT), which also operates under the supervision of the Ministry of Education, draws up development programs, makes recommendations for the various sectors, lays the groundwork for decision-making, etc. The quality of teaching and research activities, as well as that of researchers' training courses offered by the universities, is monitored and evaluated by the National Accreditation Committee. New, private (non-state) colleges were established in the transition period. Some of them (26 units) are church institutions, others (six units) are managed by various private foundations. Nearly 10 per cent of students attend private colleges.

Research activity in Hungarian universities was strengthened during the transition period, even though funding is still far from adequate (Figure 13.4). Moreover, research and teaching are so closely linked that it appears impossi-

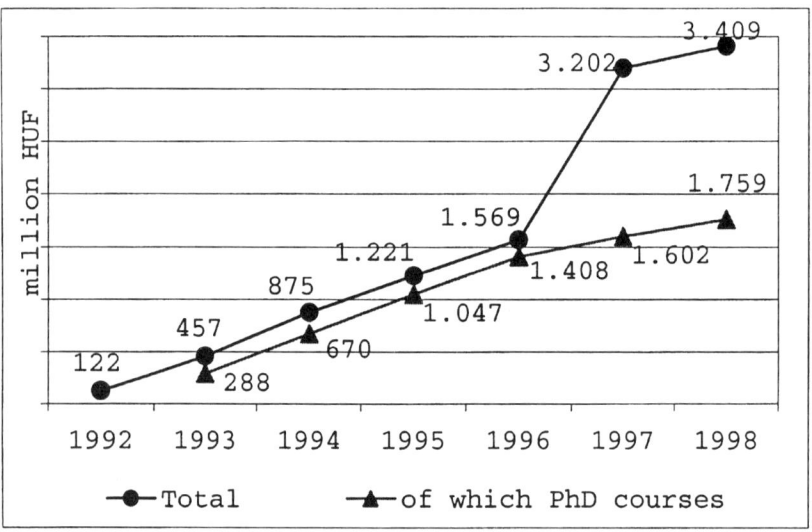

Fig. 13.4 Hungary: direct expenditure by the Ministry of Education on R&D in HE institutions Source: OMFB, 1999.

ble to separate out expenditure on research from that on teaching. Since there is a relatively large number of external sources providing research funds, teaching is in many cases being supported by research grants (Deliné Pálinkó and Dévai, 2000). Recently, there has been a discrepancy between the incentive to take on more graduate students and the aim to engage more strongly in research. In the field of higher education normative financing exists: the more students a higher education institution has, the more money is allocated for teaching responsibilities from the state budget. However, those who are also involved in research (and most lecturers and professors are involved, of course), are complaining that with too many students they are unable to do their research work, although it is strongly accounted in the evaluation process of the units. As a result, both teaching and research are endangered (Török, 1999). In the higher education sector, centralization and consolidation are now the watchwords. In recent years, a program of institutional consolidation, initiated and directed by the central government acting on the recommendation of the World Bank and funded by a loan from that institution, has been taking place within the higher education sector. Universities have merged and many colleges have been attached to them as faculties. As a result, the number of universities has been reduced by about half (as of January 1, 2000 there are twelve state universities, eleven state colleges and five state-owned art colleges in Hungary). Colleges that have not found university partners have had to merge with each other, on the basis of either geographical proximity or area of specialization. Just a few large, well-established colleges were allowed to preserve their independence and to operate as single units. It is still too early to judge to what extent this consolidation will be regarded as progressive or as damaging to the HE sector in Hungary.

It is still the case that higher education is by far the largest and, in many respects, the most active part of the Hungarian scientific community. Sixty per cent of S&T capacity (human resources) is affiliated to higher education institutions. Universities' position within the S&T system (STS) was strengthened in the transition period, since staff levels were reduced less drastically than in other sectors. It should also be noted that, of all the S&T fields, only (higher) education is directly represented in the government. The only other forum in which representatives of other leading institutions participate is the Science and Technology Policy Collegium (see above).

Following the transformation process and new legislation, the HAS became a so-called 'public body' or public-law association that is wholly independent of government. According to the civil code, a public-law association is self-governing, has registered membership and is governed by specific legislation. Such associations are also required to serve the public interest in its own sphere of activity. Anybody who has a scientific qualification (Candidate or Doctor of science, PhD) may apply for membership of the association, which is known as the Public Body of the HAS; such membership is not, of course, the same thing as membership of HAS, the country's highest scientific body. Association members have the right to send 200 delegates to the HAS General Assembly. The new legislation gave the Academy some property of its own (mainly buildings) and a guaranteed state subsidy. It is still funded by central government but enjoys special independence in respect of its economic management system. The Academy's activities and policy-making processes are more transparent than they used to be. Management and supervision of the various institutes has been largely delegated to a new body called the Council of the Research Institutes of the Academy. The HAS network now consists of

35 research institutes in addition to 125 small R&D units operating in various higher education institutions. The number of scientific staff (in full-time equivalents – FTE) was 2 297 in 1994 (19.6 per cent of all full-time scientific staff in Hungary). In the year 2000, there were 2 096 scientists and engineers employed by the HAS research network; they accounted for about 20 per cent of all full-time scientific staff in Hungary. In other words, there has been only an insignificant change in the number of researchers. The number of technical and administrative staff, on the other hand, has been reduced by about 30 per cent. With the collapse of R&D activities in most state-owned Hungarian enterprises, HAS and the higher education sector (with 4 768 FTE researchers) account for two thirds of the research staff in the country. The business sector's share in R&D employment is about 25 per cent.

The HAS research network enjoyed great institutional stability even in the early, most critical years of transformation. Skilful maneuvering by HAS presidents successfully repulsed political attacks and secured a peaceful transition for scientists involved in basic research. This was highly appreciated by the research community. In the long run, however, stability could not be regarded as the only or even the most significant value. Decisions on finance, organization and human resources had to be taken for the benefit of the most productive and eminent research units (Mosoni-Fried, 1997). A process of consolidation was set in motion in December 1997 and was more or less completed by the end of 2000. HAS received additional funds from the state budget for three years in order to implement this process of consolidation; the increase amounted to about 5 per cent of the Academy's total annual state budget allocation in the years in question. The distribution of this extra funding, about 600 million Hungarian forint (HUF) or 2.4 million EURO per annum, was approved by the General Assembly of the Academy in 1998. Life sciences were given 47.3 per cent, mathematics and natural sciences 39.5 per cent, and social sciences 13.2 per cent of the additional funds. In addition, a core staffing level was identified for each institution, with budgetary support being guaranteed only for the number of employees thus defined. Any staff employed over and above this limit have to be paid for from external (non-budgetary) sources. Average unit costs were defined by the Consolidation Committee as follows:

– 213 000 HUF/per person per year in life sciences;
– 298 000 HUF/per person per year in natural sciences;
– 176 000 HUF/per person per year in social sciences.

Certain institutional changes did take place in connection with this process, but – with a few exceptions – they did not endanger employees' academic careers. Any changes made were intended to have a beneficial not an adverse effect on the HAS. Two research institutes were transferred to the university sector. Seven institutes had to be downsized and merged with other HAS institutes. This effectively reduced the number of institutes from 44 in 1998 to 35 in 2000.

These institutional changes also affected research priorities. Three research areas have been given priority in the consolidation process: ecology, regional development and national minorities. Three small HAS institutions have been set up to study these areas.

Applied research is a missing link in the chain in Hungary. Since the decline of the branch institutes, only one organization has been established in this area, namely the *Zoltan Bay Foundation for Applied Research*, which is modeled on the German Fraunhofer-Gesellschaft. It was expected to grow rapidly into a new applied research network, geared to the high-tech sector. However, scarce resources mean the foundation has so far been able to set up only three institutes (in biotechnology, logistics and material sciences). They are funded from the foundation's own budget and external sources.

Industrial R&D is still in a state of transition. On the one hand, significant restructuring is taking place, with large enterprises, including Hungarian subsidiaries of multinational companies, playing an increasingly important role. On the other hand, the branch institutes, where industrial R&D used to be concentrated, are mere shadows of their former selves. There are many innovative small and medium-sized companies, but most of them lack R&D capacities. They are adapting to the new conditions but carry out very little R&D. Unlike in other countries, branch institutes were not automatically dissolved in Hungary in the crisis years of transformation. In the first half of the 1990s, 16 institutes in different branches seemed to have been saved by various declarations and, in particular, the 1995 Act on the Sales of Company Assets in State Ownership. Under the provisions of this act, the state retained a 50 per cent + 1 vote share in these institutes. This apparent willingness on the government's part to let these institutes survive did not in fact help very much. With a few exceptions, they have all now been shut down due to a permanent lack of both capital and a market for their products or services. Those that remain are doing mainly non-R&D activities. The Ministry of Economic Affairs still has plans for their reorganization, but no direct state assistance can be expected. It is now too late for the branch institutes, since they no longer produce high-quality R&D. However, as the industrial sector develops, a partial revival of a few branch institutes remains at least a possibility. The original idea of privatizing them was deemed a failure. To date, only three branch institutes have been privatized and preserved as R&D institutions (Plastics Research Institute, Pharmaceuticals Research Institute, R&D Company for the Automotive Industry). They were privatized in three different ways.

The Plastics Research Institute is now owned and managed by Pannonplast, one of the most successful domestic companies. This 'marriage' was a pragmatic measure, since the institute had already been working for this company and for its predecessor for many decades.

The Pharmaceuticals Research Institute has become a contract research organization owned and managed by a foreign investor, while the R&D Company for the Automotive Industry was privatized through a management buy-out and is now serving Hungary's relatively substantial automotive industry. The complete privatization of these institutes was authorized by the State Property Agency, since nobody was willing to invest in companies in which the state had a controlling power.

In-house R&D was reduced very rapidly between 1989 and 1992. Most company R&D units were closed, or simply disappeared when firms went bankrupt; about 60 per cent of companies' R&D employees had to seek alternative employment. Most in-house R&D capacities were liquidated or cut back on privatization, whether ownership fell to domestic or foreign investors (Mosoni-Fried, 1998). Despite the increasing number of R&D units (Table 13.1), therefore, the total number of scientists and engineers working in R&D declined markedly in this period (Table 13.2). Foreign firms, whether brown-field or green-field investors, have transferred technology to the host country but most have proved to be uninterested in domestic R&D. In some cases, however, positive experiences have led to R&D laboratories being preserved and integrated into the newly privatized enterprises. Fortunately, some newly founded companies, including some created when existing enterprises were split up, have managed to establish themselves as R&D-oriented firms. In addition, some large firms managed to quickly revitalize themselves after the great transition crisis between 1990 and 1992 and were not forced to dismiss R&D employees. Large state-owned companies, e.g. in the energy sector, have also preserved their R&D capacities. The most favorable situation can be found in firms in the machinery, electronic, pharmaceutical and lighting industries in particular, where R&D units have been incorporated into the foreign company's headquarters. In recent years, a few multinational companies and other foreign organizations have undertaken substantial R&D activities in Hungary. Although these activities are largely restricted to supporting single in-house research laboratories, some firms, such as Ericsson and Nokia in the telecommunications sector, for example, are operating single new R&D institutes, funding university research laboratories for graduate and postgraduate education, etc. New R&D investments are also generously supported by state grants. For example, a firm that invests at least 500 million HUF in Hungary and employs at least thirty scientists and/or engineers for a minimum of three years in a research unit may receive a government subsidy of 200 million HUF. This is

Table 13.1 Hungary: number of R&D units

	1988	1990	1993	1996	1997	1998	1999	2000
Total number of R&D Units of which:	1 323	1 256	1 380	1 461	1 680	1 725	1 887	2 020
R&D Units in HE	944	940	1 078	1 120	1 302	1 335	1 363	1 421
R&D Institutes	68	69	68	73	80	74	130*	121*
Company R&D Units	235	174	178	220	246	258	394	478
Other Research Units	75	73	56	48	50	58	.	.

* "Other research units" are included here.
Source: KSH, 1989–2001.

Table 13.2 Hungary: scientists and engineers (calculated FTE) in R&D institutions

	1988	1990	1993	1996	1997	1998	1999	2000
R&D Units in HE	5 251	5 204	4 546	3 857	4 194	4 398	4 768	5 852
R&D Institutes (incl. HAS)	6 035	5 189	3 450	3 097	3 072	3 061	4 550*	4 653*
Company R&D Units	8 504	5 681	2 637	1 955	2 394	2 725	3 261	3 901
Other Research Units	1 637	1 476	1 185	1 499	1 494	1 547	.	.
Total Staff	21 427	17 550	11 818	10 408	11 154	11 731	12 579	14 406

* "Other research units" are included here.
Source: KSH, 1989–2001.

obviously advantageous for those mostly foreign-owned firms that can afford to exploit these opportunities.

Of the many thousands of foreign-owned firms, fewer than 50 are engaged in R&D in Hungary and only 5 or 6 subsidiaries of multinationals have integrated local research laboratories into their Hungarian headquarters. According to statistical surveys, foreign firms spent a total of 6 663 million HUF on R&D in 1994, while in 1996 they spent 11 918 million HUF (about one seventh of GERD in each year). Of these, wholly foreign-owned firms spent 292 and 1 142 million HUF in 1994 and 1996 respectively (see Meskó, 1997). These figures are official, published by the Hungarian Statistical Office, but in our view they should be regarded only as indicative. Many foreign firms, as well as a lot of domestic ones, ignore the fact that reporting on R&D is mandatory in Hungary.

4. RESKILLING

In the early years of transition, it was not at all obvious that the R&D sector could be adjusted to market economy conditions. Its heavy dependence on the state budget and the Eastern European market were rather unfavorable preconditions for a successful restructuring and adjustment process. Taking stock after more than ten years, we can say that the loss in quantitative terms was great and the end result negative. In qualitative terms, however, the rapid integration of Hungarian R&D on a project and/or unit level both within and outside the European Union gives ground for regarding the overall end result as positive. Knowledge and experience accumulated under the previous political regime has been converted into new skills, at least by those who emerged from the changes as winners.

What are the factors that might endanger these positive trends? They include the following:

- scarcity of human resources: in recent times R&D has not been a competitive sector of the national economy; an ageing workforce is one of the most worrying aspects of academic research;
- low investment in R&D infrastructure in the government sector: this mainly affects buildings, scientific libraries and small-scale instruments (research facilities);
- predominance of project financing: institutions have been neglected under the current funding system;
- dominance of the state budget in financing: nationally, 53 per cent of resources flow from the state budget. In the academic sector, the figure is between 80 and 90 per cent. The domestic market for R&D remains small;

- concentration of research contracts from the business sector in a few disciplines: for example, 90 per cent of HAS research network income from the business sector is generated by just one of the 35 HAS institutions, which is involved in computer related sciences;
- in most developed OECD countries, a growing share of business sector R&D is in services. The service sector – especially knowledge-intensive business services – has been very late in developing in Hungary. It is still unclear to what extent service firms will engage in R&D;
- the business sector is being encouraged to invest in in-house R&D and networking. The larger a company is, the more certain it is that government incentives hit the target. However, since large companies with 250 employees or more currently account for less than one percent of all companies, the extent of their R&D activity remains far from satisfactory.

5. S&T POLICY AS PART OF HUNGARY'S INTEGRATION STRATEGY

Consolidation of the R&D system continues to be an important task. Now the country has entered the *third phase of institutional transformation* (Meske, 2000), the building of a qualitatively and quantitatively new STS is on the government's agenda. In the public sector, political power and decision-making are being centralized and human and financial capacities are being concentrated. *Laissez-faire S&T policy* belongs to the past, as is demonstrated by the new policy guidelines and the action plan for 2001 – 2002. The document "Science and Technology Policy 2000" drawn up as a 'white paper' by the government's Science and Technology Policy Collegium (cf. STPC, 2000), set out the following policy measures:

Additional state funds were to be made available for R&D in 2001 – 2002. The extra resources amounted to 17.5 billion HUF in 2001 and to 19.0 billion HUF in 2002 (about 25 per cent of GERD in 2000). The greatest share of this additional allocation was to be given to large projects in priority areas. Furthermore, the additional funding also made it possible to introduce as from January 1, 2001 a significant wage adjustment for research scientists and engineers in the higher education sector and research institutes. This adjustment will particularly benefit very badly paid young researchers, giving them an increase in monthly pay of between 40 and 50 per cent. This measure is helping to narrow somewhat the earnings differential between the academic and business sectors. The document also stated that the share of GERD in GDP, which stood at 0.70 per cent in 1999, was to be doubled by 2002. Within that figure, moreover, the share of business enterprise R&D expenditure (BERD) was to reach 50 per cent by the same time (the figure for the year 2000 was 37.8 per cent) (Figures 13.5 – 13.7).

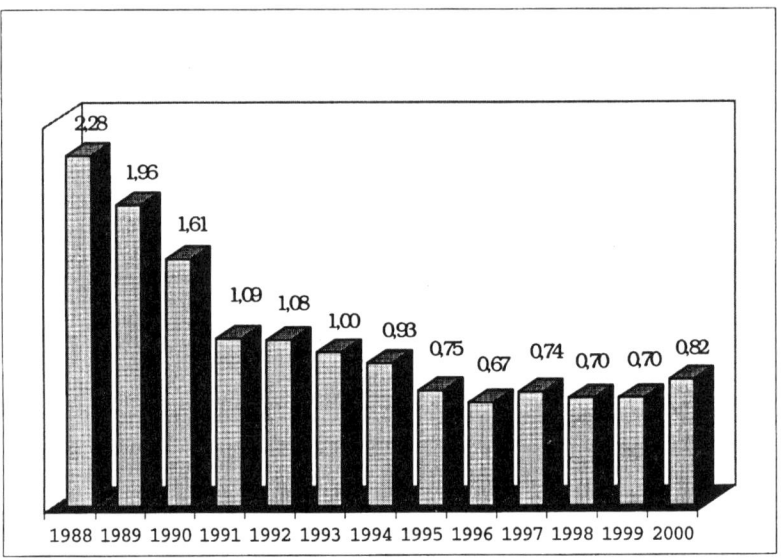

Fig. 13.5 Hungary: R&D expenditure (GERD) as percentage of GDP
Source: KSH, 2001.

National Research and Development Programs (NKFP) were drawn up. They cover five priority areas:
– improvement of quality of life;
– information and communication technologies;
– environmental protection and material sciences;
– agriculture and biotechnology;
– national heritage and social challenges (humanities and social sciences).

Other goals include increasing the number of PhD students, taking greater account at national level of the needs of industrial enterprises, developing attractive career paths for scientists and facilitating knowledge transfer between academic institutions and the business sector.

These national research and development programs are part of a broader economic development program called the Széchenyi Plan. This is a unique initiative, for the following reasons:
– it encourages and values cooperation between sectors and/or institutions;
– it gives preference to large coordinated projects;
– it is apparently linked to economic priorities;

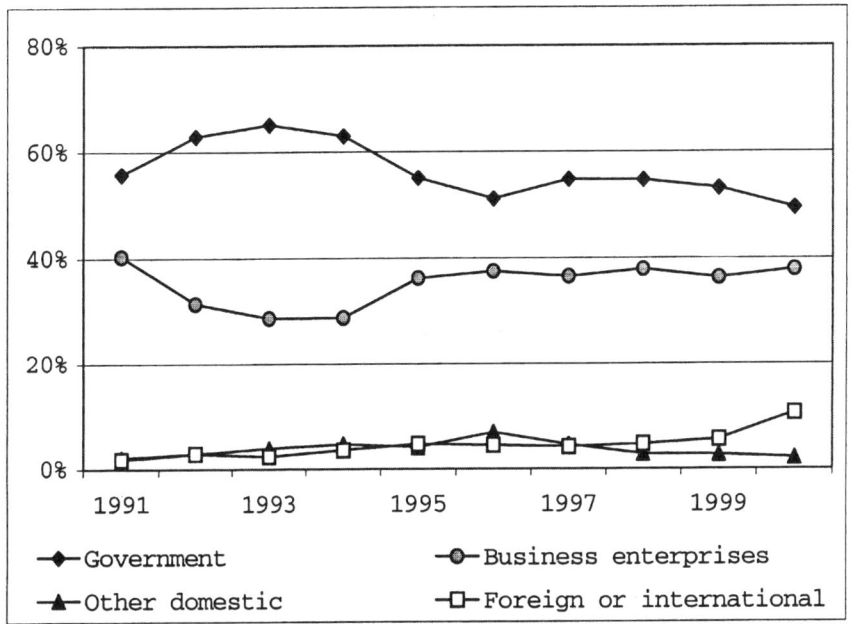

Fig. 13.6 Hungary: R&D expenditure by financial source (%)
Source: OMFB, 1999; KSH, 2001.

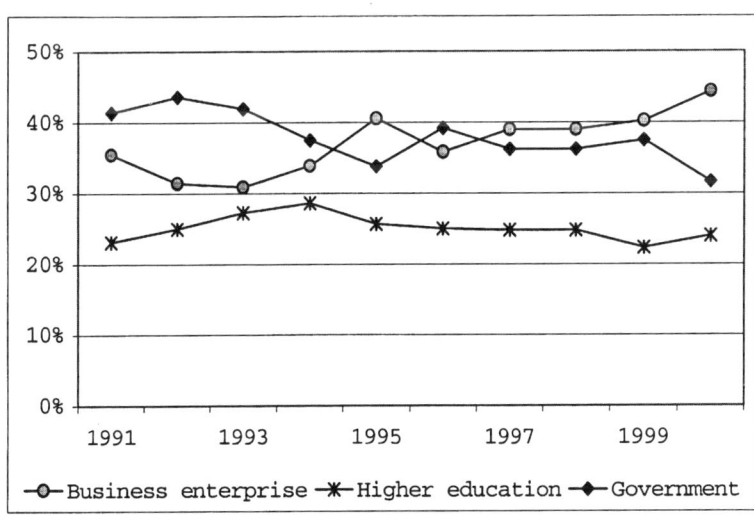

Fig. 13.7 Hungary: R&D expenditure by sector of performance (%)
Source: KSH, 1991–2001.

– it seems to provide a good basis for Hungary's participation in the 6th EU Framework Program (Siegler, 2001).

In the past 15 years of the grant system, researchers have become accustomed to very small grants. In the first four priority areas set out in the national research and development programs, the minimum cost limit for single applications is 100 million HUF (about 400 000 EURO). The state can pay 100 per cent of total costs for projects classified as basic research, 50 per cent for those classified as applied research and 35 per cent for those classified as experimental development. The balance must be found from the applicants' own resources. In the social sciences and humanities, the minimum cost limit per project is 10 million HUF. These projects can be fully funded from the state budget.

The S&T policy document also emphasizes that *there is no need for any significant institutional change in the near future*. The government has taken responsibility for national research institutes and the higher education sector and is providing support for business R&D. A growing share of research support is to be provided on a competitive basis. New support schemes are to be launched for infrastructure development. As part of these schemes, scientific libraries are to be allocated additional funds.

The nationwide *information infrastructure program* is acknowledged as the most successful development in the Hungarian R&D sector in recent years. Continuous upgrading of the well-established infrastructure is crucial. Hungary, which already participates in the European Union's framework programs, is joining the European Union Quantum/TEN 155 network project. As a result, the United State's Internet2 and NGI projects are also to be made available to the Hungarian research community. The modernization of laboratories, the purchase of instruments of all kinds and sizes, investment in machinery and construction, etc. are key questions for the whole sector.

Researchers in Hungary have been relatively free to develop *international relationships* since the 1960s. This freedom has been further strengthened by the transition to democracy. International projects open to Hungarian citizens and the various aid and other programs have played a considerable role in the survival of the R&D system. Through its PHARE, TEMPUS, COST, PECO, COPERNICUS and EUREKA programs, among others, the European Union has given significant assistance to the Hungarian scientific community, particularly in the higher education sector (Balogh, 1997; Inzelt, 1999). In addition, the renewal of agreements with large foreign scientific organizations as well as membership of CERN and of the European Science Foundation have been very valuable for Hungarian researchers. Significant progress has been made in the field of S&T cooperation with the OECD and NATO. Scientometric figures reveal some decrease in the country's scientific output and a slight increase in its quality; in many disciplines the number of citations is higher than it was 5–10 years ago. It seems that, even today, Hungary's position in science is well ahead of its economic productivity (20th and 43rd place, respectively, in the world rankings) (Török, 2000).

In the next few years the most important objectives in R&D are linked to Hungary's *participation in the European Union's Framework Programs*. The country's full participation in the 5th Framework Program (FP5) appears to be accelerating Hungary's integration into the European scientific and technological community and paving the way for the country's accession to the European Union. Cooperation in this program is expected to be highly beneficial for all concerned. Hungary is participating in 130 projects within FP5. It is worth mentioning that most participants are affiliated to the HAS research network. Applicants from Hungary have been acknowledged primarily in the fields of information technologies, life sciences, environmental and material sciences. Until recently six Hungarian centers of excellence were acknowledged by the EU. A few others are also expected to be awarded this kind of recognition in the near future.

Experience of the *consequences of dissolution and fragmentation* has encouraged organizations to explore the possibility of merger with other institutions. As part of the consolidation of the HAS research network, for example, eight research institutes in social sciences and humanities have been integrated into one large (administrative) unit called the Research Center for Social Sciences. In legal terms, each institute is independent, but they are entitled to participate in large-scale projects or programs as a single HAS institution.

The *future of the Hungarian S&T system* depends on two decisive factors. The first is EU enlargement, since Hungary's accession to the European Union will have an enormous impact on both social and economic development. Should external and internal conditions be favorable, it is likely that new EU member states such as Hungary will slowly but surely claw their way out of their peripheral situation, as Spain, Portugal and Greece did before them (Rácz, 1994). R&D is one of the main areas of activity for developing links with institutions and individuals from the most advanced countries. Scientists are well aware that integration is a lengthy process and will not happen overnight on accession (Weber et al., 1999); it is based on cooperation and the ability to make significant contributions to common goals over the long term. The second factor is macroeconomic development during the first decade of the 21st century. The risk that a dual economic system will emerge, with a strong foreign-owned and a developing but on the whole still underpowered domestic business sector, is still very real. It also exists in business sector R&D: a high share of the best Hungarian company R&D laboratories are owned by foreign firms. They are able to recruit the best scientists and engineers (including young ones), can afford the most sophisticated facilities and, moreover, since they are mostly in the mainstream of the given research field, they are often awarded relatively high domestic research grants as well. The current situation is not a truly competitive one: domestic enterprises lag behind foreign firms in many fields, including technology transfer, salaries, networking, etc. Most of the R&D embodied in Hungary's main export products, for example, is carried out either in foreign-owned companies or in research sites supported by foreign investors in Hungary. In this sense, the economic, technological, organizational and other changes currently taking place are a response to demands and opportunities created by foreign capital, foreign markets and foreign management as well as by international organizations, the EU and other international financial sources, programs, etc. "[The] transition towards [a] market economy should result in the decisive role of competition i.e. there should be a major change in enterprise behavior as far as innovation, including diffusion of new products/processes, is concerne" – wrote Havas in 1991 (Havas, 1991, p. 164). This consideration has proved correct: competition, which was not a basic feature of the planned economy system, has become one of the most significant driving force of structural and other changes in the first half of the 1990s. However, in the meantime it had become apparent that the strong competition in the domestic market – unusual in this region in former time – had yielded not only prosperous innovative firms, but had also reduced the innovative capacities of many firms, since they became unable to make progress and profit, to export their goods, etc. Innovation for these enterprises was not a question of 'behavior' but one of profitability and financial resources. Therefore competition did not strengthen innovation to the expected extent in the 1990s. Poor players, even if they would have been capable to innovate due to their knowledge base, were unable to compete with strong ones, became unable to innovate and simply were no more competitive partners, if they survived at all. It has become clear that Havas' further consideration was true that "competition is a necessary but not sufficient condition of [the] faster and wider dissemination of innovations." (Havas, 1991, p. 164). This view, expressed more than ten years ago, seems to be even more valid today. Competition plays a very important role but it is not in itself sufficient to produce real competitiveness in the market. Cooperation, knowledge and technology transfer are also needed, as well as appropriate fiscal, economic and S&T policies. This is the only way to avoid the emergence of a dual structure in both the economy and R&D.

6. MAIN INDICATORS IN THE 1990S

Due to the high rate of inflation in the first half of the decade, total R&D expenditure at current prices grew by 388.9 per cent between 1991 and 2000. In real terms, however, GERD decreased by 17 per cent over the same period. (In the first half of the 1990s, the inflation rate reached 20 – 30 per cent per annum; in 2000 it was 9 per cent.) The share of central state resources increased, while company resources declined, by 40 per cent in 1991 and 49.5 per cent in the year 2000. GERD as a percentage of GDP was 1.09 per cent in 1991 and decreased to 0.70 per cent in 1998 but increased to 0.82 per cent by the year 2000. Human capital in R&D (total number of employees in FTE) decreased by 20 per cent between 1991 and the year 2000. The number of scientists and engineers was 21 427 in 1988 (the last year of the old system) and 14 406 in 2000 (67.2 per cent). However, this figure has been increasing since 1997. In the past, reductions in human resources occurred in very different ways. In the higher education sector, for example, there was no significant change during the early years of transformation (Imre, 1998). The radical decrease occurred in 1995 when – due to financial restrictions – many lecturers and researchers were forced to leave their place of employment. HAS financial resources decreased by about 50 per cent in real terms between 1988 and 1995. This was the main reason behind the mostly voluntary mobility of scientists: 20 per cent left either for foreign R&D units or for the Hungarian business sector. In industry and agriculture, most of the research laboratories were closed down and about 70 per cent of researchers had to seek alternative employment.

In 1988 there was a total of 1 323 R&D units. These included 944 higher education units, 68 R&D institutes (HAS and branch institutes), 235 enterprise units and 75 other units. In the year 2000, there were 2 020 research units registered by the Hungarian Statistical Office (cf. Table 13.1). This relatively great difference is due mainly to the extended statistical coverage of business enterprises and the growth in the number of R&D units in the higher education sector over the last few years.

In real terms, the 1990s witnessed a significant decline in the Hungarian R&D sector. However, institutional and structural changes, together with the relatively new trends towards integration, might be seen as compensating for the deterioration in certain areas of domestic R&D. It is still uncertain, for example, whether new and positive developments in company R&D, primarily in manufacturing industry, remain limited in scope or are the first signs that an integrated R&D system, capable of competing on a European or global level, is emerging in the corporate sector. In the academic sector, the quality of higher education and the upgrading of the R&D infrastructure seem to be the most decisive factors in future development and successful international integration.

REFERENCES:

Balázs, K. (1994) *Transition crisis in Hungary's R&D sector.* Budapest: HAS, KTI (Institute of Economics).
Balogh, T. (1997) "Magyarország az EUREKA-ban" (Hungary and the EUREKA). *Magyar Tudomány* no. 11: 1346–1350.
Deliné Pálinkó, É., and K. Dévai (2000) "A felsõoktatás K+F finanszírozásának egyes kérdései" (Financing R&D in the higher education sector). *K+F a Budapesti Mûszaki és Gazdaságtudományi Egyetemen* (R&D in the Budapest Technical University). Ed. K. Dévai. Budapest: Mûegyetemi Kiadó (Publishing House of the Technical University of Budapest), 85–102.
HAS (1998) *Akadémiai Értesítö.* Newsletter of the HAS, October 16.
Havas, A. (1991) "The Hungarian Laser Industry in Transition." *Innovation, Competition, Competitiveness.* Ed. A. Inzelt. Budapest: IKU (Innovation Research Center), 147–172.
Imre, J. (1998) "S&T in Hungary: Past, Present and Future." *Transforming Science and Technology Systems – The Endless Transition? NATO Science Series 4: Science and Technology Policy – Vol. 23.* Ed. W. Meske, et al. Amsterdam: IOS Press, 69–82.
Inzelt, A. (1999) "Kutatóegyetem a finanszírozás tükrében" (Financing R&D at research universities). *Közgazdasági Szemle* April: 346–351.
KSH (1989–2001) *Kutatás és Fejlesztés* (Research and Development – Yearbooks 1988–2000). Budapest: Central Statistical Office (Központi Statisztikai Hivatal – KSH).
Meske, W. (2000) "Changes in the innovation system in economies in transition: basic patterns, sectoral and national particularities." *Science and Public Policy* 27, no. 4: 253–264.
Meskó, A. (Ed.) (1997) *Foreign Direct Investment in Hungary 1995–1996.* Budapest: KSH – National Accounts Department.
Mosoni-Fried, J. (1997) "Transformation of the R&D System in the Transition Economies: The Changing R&D System in Hungary – Country Report." Unpublished paper. Berlin: Wissenschaftszentrum Berlin für Sozialforschung (WZB).
Mosoni-Fried, J. (1998) "Structural Changes in Industrial R&D in Hungary: Losers and Winners."*Transforming Science and Technology Systems – The Endless Transition? NATO Science Series 4: Science and Technology Policy – Vol. 23.* Ed. W. Meske, et al. Amsterdam: IOS Press, 171–182.
OMFB (1999) *Research and Development (R&D) in Hungary 1999.* Budapest: National Committee for Technological Development (OMFB).
Rácz, M. (1994) *A perifériahelyzetben lévõ országok és a K+F politika* (Countries in the periphery and R&D policy). Budapest: OMFB.
Siegler, A. (2001) "Integráció és bõvítés az Európai Kutatási Térségben" (Integration and enlargement in the European research area). *Magyar Tudomány* 6: 714–728.
STPC (Science and Technology Policy Collegium) (2000) *Tudomány és technológiapolitika 2000* (Science and technology policy 2000). Budapest: Oktatási Minisztérium (Ministry of Education).

Tamás, P. (1985) "Historical Process of the Institutionalization of Science and Technology." *Science and technology policy in Finland and Hungary – A comparative study*. Ed. K.O. Donner and L. Pál. Budapest: Akadémiai Kiadó, 30 – 46.

Török, Á. (Ed.) (1999) "Alapkutatás, alkalmazott kutatás, piacgazdaság" (Basic research and applied research in market economy). Unpublished paper. Budapest: HAS.

Török, Á. (2000) "Mérhetõ-e a magyar K+F nemzetközi versenyképessége?" (Is it possible to measure the international competitiveness of Hungary's R&D activity?)*K+F a Budapesti Mûszaki és Gazdaságtudományi Egyetemen* (R&D in the Budapest Technical University). Ed. K. Dévai. Budapest: Mûegyetemi Kiadó (Publishing House of the Technical University of Budapest).

Weber, M., W. Meske, and K. Ducatel (1999) *The Wider Picture: Enlargement and Cohesion in Europe*. Futures Report Series 15. Seville: Institute for Prospective Technological Studies (IPTS).

14. ROMANIA: TRANSFORMATION OF THE S&T SYSTEM

Steliana SANDU

From the early 1980s onwards, Romania's science and technology (S&T) system struggled under the imposition of numerous and various burdens. The expanded function of the National Council for S&T, chaired by Elena Ceausescu, gave that body a political and bureaucratic supervisory role that did little to enhance its performance in solving the problems of R&D coordination. The entire research sector fell prey to politics and ideology, while forfeiting its organizational autonomy and freedom of scientific activity. The S&T system had a pyramidal structure, the base of which was made up of the huge Central Research Institutes. Thus when the reforms began, the Romanian situation was characterized by features that set the country apart from other East European states. They included:
- the organizational separation of industrial research from industry and its concentration in large, specialized institutes;
- priority-setting criteria and mechanisms oriented to national autarchy rather than to competition;
- an internal evaluation system distorted by political and ideological factors;
- an inadequately developed R&D infrastructure;
- the almost isolation of Romanian science from the international scene;
- a research – development – production cycle that never really functioned properly.

1. FROM R&D SYSTEM TO INNOVATION SYSTEM

After 1990 Romania moved within two years from an excessively centralized to an excessively liberal R&D system. The laissez-faire principle is now accepted as the prevailing philosophy in Romania, in R&D as elsewhere. In the field of R&D, a new legal and institutional framework was gradually put in place that seeks to encourage competition in R&D fund allocation, the development of in-house R&D capacities in private firms and investment in industrial innovation.

The first step in this direction was the dissolution of the National Council for S&T and the Central Research Institutes, which eliminated the top level of the former S&T system. The Romanian Academy of Sciences (AoS) was acknowledged as the highest scientific forum in Romania and as an autonomous scientific entity; its network of research institutes was rebuilt. A reorganized Ministry of Education and Science was given responsibility for domestic and international scientific affairs. The second step involved the elimination of all hindrances created by the over-centralized system by granting the research units organizational and decision-making autonomy. The third step was to grant operational autonomy to all economic units, including R&D institutes, a measure that coincided with the emergence of new enterprise structures.

Despite the political rhetoric, the role of S&T was marginalized by policies introduced between 1990 and 1993. As a result, science and technology has played little part in solving Romania's social and economic problems. Policymakers disregarded S&T and introduced radical neo-liberal measures, which were insufficiently grounded. This disrupted S&T activities and affected the quality of research for a long period. However, the scientific community, anxious to preserve its past by getting public funds for projects not associated with the economic and social needs of the country, undertook a series of *restorative actions*. The rights of reputable institutes and scientists were restored and international contacts resumed. The universities now have operational autonomy. Few academics were exclusively engaged in research, while research activity within universities was a peripheral activity. At the same time, R&D institutes faced a shortage of funds. In 1991, the state funding allocated to research amounted to only 0.18 per cent of GDP. Institutes were threatened with liquidation and it became conceivable that R&D would disappear altogether from Romania.

R&D, HE and Academy of Science institutes all suffered from inadequate and outdated infrastructure, while research was hindered by the fact that teaching staff were overburdened. Industrial R&D has undergone comprehensive structural changes since 1990. Industrial R&D institutes went through a brutal process in order to achieve autonomy, most of them being forced to become commercial companies, while others were assimilated into industrial companies. The fall in demand for domestic and foreign R&D, coupled with the slow pace of privatization and the lack of incentives to invest in the modernization of production processes, inhibited radical reform in industrial research. A special fund was set up, financed by state and private companies, with the aim of providing financial support for priority research programs and eventually eliminating subsidies and the restrictions on institutions associated with public funding.

After 1994, the National University Research Council and Academy Grant Council were set up with the twin aims of stimulating research quality through selective competitive funding and encouraging young researchers.

At the end of 1992, the Ministry of Research and Technology (MRT) was established in order to coordinate R&D activities with a view to stopping the 'uncontrolled restructuring' of R&D – 40 per cent of personnel left the sector between 1990 and 1992 – and maintaining technological development. Industrial Liaison Offices (ILOs) and related local organizations, which support innovation and technology transfer, have become established in Romania since 1992 through a combination of bottom-up initiative and top-down strategy. The Ministry of Research and Technology has provided guidance and funding to many of these initiatives. The emphasis has been more on the commercialization of research results rather than on the identification of, and response to, company needs for technology and technical improvement (PHARE, 1996). In 2001, only some of the 34 ILOs initially established were still operating, due to a lack of state support.

A major shortcoming of the reforms in Romania is that no specific bodies were put in place with responsibility for drawing up S&T strategy and policy at national level and coordinating activities and organizations.

The excessively simplistic approach to innovation that has been adopted, with attention being focused only on the application of scientific research results, as well as the almost general tendency to ignore the fundamental role of innovation in modernizing economic structures in Romania, has led to the importance of innovation as a factor in economic development being neglected.

Considerable efforts need to be made to change the technological structure of the economy. Increasing the national capacity to produce and absorb new technologies is a prerequisite for compatibility with production systems in the EU, which Romania seeks to join, and for deriving economic benefit from EU membership whenever it comes about.

The first step towards the creation of a Romanian innovation system was to provide institutional support for the stimulation of demand for innovation by laying down criteria for the rational selection of industrial companies. The institutional changes introduced to date have basically been aimed at changing the status of the various elements of the science and technology system but have taken little account of the need to create an informal, systemic and coherent framework within which the non-linear model of innovation can be applied. The Ministry of Research and Technology (MRT), as the main decision-making and strategic body in this field, has devised and partially implemented innovation strategies; this process began in 1993, but the strategies have been reviewed periodically since then, with political criteria unfortunately being brought into play, following changes among senior MRT staff.

The MRT drew up short-term strategies for R&D in each branch. New institutes were spun off from existing institutes, which were also reorganized and restructured. The privatization process had already begun in 1993 with the sell-off of approximately 5 per cent of all existing R&D units and the establishment of new private R&D-oriented companies. This process of R&D restructuring continued in 1994 with assistance from the World Bank and EU organizations delivered through special programs (e.g. PHARE) and guided by several strategies inspired by Western European models. The Romanian authorities made efforts to adapt the R&D infrastructure and to introduce policies similar to those of OECD countries.

A national R&D reform program was developed for 1995. It had the following objectives:

– reform of the system of R&D public financing through the introduction of a competitive system of fund allocation based on evaluation and selection procedures; in 1995, 25 per cent of total R&D funds were allotted in this way. The first evaluation of research institutes was published in 1995. In addition to such retrospective evaluation, some institutes involved in the PHARE program were audited and potential projects were subjected to preliminary evaluation in order to select those deserving grants and core budget funding;
– fund allocation to be based on priority national programs;
– co-financing of technology transfer projects, with about 11 per cent of overall R&D budget funds being used for co-financing (up to 20 per cent of the costs required for transfer projects and the implementation of R&D results);
– the creation and development of an experimental structure for the transfer and implementation of research results as well as logistical support for the dissemination of scientific and technological information on a regional level via newly established regional invention centers and the establishment of an experimental innovation network.

Since no fundamental changes resulted, a new R&D funding strategy was devised by the new Minister of Research and Technology from January 1997 onwards. The aim of this strategy was to change the institutional bodies and funding mechanisms and to involve civil society in the management of research funds. All the public financial resources required to implement the R&D and innovation policies were supposed to be concentrated in the MRT budget. They were to be distributed through two funds – the R&D Fund and the Incentives for Innovation Fund – which were to be set up once governmental approval had been given. This strategy was not implemented due to the removal of the minister who had drawn it up.

The lack of a conceptual framework within which innovation can be treated as a dynamic process of continuous learning, including not only the creation of new knowledge and ideas but also an element of technological innovation produced outside the research institutes, has led numerous political decision-makers and economic analysts to consider innovation as a result of the play of market forces, in which the distinction between suppliers and recipients is easily drawn and definitive. This point of view is certainly at odds with the modern innovation system, in which the organizational structure is basically the network and not the company and processes are governed more by cooperation than by competition, with decision-making emphasizing consensus rather than conflict.

The dimensions and nature of the cooperation between the actors involved in the innovation process have not yet been shaped by an awareness of common interest, as the race for increasing industrial competitiveness goes on. Adoption of the non-linear model of innovation affects the architecture of the existing system and creates new formal and informal connections between its components, linking areas of activity that have hitherto been strictly separate.

At the stage it has now reached, Romania has put in place and has some experience of new institutional forms intended to strengthen relationships between scientific research and the market. Enlargement of the communication channels is the most significant problem faced in the creation of an innovation system in Romania, and it cannot be solved unless there is real interest on both sides. The interaction between science and the market works as long as enterprises are encouraged to innovate and the potential demand for new technologies is stimulated. It works, in other words, to the extent that innovation can be turned into an accessible instrument of market competition. Although setting up a non-linear innovation system involves the creative use of new technologies, each scientific research unit constitutes an important link within the network.

2. THE R&D SYSTEM AND ITS STRUCTURAL SHIFTS

The scientific element of the national innovation system in Romania consists of the following public and private research and education establishments:
- national R&D institutions;
- R&D units organized as public institutions and controlled by the central public administration;
- R&D units and structures in the universities;
- scientific research units operated by the Romanian Academy, organized as public institutions;
- research units of the branch academies operated as national R&D institutes or as public institutions;
- commercial R&D companies, R&D units and departments within commercial companies or autonomous state-owned enterprises in public-interest sectors such as electricity, gas and mining, which have some financial au-

tonomy and are engaged in production, as well as R&D units subordinated to such state-owned enterprises;
– museums and other units with an independent legal status.

Despite the fact that this structure is considered to be one in which research is guided by the market, the overall coherence of the system is impaired by two important deficiencies. Firstly, basic research units are highly specialized, which means that they interact only with similar units or within very restricted networks. Secondly, in the absence of a cooperation strategy based on common objectives, the links between basic and applied research units, on the one hand, and users, on the other, have no consistency and prioritize short-term objectives. This last characteristic is further reinforced by the lack of strategic cooperation between the main coordinating institutions in science and education, namely the Romanian Academy of Sciences, the Ministry of Research and Technology and the Ministry of Education. The Ministry of Education and Research, established in 2000 following the merger of the two ministries, is a new and auspicious political initiative that could improve cooperation among the main actors in the R&D system.

The institutional R&D system created in Romania after 1990 has been characterized by extreme fragility because of the absence of clear guiding principles governing the overall approach to R&D and its institutions and functional mechanisms. The transformation and management of the S&T system have been considerably affected by the vicissitudes of Romanian politics, with 6 ministers having been in charge over an 11-year period, each with markedly different approaches to the organization and management of S&T.

Nevertheless, the institutional structure has seen some fundamental changes. The number of units engaged in R&D activity (including industrial firms) increased from 369 in 1990 to 591 in 1994 and 645 in 1997 (see Table 14.1). These increases are attributable to the division of units following their conversion into commercial companies, the re-establishment of the Academy network and the establishment from scratch of private research units (cf. Table 14.2). Since 1998, the trend has been downward, resulting in a decrease from 643 in 1998 to 601 in 2000, especially in the enterprise sector. The number of employees decreased from 148 513 in 1990 (100 per cent) to 117 046 in 1994 (79 per cent) and 62 572 in 2000 (42 per cent) (see Table 14.1), mainly through voluntary departures caused by low incomes, career uncertainty, migration abroad and a lack of effective recruitment. Between 1995 and 1999, the number of personnel decreased by a further 37 874 in the public sector and increased in private and mainly privately-owned R&D companies (taking their share from 5 to 19 per cent) (cf. Table 14.2). Whereas in some social science fields the number of researchers increased, in technological research the employment level dropped drastically.

R&D expenditures (mostly for technology development and design activity) reached a peak of 2.6 per cent of GDP in 1989. Economic crisis (1990–1991) and the very slow upturn beginning in 1993 led to a sharp decline in R&D expenditure in real terms. Its share of GDP was 0.82 per cent in 1993 and 0.37 per cent in 2000 (Table 14.3). Insufficient funding for R&D, combined with a reduction in R&D investment, has been a permanent state of affairs since 1990, despite the sharp decline in the number of R&D personnel since 1989.

Since 1994, there has been a single and relatively permanent institutional body, namely the Consultative College for R&D and its specialized committees. Its main function has been to determine thematic priorities for the 22 broad areas covering the whole spectrum of Romania's research and financial resources allocation; this has been done on a competitive system, with projects being put forward by researchers from all institutions, regardless of institutional structure or form of ownership. Wide access to this competitive funding allocation can be considered an important step in making the Romanian financing system compatible with that of the EU. However, the extremely large number of research institutes in Romania which in 2000 had 23 179 researchers as potential suppliers of projects, combined with the extremely low level of financial resources (0.37 per cent of GDP in 2000) and a wayward allocation system, has eroded budgetary allocations and led to inefficient use of funds (see Table 14.3).

Within these general tendencies, there have been some shifts in the various sources of funding and the activities they are used to finance (cf. Table 14.4). Features specific to the main R&D sectors are outlined in the following section.

The *universities* in Romania lost their independence after 1948; their research institutes were transferred to the AoS. After 1989, increased demand for university places was met by increasing the number of higher education (HE) faculties. The establishment of private universities began after legislation was passed to make this possible in 1990. At the beginning of the 1992–93 academic year, there were 129 HE institutions in Romania, of which 62 were public and 67 private (the latter accounting for 26.8 per cent of students). In 1996 there was a total of 102 universities, of which 44 were private, with 485 faculties, of which 161 were private. In 1999 there were 121 HE institutions with 632 faculties, of which 221 were private. Most of the students in private HE are candidates who failed the entrance examination for public HE. From the very beginning, this situation has been detrimental to the creative scientific potential of private university graduates. The private universities are staffed partly by former and present state university professors and partly by scientists

Table 14.1 Romania: R&D units and personnel (end of year)

	1990	1991	1992	1993	1994	1995	1996	1997	1998	1999	2000
a) Number of Units											
TOTAL of which:	369	382	423	617	591	615	616	645	643	626	601
– enterprise sector				460	452	454	455	496	493	473	439
– governmental sector				120	105	120	122	109	114	109	110
– higher-education sector				37	34	41	39	40	36	44	52
b) Number of Personnel											
TOTAL of whom:	148 513	132 635	123 064	118 329	117 046	105 195	104 185	92 822	89 797	76 744	62 572
– enterprise sector				90 876	90 272	78 308	77 882	68 702	62 832	53 216	39 054
– governmental sector				24 224	24 091	23 487	21 351	18 493	19 121	17 941	17 563
– higher-education sector				3 229	2 683	3 400	4 952	5 627	7 844	5 587	5 955

Source : NIS, 1996, p. 301; 1997, p. 299; 1999, p. 273; 2000, p. 234; NIS, 2001, p. 7.

Table 14.2 Romania: R&D units and personnel by form of ownership

	1995				1996					1997					1998				1999		
	TOTAL	Type of ownership			TOTAL	Type of ownership				TOTAL	Type of ownership				TOTAL	Type of ownership		TOTAL	Type of ownership	
		Public	Mixed	Private		Public	Mixed	Private			Public	Mixed	Private			State majority	Private majority		State majority	Private majority
a) Number of units																					
TOTAL of which:	615	548	14	53	616	546	13	57		645	543	53	49		643	490	153	626	444	182
Enterprise sector	454	390	14	50	455	388	13	54		496	397	53	46		493	342	151	473	294	179
governmental sector	120	120	–	–	122	122	–	–		109	109	–	–		114	113	1	109	107	2
higher-education sector	41	38	–	3	39	36	–	3		40	37	–	3		36	35	1	44	43	1
b) Number of personnel																					
TOTAL of whom	105 195	100 333	2 085	2 777	104 185	98 891	2 457	2 837		92 822	82 900	6 916	3 006		89 797	77 219	12 578	76 744	62 459	14 285
Enterprise sector	78 308	73 664	2 085	2 559	77 882	72 813	2 457	2 612		68 702	58 849	6 916	2 937		62 832	50 346	12 486	53 216	39 019	14 197
governmental sector	23 487	–	–	–	21 351	21 351	–	–		18 493	18 493	–	–		19 121	19 036	85	17 941	17 871	70
higher-education sector	3 400	3 182	–	218	4 952	4 727	–	225		5 627	5 558	–	69		7 837	7 837	7	5 587	5 569	18

Source: NIS 1997, p. 300; 1999, p. 274; 2000, p. 235.

Table 14.3 Romania: R&D expenditure

	1990	1991	1992	1993	1994	1995	1996	1997	1998	1999	2000
TOTAL (billions of Lei; current prices)	15.6	35.0	92.3	426.7	862.8	1 284.3	1 901.8	3 610.9	4 669.5	5 926.7	8 010.3
– Current Expenditures of which:	14.4	32.6	89.8	415.5	834.1	1 234.5	1 799.9	3 419.2	4 444.5	5 664.4	7 563.2
+ labor force	8.3	17.3	44.4	148.7	352.8	567.9	434.2	798.3	1007.1	1200.4	.
– Investments	1.1	2.4	2.5	11.1	36.2	59.3	101.9	191.7	224.9	263.3	447.1
TOTAL (in millions of USD)	667.1	437.6	288.2	547.0	516.0	626.2	616.0	490.0	526.1	386.5	369.3
– Current Expenditures of which:	620.0	407.0	280.0	532.0	494.6	597.3	583.8	464.0	500.7	369.4	348.7
+ labor force	355.0	215.0	139.0	191.0	210.0	276.0	140.8	108.3	113.5	78.3	.
% of GDP –Total R&D expenditures	(2.6 in 1989)	.	.	0.82	0.68	0.80	0.71	0.58	0.49	0.41	0.37*
– Enterprise sector	.	.	.	0.64	0.55	0.62	0.52	0.47	0.38	0.30	0.26
– Government sector	.	.	.	0.16	0.11	0.16	0.17	0.09	0.08	0.08	0.07
– Higher education sector	.	.	.	0.02	0.01	0.02	0.02	0.02	0.02	0.03	0.04

* Provisional Data
Source : NIS 1996, pp. 300, 301; 2000, pp. 234, 248; NIS, 2001, p. 7; own estimation in USD.

Table 14.4 Romania: structure of R&D expenditure (%; total expenditure = 100)

R&D Expenditures	1993	1994	1995	1996	1997	1998	1999	2000
a) by Sector of Performance								
Enterprise	84.5	85.6	84.9	82.8	82.0	76.7	77.0	69.4
Government	14.1	13.3	13.9	15.8	16.7	18.7	20.4	18.8
HE (higher education)	1.4	1.1	1.2	1.4	1.3	4.6	2.6	11.8
b) by Type of Research								
Basic Research	20.0	16.7	13.3	15.0	13.9	17.8	18.0	17.5
Applied Research	72.1	74.9	63.9	68.5	66.7	64.3	63.4	61.7
Experimental Development	7.9	8.4	17.8	17.5	19.4	17.9	18.6	20.8
c) by Source of Financing								
Public Funds	71.5	67.8	57.4	54.8	42.4	52.9	46.7	40.8
Enterprises' Funds	16.5	19.9	22.8	22.7	17.0	13.4	19.2	17.4
HE Funds	0.21	0.16	0.37	0.2	1.5	2.9	0.05	0.3
Foundations and other funds	2.43	3.04	3.33	2.8	3.2	1.6	4.75	3.4
Own Sources of Research Institutes	9.06	8.7	13.0	16.8	33.0	27.4	26.8	33.2
External Funds	0.4	0.4	3.1	2.7	2.9	1.8	2.5	4.9

Source : NIS 1996, p. 310; 1997, pp. 312, 313; 1999, pp. 284, 292; 2000, pp. 234, 248; NIS, 2001, p. 43; own estimation in percentages.

with full-time jobs, either at state universities or Academy institutes, who are seeking to top up their meager incomes. Some private universities are of low quality, and some were not accredited when evaluated by the new Romanian Accreditation Committee. Research in private HE is in its infancy.

The infrastructure in some parts of the public HE system is still inadequate and outdated. Few members of academic staff are exclusively engaged in research and research activity is a peripheral concern in universities, often being a hobby of some well-known professors who are also members of the Academy or directors of institutes. There are more AoS researchers teaching in HE institutes than there are teachers carrying out research in scientific research institutes. Nevertheless, scientific cooperation between AoS and HE is growing and is an important element in the reform of R&D. Because of the decline of the Romanian economy, the funds allocated to research in HE dropped significantly between 1989 and 1993. In 1994 the Ministry of Education set up the National University Research Council with the aim of raising research quality through selective competitive funding and encouraging university research. However, this aim was never really fulfilled because of the country's economic crisis.

Academies of Sciences have a long tradition in Romania. The Romanian Academy of Sciences (AoS) was founded over 125 years ago and has always had its own network of research units in diverse disciplines alongside the Academies of Agriculture and of Medical Sciences. The AoS was granted autonomy again in 1990 and remained a learned society. It recreated a network of institutes similar to the one that had been in existence in 1948, but also reclaimed the research institutes that had been under its umbrella between 1948 and 1974. Moreover, it expanded its portfolio to take in other institutes attached to ministries. Between 1990 and 1993, the number of research institutes and centers in the AoS increased from 54 to 66 and the staff from 4 481 to 5 092 people. At the beginning of 1994, the Academy network numbered 73 institutes, centers and teams. Most of the units were relatively small, with a staff of fewer than 90 employees. By the year 2000, the number of institutes had fallen to 66 and the number of employees to 3 646. The two branch academies are also in the process of reorganizing and restructuring their institutional network and financing mechanisms. While the AoS is entirely autonomous and publicly funded and its associated institutes are directly subordinate to it, these other academies merely coordinate the scientific activities of relevant institutes, which, administratively and financially, are subordinate to other ministries.

The AoS still has a weak relationship with industrial or other sponsors and now finds itself increasingly unable to support its large network. The humanities and social sciences predominate (62 per cent) and mathematics and chemistry are also well represented. Many units suffer from a shortage of facilities and equipment.

Since the Academy can hardly provide adequate conditions for research, it has become unattractive for young specialists, who tend to choose careers either in more lucrative fields or in HE, where they are able to hold more than one job in the numerous private or state universities.

Basic and applied research activities are financed from the state budget. Although the funds allocated are somewhat meager, they are paid on a regular basis. The institutes are able to increase their income by contracting with local economic entities (10 to 25 per cent of their total funds) or through various types of domestic and foreign grants, including those awarded by the Academy Grants Fund, set up in the spring of 1995. The AoS developed a long-term research strategy for the period up to the year 2010; it is seen as an instrument for coordinating the activities of research institutes and encouraging the efficient use of resources.

Industrial R&D has undergone thoroughgoing changes in its management and structure since 1990. With the formation of commercial companies in industry, some R&D institutes were allocated to these enterprises as subunits and integrated into their economic activities, completely transformed into independent commercial enterprises or given a hybrid status as 'research-development-design institutes'. In all cases, the activities of the R&D institutes were changed or redirected towards commercial goals and technological services intended to generate instant earnings. The government essentially left these new units to their own devices, on the assumption that the market would take care of their development. However, during the economic transition, the demand for domestic and foreign R&D decreased. The financial distress of state-owned industrial enterprises, coupled with the lack of incentives for investments in the modernization of production processes, prevented radical reform of industrial research. The fiscal incentives for enterprises to invest in R&D were low. Hiding behind notions of decentralization and autonomy, the state adopted a minimalist approach to its role in research financing, which was limited to the inadequate resources of an extremely small budget. Most of the R&D institutes were threatened with liquidation. Accordingly, a 1 per cent tax was imposed on the turnover of all state-owned economic units in the main branches, and a special fund for research financing was thus formed (replaced in 1995 by the State Budget for Research).

The consequences of this policy were contradictory and industrial R&D has still not recovered from its impact. Many R&D units, having been converted into commercial firms, are in a precarious financial state, forced to juggle R&D, technical services and small-scale production in search of income. Taken together, they constitute neither an efficient, effective industrial innovation support system nor an R&D infrastructure able to address longer-range technical problems of economic importance. Many industrial research institutes, supported until 1989 from industrial funds, have abandoned research activity altogether and embarked instead on commercial activities, stimulated by the very permissive legal framework for entrepreneurial activity within R&D institutes. Turned into business firms overnight, the industrial research institutes plunged into competition with other manufacturing and service firms. Consequently, the number of specialized R&D staff employed fell by 42 per cent between 1990 and 1995 and by 50 per cent between 1996 and 2000. The average age of research staff has increased alarmingly: in the year 2000, 62.8 per cent of research staff were over 40 (NIS, 2001, p. 38).

The scarcity of public funds after 1995 has forced R&D institutes to generate additional income from small-scale production departments or other production activities.

Since there were no strategies for promoting excellence, prestigious research institutes implemented the most unusual measures in order to survive from one day to the next. Thus the units vary widely in status and include those recognized as institutes subordinate to the MRT, independent commercial firms able to engage in small-scale production and other income-generating activities as well as obtain government contracts, public institutes without any ministerial affiliation and branch institutes affiliated mainly to the ministries of industry, public works, education, environment, health and agriculture. A new and auspicious trend is the establishment of industrial in-house R&D, either by transferring some departments from research institutes to industrial enterprises or by establishing new research departments within viable firms. There are also small independent private institutes. In 1995, there were 454 industrial R&D units, 64 of which were private. In 2000 there were 439 industrial R&D units, of which 201 were largely privately owned.

According to statistical data, the share of the private sector in R&D activity has been increasing since 1990, but so far it remains modest; in 1999, the private sector accounted for 18.6 per cent of total R&D employment (cf. Table 14.2).

Privatization of some publicly owned R&D units has been underway since 1995. Of all the R&D units that receive funding from the MRT, the State Ownership Fund has identified forty as candidates for privatization and some of them have already been privatized. Most of these are commercial firms, which have become virtually self-financing. An unknown number of industrial R&D units would also be potentially viable business ventures as contract research organizations or private technical consultants. A small but significant number of R&D units have been successful in amassing private resources from consultancy and small-scale production. Compared to other state-owned enterprises, the R&D units have low levels of debt, valuable property in major urban areas and significantly depreciated physical assets. The critical bottlenecks are lack of capital, lack of ability to manage technological investments and lack of marketing skills. Policy initiatives that might help to clear these bottlenecks could include low-interest loans and very vigorous support for market development. The 'MEBO' (management and employee buy out) privatization technique, which was used in Romania for some years, did not provide enough working capital to enable many newly privatized industrial R&D companies to become viable.

Experience in the past 10 years has shown that it is easier to reform the institutional system of academic and basic research than to turn industrial R&D into a self-sustaining system. The lack of a clear framework for developing autonomy and efficiency, a tradition of political and bureaucratic micromanagement of R&D activities and the absence of an overall strategy for restructuring industry are the main problems that had to be faced in reforming industrial R&D. The revitalization of industrial R&D institutes will be a complex process that will seek to revive industrial innovation capabilities; it will involve initiatives to improve not only the supply of R&D services but also the demand for them.

Romania has not yet completed the transition to an STS with competitive grant-based allocation of resources for fundamental and strategic research, on the one hand, and demand-oriented and user-financed provision of R&D services, on the other.

Until 1995, policy and institutional arrangements encouraged supply-side approaches to product innovation on the part of R&D institutions. Separate from enterprises, though largely financed by them through a special levy on turnover until 1995, many Romanian institutions of applied R&D have aspired to do what they have learned to do. The principal connection between the industrial R&D system and the enterprise sector was a system of contracts for research products between the R&D units and the MRT. Research results, in the form of specification sheets for new products and processes, were collected in catalogues and distributed to industry. The result was a considerable potential supply of results that could find few takers in the domestic industrial sector. In all this, little progress has been made in reforming industrial R&D activities. From 1996–1997 onward, the MRT has increased the responsiveness of the public R&D system to the needs of enterprises by expanding the representation of users in the national policy-making bodies that control the uses to which funding from priority programs is put.

The balance between the various sources of R&D funding was modified by increasing the share of funds from enterprises and reducing that from the state budget. In addition to their public funds, state and private companies are allowed to spend part of their revenue (in 2000 up to 33.2 per cent) on R&D activities at their own discretion. As in-house R&D is still limited, most of these enterprise funds are spent on contract R&D in different branches of industrial research. Many small private firms declaring an interest in doing R&D have emerged spontaneously. In many cases R&D is not their main activity, which is usually consultancy and other services. Some of the managers were formerly researchers in public R&D institutes, while others are still formally employed in a state research institute.

3. PRIORITY-SETTING AND FINANCING — AN UNRESOLVED PROBLEM

Unlike the EU member states or some of the associated countries, Romania has difficulties in implementing government decisions on science and technology policy. There are no dedicated institutions (such as research councils or R&D forecasting institutes) able to provide scientifically substantiated studies of international developments or technological forecasts. At the same time, it must be noted that institutions that were supposed to play a major part in determining science policy on the basis of economic and social priorities, such as the Inter-Ministry Council for Science and Technology (established in 1996 to draw up R&D policy with the participation of the relevant ministries, academies, professional organizations and users of R&D results and still formally in existence but no longer operational), have not functioned in practice (Sandu, 2001).

The problem of the existing oversized research units could be satisfactorily solved not by administrative decisions but by applying criteria aiming to optimize their size according to their activity profile and the compatibility of their R&D results with the demands of the market system.

New technology users, who constitute the 'innovation market', are predominantly large enterprises implementing complex technologies. Regardless of whether their technological level is low, medium (metallurgy, chemical industry, automotive industry) or high (electrical machines and communications equipment, aircraft, office-related industry), enterprises require not only imports and foreign licenses but also continuous collaboration with national research if they are to remain competitive. The relationship between enterprises and the specialized research institutes must be based on mutual interest. Enterprises have to be aware that exposure to competition imposes quality standards on products and processes, the design and production of which absolutely require scientific research and technological development. For their part, research units have to reorganize their portfolio of interests in accordance with market demands.

Communications between the two parties are difficult in Romania because of differences in technological capacity and know-how and a reticence that is psychological in origin. Over and above these difficulties, there is also less interest among researchers and potential beneficiaries in applying research results. On the other hand, the instability of the legislative framework and the lack of any institutions and financing mechanisms designed to favor competition and scientific performance have led to considerable resources being wasted in funding projects developed on the basis of institutional, political and client-focused criteria.

The inertia of the industrial structure against the general background of a painfully slow privatization process led to the development of strategies intended to provide further support for industrial research, which continues to dominate the R&D system and consume a significant share of state funding (cf. Tables 14.4 and 14.5). Thus between 1995 and 1999, enterprises accounted for more than 70 per cent of total current R&D expenditure (representing 94.8 per cent of total expenditure; the other 5.2 per cent is investment). Independent industrial R&D institutes are business-oriented, but in terms of funding sources they are neither self-supporting nor entirely dependent on government. Although R&D institutes are now market-oriented, they have yet to become integrated into firms (Radosevic and Lauriol, 1998).

Scarcity of R&D funding was a permanent feature of the R&D reform, and the most striking difficulty faced by the R&D institutes was the severe diminution of R&D investment expenditure. The share of R&D expenditure within GDP decreased from 0.82 per cent in 1993 to 0.37 per cent in 2000, in particular in the enterprise sector (from 0.64 to 0.26 per cent) (cf. Table 14. 3).

Core institutional funding has gradually diminished in favor of grants or other forms of funding based on competition, and even because of the lack of a national R&D strategy and priorities, along with an inadequate institutional and legal framework. Increasingly scarce resources are not always allocated on the basis of scientific or economic performance. There have on occasions been difficulties with the peer review system of evaluation and the weak links between user demand and research providers.

The fiscal incentives for enterprises to invest in R&D are weak and policy and institutional arrangements still encourage supply-side approaches to product innovation on the part of R&D institutions (Sandu, 1998).

Because of Romania's legacy in this field and the conditions specific to the transition, the problem of priorities could not until recently be considered to be a major concern for policy-makers. From 1990–1992, the disappearance of internal demand for applied research and of funding sources created a confused situation that ended with the transformation of most technological research institutes into commercial firms. Between 1992 and 1994, the priority was to

Table 14.5 Romania: structure of current R&D expenditure in enterprises by CANE* activities (%; total expenditure = 100)

		1995	1996	1997	1998	1999	2000
1.	Agriculture, Forestry and Fishery	12.5	13.7	13.05	12.6	11.0	12.0
2.	Mining and quarrying	5.7	5.98	5.09	5.42	6.9	5.7
3.	Manufacturing	58.7	59.87	62.6	66.5	66.9	69.9
	Food, beverages and tobacco	0.7	0.84	0.82	0.94	0.13	0.1
	textiles, leather and footwear	1.5	1.85	2.12	2.91	2.03	1.5
	Wood processing (excluding furniture)	0.08	0.13	0.23	0.19	0.11	0.1
	Pulp, paper and cardboard	0.65	0.54	0.73	0.64	0.58	0.7
	crude oil processing, coal cooking	0.97	0.33	0.87	0.89	0.16	0.4
	rubber and plastic processing	0.38	0.56	0.62	0.54	1.17	0.7
	chemistry and synthetic fibers	10.2	8.04	8.09	6.33	6.93	5.4
	other non-metallic mineral products	1.66	2.56	2.47	2.0	2.97	2.2
	Metallurgy	4.98	4.80	5.0	6.28	6.53	11.4
	metallic construction, machinery and equipment	36.7	39.4	40.63	44.93	45.37	46.8
	furniture and other non-classified activities	0.89	0.77	0.94	0.82	0.82	0.5
4.	Electric and thermal energy, gas and water	11.2	12.1	13.95	9.86	9.72	8.2
5.	Construction	2.42	2.3	1.75	1.29	1.59	1.3
6.	Other activities	9.48	6.02	3.6	4.22	3.84	2.9

*Classification of Activities of National Economy
Source : NIS 1997, p.310; 1999, p.286; 2000, p.244; NIS, 2001, p. 37; own estimation in percentages.

'rescue' the technological research potential by setting up the Special Fund for R&D, into which enterprises paid compulsory contributions. Under these conditions, a large number of applicants investigating an extremely wide range of topics was financed. Funds were granted for more than 4 000 projects each year, many of which had no direct connection with the needs of the companies that were paying for them.

Starting in 1994/1995, the idea that funds should be allocated according to targets and priority programs was advanced with the launch of the National Program of Research-Development-Innovation 'Horizon 2000' (R-D-I Program). This program was conceived for the purpose of financing "programs of an interdisciplinary and inter-sectoral character intended to promote partnership for the solution of complex problems" (ANSTI, 1998, p. 1). Actually, what was financed by means of this program was, in the opinion of the former president of the National Agency for Science, Development and Innovation, "everything that Romanian science could provide" (ANSTI, 1998, p. 3).

Thus in 1998, for instance, funding was provided through the 22 specialized committees for 8 286 research projects, operational programs, regional and interdisciplinary programs, carried out within hundreds of national institutes, institutes of the Romanian Academy, units in HE and non-governmental organizations, as well as public and private commercial companies.

A further step towards defining priorities on the basis of major economic and social development targets was taken in 1999, when the priority national projects RELANSIN, CALIST, INFRAS and CORINT were launched as part of the new National Research, Development and Innovation Plan (R-D-I plan). These projects were subsequently completed and another 10 priority programs were added in 2001 in various areas, including agriculture and food, biotechnologies, nanotechnologies, aerospace, economic and social sciences etc.

The aim of these programs was to increase the impact of R&D activities on the economy and society, with a view to:
- promoting economic recovery and sustainable development;
- strengthening innovation in order to increase the quality and competitiveness of Romanian products and services in domestic and international markets;
- focusing the country's scientific and technological resources on the extension of the national scientific, technological and innovation base, and
- bringing the national legislative, institutional and procedural framework into line with that of the EU in order to implement the partnership for integration quickly and efficiently.

The majority of pubic funds have hitherto been allocated through the R-D-I plan. An analysis comparing the structure of the funds allocated on a competitive basis by the specialized committees with the branch structure of industrial output and the total export share of particular industrial branches shows that the fund allocation priorities for industrial R&D do not accord with current trends in industrial development.

Real prioritization in the R-D-I field is still in its infancy; indeed, until the year 2000 this problem was not even raised when the steps to be taken with regard to European integration were being planned. Despite its extreme importance and urgency, the problem of establishing priorities is currently being settled only at the formal level because of the severe cuts in R&D expenditure as a share of GDP over the last 5 years. The institutions, mechanisms and resources required to implement priorities have still not been put in place.

The fragmentation of the R&D system has thwarted the establishment of priorities at the national level. Each part of the system has sought to set its own priorities and to obtain the resources for their implementation as much as possible from public funds. The extreme diversity of research topics and institutions and a lack of involvement on the part of branch ministries and the users of research results have been a further obstacle to the establishment of priorities in fields of major interest. Moreover, there has long been considerable confusion about the trends that have emerged following the reorganization of the main branches of the economy.

4. EUROPEAN INTEGRATION – A CHALLENGE TO THE ROMANIAN R&D SYSTEM

Preparing Romania for EU membership is a complex process, with one of the primary objectives being to promote policies compatible with EU mechanisms in the area of R&D. S&T policy occupies a distinct place within the national medium-term development strategy, which incorporates the main objectives and policies required for Romania to be able to fulfill the essential conditions for EU membership by 2007.

The R&D chapter within the 'acquis communautaire' is one of the chapters that have been temporarily closed; this marks a necessary but not sufficient step towards harmonizing the operational mechanisms of the Romanian R&D system with those of the EU (MER, 2000). Complex comparative analysis of the R&D system in Romania with that of the EU countries has revealed significant differences, on multiple levels, between its configuration and sustaining mechanisms, on the one hand, and between the good intentions underlying the R&D-related objectives listed in the official documents on Romanian membership and the country's medium-term development strategy and the reality of the Romanian situation, on the other.

In addition, further sustained effort is required in order to remedy the shortcomings already identified (deficiencies in the institutional and legislative system, problems with establishing priorities, funding difficulties etc.) in order to create the conditions for the efficient integration of Romania into the EU. The revival of the Romanian economy, based on the exploitation of competitive advantages, depends to a great extent on these shortcomings being remedied.

The problem of priorities, for example, is crucial to the future configuration of the R&D system, since priorities represent a distillation of national strategies and policies in this field. They provide a basis for the allocation of funds to R-D-I and they must be fixed in accordance with the priorities of the economy and social development, which are strongly influenced by science.

The main medium-term objectives for R&D in Romania are in line with the instructions included in a special document that defined the concrete actions and instruments needed to transform the European Research Area concept into reality (MER, 2000; EU, 2000).

The theme of establishing priorities within R-D-I took on greater significance when Romania was invited to start negotiations on its accession to the EU. The national strategy for medium-term development that was drawn up in this context lays down a series of priority objectives for R&D, in particular development of the capacity to generate scientific and technological knowledge and the integration of the country's specialized units into international networks and programs (Government, 2000).

The aim is to improve the quality and efficiency of R&D units by applying standard international procedures for the evaluation of the units, activities and staff, improving selection procedures, management practices and the guidance given to R&D units on their way towards the market, developing the ability to diffuse scientific and technological knowledge, expanding marketing services and links with industry, and so on.

Romania's position with respect to integration into the European Research Area was upheld in 2000 by a series of official documents reflecting acceptance of the 'acquis communautaire' as it relates to science and research. These documents specify a series of priorities for the development of the legislative, financial and organizational foundation required for participation in the European Community's Framework Programs, of research infrastructures and of human resources in science, technology and innovation in accordance with the European pattern of scientific careers.

In order to be able to participate in Community programs intended to strengthen SMEs' innovative capacities, Romania will take action to:
– encourage, through specific national programs, collaboration between R&D units and companies;
– stimulate company research, especially in high-tech areas;

– put in place programs leading to the establishment of an information, documentation and technological support network for SMEs, and
– promote technology transfer.

The goal of integration into the European Research Area is forcing Romania to make a sustained, long-term financial effort in order to achieve two major objectives. The first is to increase Romania's ability to enter into partnerships with European researchers, the second to improve the quality and effectiveness of the country's participation in the R&D Framework Programs at EU level.

At present, due to the exigencies imposed by Romania's decision to seek EU membership, there is an acknowledged need to undertake general institutional construction and to adopt legislation specific to the field.

To this end, the government that came to power in December 2000 has included support for R&D among its strategic options, with a view to facilitating the transition of the Romanian R&D system to a European type of operational structure, increasing competitiveness and capitalizing on the country's R&D potential (Government, 2000a). Efforts are also being made to improve institutional management, increase the competence and efficiency of R&D activity and decentralize program management.

These ambitious objectives can be attained through further institutional and legislative reform. The new law on research now being debated in Parliament calls for the setting up of 4 new institutions, the most important of which is the National Council for Science and Technology Policy, charged with drawing up the national R&D strategy and answering directly to the Prime Minister (EU, 2000a). Another aim is to set up a consultative committee on research, development and innovation, on which the scientific community, the ministries and the important economic agents would be represented.

The lack of a clear framework for developing autonomy and efficiency, a tradition of political and bureaucratic micro-management of R&D activities and the absence of an overall strategy for restructuring industry are the main problems that had to be faced in reforming industrial R&D. The revitalization of industrial R&D institutes will be a complex process that will seek to revive industrial innovation capabilities; it will involve initiatives to improve not only the supply of R&D services but also the demand for them. The next step in the transition of the S&T system will be to a competitive grant-based allocation of resources for basic and strategic research, on the one hand, and demand-oriented and user-financed supply of R&D services, on the other.

REFERENCES:

ANSTI (1998) *Evaluation Report of the National R&D Program 'Horizon 2000'*. Bucharest: National Agency for Science, Technology and Innovation (ANSTI).
EU (2000) "Innovation in a knowledge driven economy – Commission Communication." *Innovation and Technology Transfer* special edition, November: 12 – 24.
EU (2000a) *Regular Reports from the Commission on Progress towards Accession by each of the candidate countries – Romania*. November 8th. (Internet europa.eu.int/comm/enlargement/report_11_00)
Government (2000) "Strategia Nationala de Dezvoltare a Romaniei pe Termen Mediu" (National Strategy of Romanian Medium Term Development). *Romanian Government Bulletin* March: 31.
Government (2000a) "Programul de Guvernare pentru perioada 2001 – 2004" (Governmental Program for 2001 – 2004). *Official Gazette of Romania* no. 700, December: 32 – 34.
MER (2000) *Romanian Position Paper on the EC Communication – Making a Reality of the European Research Area*. Bucharest: Ministry of Education and Research (MER).
NIS (1996 – 2000) *Romanian Statistical Yearbooks*. Bucharest: National Institute of Statistics (NIS).
NIS (2001) *Research and Development Statistics Series*. Bucharest: NIS.
PHARE (1996) *Assessment of Industrial Liaison Offices and Local Technology Transfer Organizations*. Final Report in the framework of the PHARE Program for the Restructuring of the Science and Technology System in Romania. Sussex: Science and Technology Policy Research Unit (SPRU)/Ernst & Young.
Radosevic, S., and L. Lauriol (1998) "Measuring S&T activities in the former socialist economies of Central and Eastern Europe: Conceptual and Methodological issues in linking past with present." *Scientometrics* 42, no. 3: 273 – 297.
Sandu, S. (1998) "Industrial R&D in Romania." *Transforming Science and Technology Systems – the Endless Transition? NATO Science Series 4: Science and Technology Policy – Vol. 23*. Ed. W. Meske, et al. Amsterdam: IOS Press, 244 – 253.
Sandu, S. (2001) *Inovare, Competenta tehnologica, Crestere economica* (Innovation, Technological Capability and Economic Growth). Bucharest: Expert Publishing House.

15. BULGARIA: THE LONG ROAD TO A NEW INNOVATION SYSTEM

Kostadinka SIMEONOVA

At the end of World War II, Bulgaria was one of the least industrialized Eastern European countries. The country lacked the scientific traditions that exist in most CEECs. Until the 1960s, therefore, state policy was concerned primarily with the (quantitative) strengthening of higher education (HE). Generally, the system that was put in place followed the 'Soviet model'. In 1971 the notion of integrating science, technology and HE became the guiding principle of science policy. The integration of the Bulgarian Academy of Sciences (BAS) and the University of Sofia led to the establishment of the 'United Centers for Science and Higher Education' (UCSHEs), which replaced the traditional departments of the BAS. The main task of these units was to draw up joint research and teaching plans, involving the best BAS scientists in HE and, in turn, university staff and students in research. The UCSHEs survived until 1988, having been deemed inefficient. In order to improve research activity in the HE sector in the early 1970s, so-called research sectors (RS) were set up to organize and facilitate contracts between university staff, branch institutes and industrial firms. There were 21 RS by 1991, by which time they accounted for 15 per cent of all HE employment. Several new forms of organization emerged. In 1988 there were 42 small, innovative firms in HE and 11 in the BAS; technological and transfer centers were set up in the mid-1980s to unify research, innovation, transfer and education. Nevertheless, the innovation system suffered the usual shortcomings, and in the 1980s the absorption of R&D results diminished.

1. ECONOMIC AND POLITICAL REFORMS

Economic and political reforms in Bulgaria began in 1989, simultaneously with those in the other CEECs. However, they fail to get beyond the initial stage because of a lack of political and social consensus about the reforms. The period 1990–1992 was characterized by sharp criticism of science and a number of initiatives to restructure its 'totalitarian' institutional system. The general political instability in the country (seven governments and five parliaments between 1990 and 1997) gave rise to discontinuity in economic and legislative measures. New laws were passed on the autonomy of HE (1990) and of the BAS (BAS, 1991). In 1992 a special 'De-communization Law' was adopted, according to which some categories of HE and research staff that had been involved in ideological and political activity in the previous regime were deprived of the right to participate in the leading scientific bodies for five years (Simeonova, 1995). The law was abolished in 1995.

The period immediately following the changeover of power in 1989 was characterized mainly by a restrictive science policy and a series of government measures to reduce the scientific potential and cut state funding of science. As this restrictive policy continued, most of the important scientific facilities were restructured. The Medical Academy was dissolved, the Agricultural Academy was first evaluated, but later on dissolved as well. The most comprehensive changes occurred in the industrial R&D sector. Following the approach to restructuring taken by most post-communist countries, the autonomy of R&D institutions increased considerably (Simeonova, 1998).

The temporary amalgamation in 1993 of the Ministry of Science and Education with the Ministry of Culture led to this area of politics being extremely marginalized for six months and to the delay of further reforms. At the beginning of 1995 a new *Ministry for Education, Science and Technology* (MEST) was set up in order to improve the linkages between the different parts of the 'innovation chain'. For the same reason, the special Fund for Structural and Technology Policy (FSTP) was transferred from the Ministry of Industry to the MEST, but many experts considered links with industry more important than those with HE institutions and the BAS. Because R&D activity was to be part of the general economic restructuring, reforms were delayed. Until 1997 there was no comprehensive S&T policy in Bulgaria; indeed, between 1990 and 1997, S&T had been the subject of critical public debates and conflicting approaches.

The changes that took place in S&T policy in Bulgaria between 1997 and 2000 support the conclusion that the general political situation has played, and indeed continues to play, an important role in the development of a new innovation system in the country. The changes took place following the 1997 parliamentary elections, after which Bulgaria entered a new phase of reforms marked by relative macroeconomic stability. GDP increased in 1998, 1999 and 2000 by 3.5, 2.4 and 5.8 per cent respectively (cf. Table 15.1). The newly established Currency Board played an indispensable part in achieving financial stability in the country. Foreign direct investment (FDI) increased faster than in previous years and in 1999 reached 1.78 billion USD (NSI, 2001, p. 102). Salaries also increased, although they are still comparatively low in absolute terms. Inflation dropped from 578.6 per cent in 1997 to 4.8 per cent in 2001.

In 1997 – 1999 about 57 per cent of GDP was created in the private sector, the biggest share in the service sector. By the end of 1998, 31 per cent of

Table 15.1 Bulgaria: GDP and GERD

Year	1989	1990	1991	1992	1993	1994	1995	1996	1997	1998	1999	2000
GDP index (previous year = 100)		90.9	83.3	92.7	97.6	101.8	102.1	89.1	93.1	103.5	102.4	105.8
GDP per capita (USD)	2 454	1 922	943	1 008	1 276	1 147	1 537	1 408	1 224	1 484	1 510	1 459
GERD/GDP (%)	2.63	2.38	1.53	1.64	1.18	0.88	0.62	0.52	0.52	0.59	0.59*	0.55*
Budget expenditure/ GDP (%)	1.09	0.79	0.47	0.45	0.38	0.31	0.24	0.18	0.35	0.41	0.41	0.38

* Preliminary data.
Source: NSI Data base for the respective years.

state property was privatized; in 1999 this figure had reached 56 per cent. The private sector accounted for 63.3 per cent of total employment in 1999.

However, regional instability and the conflict in Kosovo had a negative impact on the country's economic recovery, which slowed down in 1999. By the end of 1999 15.6 per cent of the economically active population was unemployed. In 2000 this figure was still increasing, surpassing on average 18 per cent. In some regions the share of unemployed is very high (25 – 30 per cent) due to the unfinished restructuring. This is why the regional context plays a particular role in Bulgarian restructuring and integration into the EU.

An important achievement in this respect is the Stability Pact (SP) for South Eastern Europe, which is at present becoming a decisive determinant of regional development.

On June 10, 1999, in response to a call from the European Union, a stability pact for South Eastern Europe was adopted in Cologne, Germany. Participating countries in the region committed themselves to continued democratic and economic reforms. The aim of the SP is to strengthen countries in the region in their efforts to foster peace, democracy, respect for human rights and economic prosperity in order to achieve stability throughout the region.

Among the objectives of the SP is the creation of vibrant market economies based on sound macro policies, markets open to greatly expanded foreign trade and private sector investment. The fostering of economic cooperation in the region and between the region and the rest of Europe and the world has become an activity of the utmost importance.

The positive impact of the SP is related to the mobilization of international donors. Special attention is to be given to projects that involve two and more countries in the region. Of all three 'working tables', which are the main structural element of the SP, the working table on economic reconstruction, development and cooperation brings bigger opportunities for R&D development in the areas of cross-border transport, energy supply and savings, infrastructure, promotion of the private sector business and environmental issues. The most important infrastructure projects planned in Bulgaria are the second Danube Bridge, at an estimated cost of 270 million EURO, the reconstruction of Sofia airport (60 million EURO) and the redevelopment of the Danube river ports.

Between 1997 and 2000, the government adopted a more applied science policy, in line with the SP. The reaction to this development was very rapid. Within two weeks, a consortium of 15 Bulgarian organizations, including R&D institutions, was set up in order to participate in the construction of the Danube bridge. For its part, the government sees this initiative as a model for all R&D organizations looking for ways to extend their contract activity and become less dependent on the state budget. The pre-feasibility studies related to this project cover waste and a range of geological, seismological, environmental, economic and other issues. Many of the BAS institutes have to shift to applied research related to the country's infrastructure projects. Another positive effect is the strengthening of regional cooperation in R&D, as reflected, for example, in the agreement on bilateral cooperation between the BAS and the Macedonian Academy of Sciences and Arts signed at the beginning of the year 2000.

Although some government officials pointed in that year to the limited role of the SP as a regional initiative and the possible disadvantages for Bulgaria that might result from the region's instability, the SP remains an important instrument in the facilitation of the enlargement process. Political and financial stability is the most important factor in the adoption of a new approach and an important precondition for continuity in S&T policy.

In 1997 the MEST was reorganized again into the Ministry of Education and Science (MES), while responsibility for technology policy was transferred back to the Ministry of Industry. However, the Fund for Structural and Technology Policy remained in the MES and the department responsible for administering it has been very active in putting it to use.

There were positive changes in Bulgarian S&T policy between 1997 and the year 2000 and it would seem that this policy area is no longer marginalized. The most important factor driving these changes is EU enlargement and the negotiations around EU membership that began in 1999. Many positive developments would not have been possible without the new government's commitment to meeting the criteria for accession. This underlines the arguments on the role of the general political context in S&T policy and in strengthening the links between R&D and the industry. The following developments merit particular attention:

1. An attempt was made in 1999 to 'revitalize' the program-based approach to R&D management by putting in place a national program for technological development (NPTD). Unfortunately, it was not implemented.
2. S&T policy became more firmly institutionalized with the establishment of a Council for Science and Technology (CST) at government level in 1998. The CST functions as an advisory body to government; all ministers with responsibility for R&D units, as well as the scientific community, are represented on it. The members are appointed by the Prime Minister and include university rectors, presidents of the academies, etc. This function had previously been limited to the MES.
3. The government has initiated a thoroughgoing evaluation of the country's R&D potential as a prelude to mobilizing it to assist the ongoing reforms and strengthen Bulgaria's export potential. In 1997 two important PHARE projects on the establishment of science parks in Bulgaria and the development and management of a science and technology development plan

were carried out under the supervision of the MES. The two projects were funded as part of the PHARE program and involved close collaboration between Bulgarian research teams and foreign consulting firms. This helped to ensure that the projects were well designed and the results reliable.

The PHARE projects on the evaluation of Bulgaria's S&T potential were closely linked to an evaluation of the country's export structure and the competitiveness of its industries in international markets. These links were considered crucial to the development of the NPTD. The decision was made to establish science parks as a means of restructuring the branch research institutes. The setting up of science parks (known as technoparks) is an element of the government's strategy and reflects the assumed importance of certain branches for the recovery of Bulgarian industry.

4. The impact of the Currency Board on S&T policy resulted in institutional changes being made to the funding of R&D. In March 1999 the Fund for Structural and Technology Policy and the National Research Fund were abolished as independent funding bodies and their functions transferred to the MES. Although the volume of expenditure allocated via the grant system is still determined by the Budget Act, the research committees' autonomy in allocating resources was reduced, with such allocations becoming dependent on other MES expenditure and priorities.

5. The restructuring of R&D institutions is still in progress, as can be illustrated by the changes in the Agricultural Academy. In December 1999 a law was adopted that provided for the establishment, within the Ministry of Agriculture and Forestry, of a National Center for Agricultural Sciences; the same act dissolved the Agricultural Academy.

6. Much attention has been paid to improving the dissemination of information between the producers and users of S&T information. In this respect two projects, completed in 1998 and coordinated by the MES, are important. The first is the TEMPUS JEP project, which resulted in the installation of a modern computer and communications system for three main universities in Sofia. The second is the PHARE R&D Networking Program, which led to the establishment of a national academic information system accessible to HE institutions throughout the country.

A special unit was set up at the MES to provide information and support to Bulgarian scientists participating in EU programs. Its activities include regular publications and seminars.

7. The legislative changes are enshrined in a variety of government decrees, of which the following are the most significant:

– the decree establishing the Inter-ministerial Space Research Committee;
– the act establishing the National Antarctic Program;

- the decree enabling Bulgaria's participation in the EU Fifth Framework Program and EURATOM and approval for the establishment of the Council for Accession EU-Bulgaria, passed on July 15, 1999. According to these decisions, 33 per cent of the country's financial contributions are provided from the PHARE program. The remaining money will come from the state budget;
- the amendment of the law governing intellectual property rights;
- the law governing the development of high technology in Bulgaria (submitted in 2000 and still under consideration).

The Law on the Promotion of R&D Activity in Bulgaria is one of the new laws that have been submitted to the government for consideration. It concerns such crucial problems as funding for R&D (and in particular share of GDP to be allocated to R&D), the shift from core institutional funding to project-based funding and the development of preferential tax policies in the sector.

2. CHANGES IN THE S&T SYSTEM BY SECTOR

When the reforms began in 1990, gross domestic expenditure on R&D (GERD) in Bulgaria was in excess of 2 per cent of GDP per annum. In that year, there were 112 158 employees in the R&D sector (2.7 per cent of the total working population), of which 31 707 were scientists. The total volume of capital investment in the R&D sector exceeded one billion Lev – approximately 1 per cent of the total capital investment in industry. The human and material capital was concentrated in 500 R&D organizations (Simeonova et al., 1995).

Because the political situation was unstable and there was dissension among and within the large scientific organizations and communities, institutional changes in science occurred in different periods and tended to be fragmentary. They did, however, lead to considerable changes in the type and number of R&D organizations in the various sectors and to a reduction in personnel. This was a result above all of the reduction in R&D resources. In 1992 research expenditure amounted to a total of 3 103.8 million Lev, of which 50 per cent was in industrial research, 41 per cent in scientific institutions of the academies and 8 per cent in HE (MES, 1992). Scientific expenditure amounted to 2.63 per cent of GDP in 1989 and had fallen to 1.64 per cent in 1992; budget expenditure fell from 1.09 to 0.45 per cent of GDP in this period (cf. Table 15.1).

These changes most affected industrial research, i.e. the research institutes subordinated to ministries and industrial enterprises. In 1992 alone, 66 scientific institutions and companies were dissolved, of which 27 per cent were taken over by other institutions as sub-units. By 1993 R&D personnel was reduced to 55 per cent of its previous level (1989); the number of 'scientific workers' (the classification used at the time) in branch R&D had declined by 48 per cent and in the BAS by 7 per cent, while in HE it had increased by 13 per cent (cf. Tables 15.2 – 15.5).

Table 15.2 Bulgaria: R&D organizations by sector*

	1994	1995	1996	1997	1998	1999	2000
R&D organizations in total	450	436	474	458	447	436	410
Enterprise sector	118	111	115	159	143	117	103
Government sector, of which:	227	219	228	208	207	222	207
Agricultural Academy	72	72	72	80	52		
BAS	81	72	78	79	68		
HE sector	100	102	98	86	88	87	91
Private non-profit sector	5	4	33	5	9	10	9

* In 1994 some new statistical indicators were introduced that are not always comparable with previous years. The 'branch sector' was replaced by the category 'enterprise sector'. It includes R&D units in industry and other branches of the economy (trade, transportation, construction, forestry, agriculture, etc.), but covers only part of the former sector–the rest is now classified as part of the 'state sector', comprising all state institutes within the structure of the ministries. The state sector in Bulgaria also includes the BAS institutes and those of the Agricultural Academy.
Source: NSI Data base for the respective years.

Table 15.3 Bulgaria: R&D staff and scientists by sector, 1989 – 1993

	Scientists*								R&D staff
	BAS		HE		Branch**		Total		Total
Year	Persons	%	Persons	%	Persons	%	Persons	%	
1989	4 716	14.9	13 576	42.9	13 319	42.2	31 611	100	98 338
1990	4 733	14.9	14 880	46.9	12 091	38.2	31 707	100	112 158
1991	4 616	15.9	14 862	51.1	9 582	33.0	29 060	100	88 733
1992	4 407	16.3	15 385	57.8	6 873	25.9	26 598	100	57 655
1993	4 315	16.7	15 546	59.1	6 341	24.2	26 284	100	54 324

* Equivalent to 'scientific workers'.
** Until 1994 – includes all branch R&D organizations, branch academies and industrial units; since 1994 this category no longer exists.
Source: NSI, 1991, p. 7; Eurostat, 2000.

The *higher education sector* was regarded as the most appropriate one for the new democratic and market reforms, in line with the Western model of science. Many academics were directly involved in political activity (as members

Table 15.4 Bulgaria: scientists* by sector, 1994–2000

	Total	State	HE	Enterprise	PNP
1994	25 616	7 192	16 102	2 298	24
1995	25 577	6 923	16 587	2 029	38
1996	25 853	7 381	16 711	1 729	32
1997	25 871	7 368	16 542	1 951	10
1998	25 192	7 363	16 437	1 366	26
1999	23 906	6 713	16 272	910	11
2000	22 815	6 054	15 921	831	9

* 'Scientists' according to the definition of 'scientific workers' in use since 1970. It is still applied by the NSI to denote all staff appointed in scientific positions.
Source: NSI Data Base for the respective years.

Table 15.5 Bulgaria: BAS and HE staff

	BAS		HE	
Year	Total staff	Researchers*	Total staff	Researchers*
1994	9 846	4 015	11 579	7 998
1995	9 454	3 972	11 214	7 598
1996	9 232	3 978	10 800	7 481
1997	8 782	3 922	4 821	3 989
1998	8 664	3 897	4 449	3 808
1999	8 431	3 748	4 020	3 244
2000	8 271	3 664	3 870	3 034

* Researchers: According to the Frascati Manual (OECD, 1994), all staff actually participating in R&D regardless of formal position if more than 10 per cent of working time is used for research (head count).
Source: For BAS: BAS, 2001, p. 121; for HE: NSI, 1990–1995; NSI Data Base, 1998–2000.

of parliament, in the MES, etc.). Consequently, they were in a good position to defend the interests of the universities when the reforms began. This sector has, as a result, constantly increased its share of R&D activity. The universities now have a greater diversity of programs and courses and employ more teaching staff. In the early stages of the reform process, universities became autonomous institutions. Subsequent reforms oscillated between merely cosmetic changes, such as renaming of specific chairs, and the scaling-down of some laboratories and the extension of teaching into areas, such as law, economics and management, related to the country's new economic situation. The greatest tensions arose around the legitimization of the newly founded universities (approximately ten in number): by 1994 only three of them had been accredited, but two years later, in 1996, accreditation was extended to all new universities. Generally, these new universities are considered to be chaotic institutions, engaged in activities unrelated to the economy of the regions in which they are located or to the country's labor market and labor force requirements. The uncontrolled commercialization of university education was not accompanied by improvements in the HE infrastructure or teaching quality.

In order to deal with the shortcomings of HE institutions, an amendment to the Higher Education Act was passed by Parliament in July 1999. The new accreditation procedure, which laid down strict requirements for establishing faculties and chairs, partly reduced the autonomy of HE. The National Agency of Accreditation was given additional duties and responsibilities and its supervision was transferred from the MES to the Prime Minister's office. Accordingly, state funding of HE became more dependent on the results of accreditation.

Another important change was the abolition of a practice whereby, between 1990 and the 1998/99 academic year, the universities took on additional students on a full fee-paying basis, thereby exceeding the quotas laid down by the state. Since the 1999/2000 academic year, all students have been required to pay tuition fees, amounting to 30 per cent of their institutions' educational costs.

In general, the 1997–2000 period was characterized by attempts to improve HE's regulatory base and to strengthen state control of the quality of teaching programs and staff. Research activity, as measured by the number of finished projects in the HE sector in 1997 and 1998, indicates that the downward trend continued: the total number of finished projects diminished by 4 per cent in 1997, and by 5.8 per cent in 1998. In 2000 there were 950 research projects, which was 92 per cent of the 1999 level (1 029). The recently introduced indicator of HE staff also confirms the reduction of R&D personnel in the sector (cf. Tables 15.6 and 15.7).

Table 15.6 Bulgaria: researchers by sector (in FTE)

Year	HE	State	Enterprise	PNP	Total
1994	3 460	7 243	1 882	23	12 608
1995	4 905	7 198	1 865	22	13 990
1996	4 847	8 007	1 752	145	14 751
1997	2 654	7 425	1 771	130	11 980
1998	2 576	7 501	1 783	112	11 972
1999	2 212	7 060	1 249	59	10 580
2000	1 886	6 417	1 139	37	9 479

Source: NSI, 1994, p. 25–31; 1995, p. 25–31; NSI Data Base, 1996–1999.

Table 15.7 Bulgaria: researchers with scientific degrees by sector (unit: head count)

Year	HE	State	Enterprise	PNP	Total
1997	2 393	5 340	1 061	111	8 905
1998	2 045	4 897	821	128	7 891
1999	1 466	4 533	595	72	6 666
2000	1 090	3 855	270	44	5 259

Source: NSI Data Base for the respective years.

The number of scientists also fell, with the biggest decline in technical sciences in HE (4 per cent). Positive changes can be observed in agriculture, where the number of scientists increased by 26.5 per cent in 1996–1997. This trend was maintained in the following years.

The number of students also grew between 1989 and 1998 by 93.2 per cent (133 184 in 1989/90, 196 046 in 1994/95, 257 929 in 1998/99). However some decrease was evident in the 1999/2000 academic year (250 417). This may possibly become a trend as in the 2000/2001 academic year the number of students dropped again, down to 235 807, due in part to negative demographic processes in the country and in part to the introduction of tuition fees. It is expected that the new legislation on student loans submitted to Parliament in 2001 will ease the situation and help to reverse the trend. Disregarding the initial rhetoric, universities and other HE institutions have turned their efforts to increasing their teaching provision, since the market for educational services has grown rapidly in recent years. The (decreasing) budget allocations are used mostly for teaching, not for research. Relations with the BAS research community are limited to teaching activities. Research is carried out in those 'research sectors' that still remain; they operate on a contract basis and are literally struggling for survival, offering mostly services. Their activity was quite intensive in the past, although publication output was comparatively modest and of secondary importance in the staff evaluation criteria.

The financial situation of the HE institutes has been the subject of critical discussions in the light of their extra income from tuition fees, which are considered relatively high. In some universities the practice of renting parts of the infrastructure, especially buildings, is a substantial additional source of funds.

The *academy sector* in Bulgaria comprised three academies: the Bulgarian Academy of Sciences (BAS), the Medical Academy (closed in 1991) and the Agricultural Academy (closed in 1999). The two branch academies were subjected to a 'top-down' restructuring. For its part, the BAS adopted a 'bottom-up' approach and initiated the reform itself. In accordance with the 1991 BAS

Act, a new general assembly was elected by the research community. The BAS remained the association of leading scientists and its institutes were transformed into a state-funded National Research Center. Additional sources of funding include contracts, commercialization of results and donations. The autonomy of the institutes increased. This policy was designed to reduce staff numbers and to drive some institutes from the academy. The number of staff was reduced by 6 452 people (44 per cent) between 1990 and 2000; the number of researchers decreased by 1 399 (28 per cent). The period between 1990 and 1996 saw greater weight being given to basic research. Between 1980 and 1991, small firms (about 65 in 1991 alone) and other units were set up with a view to commercializing research results. Their activity was later considered to be inappropriate for the academy and many were withdrawn from the academy umbrella or abolished altogether. In 1992 eight research units were expelled from the BAS for various reasons (Simeonova, 1995).

The grant paid from the state budget comprised about 79–80 per cent of all academy income in 1992/93; in 1994 this proportion dropped to 71 per cent. At the same time the income from contracts and royalties increased by 3 and 155 per cent respectively; following the critical drop in the number of contracts in 1992, the institutes became more adaptable. The share of extramural income for the various institutes lies between 16 and 22 per cent, the highest share being in institutes engaged in physics and chemistry. The income from contracts with private firms is insignificant – under 1 per cent of total revenue. The number of contracts with foreign organizations increased by a factor of 9.7 between 1992 and 1998 (Yuhnovski, 2001). Income from international organizations increased nominally as a share of total income. Income from donations is low and grants from the National Research Fund are also not substantial (about 3–4 per cent of annual income). The largest items in total costs are salaries and social security contributions. Capital investment is diminishing constantly (from 3.2 per cent in 1992 to 1.6 per cent in 1994), which is obviously damaging to the research infrastructure, already outdated in socialist times.

In recent years, the BAS has adjusted its policy to match the new opportunities, shifting its fundamental orientation towards more applied research. One of the reasons is the increase in international cooperation, including 35 bilateral agreements and Bulgaria's participation in the Fifth EU Framework Program. From another perspective, this policy shift is the result of an extended dialogue with different state bodies and the policy of improving transfer activity and developing high-tech parks in the country. Special agreements were signed with different ministries. More attention is being paid to improving cooperation with other organizations in the national R&D system. By the year 2000, the links between the BAS and the universities in the areas of teaching and training, which were rather loose between 1990 and 1997, had become stronger and more effective. More than 18 bilateral agreements have been concluded with HE institutions, taking the number of academy researchers giving lectures in HE institutions from 404 in 1991 to 672 in the year 2000. In the academic year 1999–2000, 6 100 students had access to the research infrastructure of the BAS institutes and the number of master's degrees completed in the BAS doubled between 1993 and 1999 (Yuhnovski, 2001; Tzonev et al., 2001).

The Agricultural Academy had 62 units in 1992, 82 units in 1993 and 72 units in 1994 and 1995. While the budget subsidy was the main source of funding until 1993, its share fell in subsequent years to only one third. In 1999 the Agricultural Academy was abolished. The National Center for Agricultural Sciences was set up in response to the restructuring policy being implemented by the Ministry of Agriculture and Forestry. This measure was in line with the general policy, pursued since 1997, of increasing state supervision of and influence over the R&D system. In contrast with the EU member states, academic research in Bulgaria falls within the *government sector*, which employed the majority of R&D personnel in 1998 (Laafia, 2000).

The relative weight of the former *branch sector* is illustrated by its share in the R&D potential at the beginning of the reforms, when it accounted for 64 per cent of R&D organizations, 70 per cent of total R&D expenditure and 70 per cent of all R&D staff. The share of scientists was about 40 per cent and the qualification level relatively high – over 50 per cent of the senior researchers in the country and about 25 per cent of PhD holders were engaged in branch R&D.

The organizational structure of the branch sector comprised two types of units. First, there were 65 large research institutes, carrying out applied research in a given branch or inter-sectoral field. Second, there were the experimental development bases, consisting of in-house units whose task it was to adapt scientific results for industrial application in the individual enterprises.

The branch structure of industrial research reflects state policy of the 1970s and 1980s. Industry's priorities are reflected in the research potential of the different branches, the largest being mechanical engineering and electronics (50 per cent), followed by agriculture (15 per cent). The dominant share – about 75 per cent – of all innovation activities concerned the results produced by the R&D organizations themselves, with 25 per cent of activity involving the transfer of foreign results through the purchase of licenses, know-how, technical documentation, etc.

The restructuring of industry and the decrease in R&D activity have changed the structure of that activity. In 1998 in the business enterprise sector, the number of R&D personnel (full-time equivalents) was highest in the manufacturing of coke, petroleum, chemicals, rubber and plastics (33.2 per cent).

In 1990 the conditions under which the industrial branch units operated changed dramatically, as did their activities. Of all the R&D organizations shut down between 1990 and 1992, units in the industrial sector accounted for the greatest share. The remaining organizations were granted autonomy from their supervisory ministry or state enterprise; they became legal entities in their own right, which gave them greater control over their decision-making and financial and human resources. According to a survey on industrial research conducted in 1991, only one research organization remained dependent on its supervisory firm.

Financial stagnation and reduced demand for industrial R&D were the main characteristics of the entire period up to the year 2000. In order to be able to retain their staff, most of the remaining units shifted from R&D to other activities, such as trade, renting flats and infrastructure, none of which were commensurate with their personnel's qualifications and abilities. Because of staff changes, most institutes have closed their scientific and expert committees. Low salaries discourage researchers from staying in their units and the best of them have left for other jobs. The number of scientists fell to 6 873 in 1992 (52 per cent of the 1990 level) and to 5 862 in 1995 (45 per cent of the 1990 figure). This trend has continued, as data gathered on the 'enterprise sector' since 1994 indicate (Tables 15.6, 15.8 and 15.9). The number of units did grow as a result of split-offs and the foundation of new units, but the number of staff declined further. In 1993 there were 11 196 employees, 3 091 of whom were researchers, but in 1996 only 4 155 remained, 2 020 of whom were researchers. This trend was maintained in the following years – in 2000 there were 1 225 researchers in the sector. This decline amounted to a 60 per cent reduction in the number of researchers from 1993 to 2000. Nevertheless, the share of this sector in all researchers increased from 8.3 per cent to 11.6 per cent between 1993 and 2000 (cf. Tables 15.8 and 15.9).

Having lost practically all their orders from firms and their support from the state budget, branch units adopted different strategies in their attempt to cope with financial and staffing problems.

The first of these was to seek new customers and sources of finance. The possibilities were very limited. State-owned industry was in collapse, the private sector was weak and, in terms of trade, mostly concentrated on international links. The major CMEA market had disappeared. As a result this line of adjustment was not realistic for most industrial units at that time. Very few of them were able to keep up contract activity abroad.

The second was to reduce the number of researchers and technicians. This turned out to be the basic element of the coping strategy, with staff numbers

Table 15.8 Bulgaria: R&D personnel total and in the enterprise sector, 1993–2000

	1993	1994	1995	1996	1997	1998	1999	2000	% to 1993
Total	54 561	31 924	30 663	31 942	21 908	21 766	18 451	16 853	30.9%
per 1 000 labor force	14.3	8.8	4.9	5.2	4.1	3.9	3.6	3.2	22.4%
Enterprise sector of which:	11 196	5 055	4 423	4 155	3 909	3 751	2 806	2 273	20.3%
Industry	7 590	3 786	3 338	2 819	2 390	2 417	1 807	1 258	16.6%
Construction	887	208	105	271	166	121	56	67	7.6%
Agriculture	1 329	178	180	169	301	59	171	165	12.4%
Forestry	48	14	23	22	–	–	–	–	–
Transportation	845	541	486	428	398	354	248	160	18.9%
Communications	313	275	252	234	352	520	269	–**	–
Trade	22	37	39	34	42	16	–	–**	–
Others*	162	16	0	185	260	264	255	207	(127.8%)

* since 1996 this category includes 'business services'.
** Confidential data.
Source: for 1993–1996: NSI, 1994, p. 19; 1995, p. 22; NSI Data Base, 1996; for 1997–2000: NSI Data Base, 1997–2000.

Table 15.9 Bulgaria: researchers total and in the enterprise sector, 1993–2000

	1993	1994	1995	1996	1997	1998	1999	2000	% to 1993
Total	37 179	18 497	17 523	18 579	14 573	14 045	12 335	10 527	28.3%
per 1 000 labor force	9.8	5.1	4.9	5.2	4.1	3.9	3.6	3.2	32.7%
Enterprise sector of which:	3 091	2 298	2 092	2 020	2 224	2 004	1 435	1 225	39.6%
Industry	2 051	1 738	1 629	1 417	1 408	1 245	964	643	31.4%
Construction	303	132	52	121	110	80	50	41	13.5%
Agriculture	352	56	68	63	107	33	84	75	21.3%
Forestry	10	5	6	5	–	–	–	–	–
Transportation	242	214	209	167	193	170	71	44	18.2%
Communications	101	131	103	100	189	301	79	–**	–
Trade	3	20	25	26	31	3	–	–**	–
Others*	29	2	–	121	186	172	187	155	(534.5%)

* since 1996 this category includes 'business services'.
** Confidential data.
Source: for 1993–1996: NSI, 1994, p. 19; 1995, p. 22; NSI Data Base, 1996; for 1997–2000: NSI Data Base, 1997–2000.

falling by 1996 to about a quarter of the 1990 level. Some staff moved to other activities, some became unemployed; others were forced into early retirement.

The third strategy was to maintain the remaining small group of employees on a drastically reduced salary, with equal pay for all job categories, and to wait for 'better times'. This had the advantage of retaining the infrastructure, which was to be used to offer different services.

However, after the initial shock, the R&D units' share of activity in the national STS continued to decline. 'Business services' in the enterprise sector became a relatively important item, with 13 R&D organizations (11.3 per cent of the branch sector); R&D units in industry underwent more extensive restructuring and reorganization than in other sectors of the economy. The privatization of research organizations is proceeding faster in industrial R&D than in the other sectors. Between 1996 and 2000, 150 branch R&D units were liquidated or privatized (Stoynov, 2001). According to the Ministry of Economy's database, in 1997 the total number of units with private shares was 32, which is 30.8 per cent of all industrial R&D organizations. Their distribution by branches is 21.8 per cent in chemical industry, 43.8 per cent in the mechanical engineering industries and 34.4 per cent in the electrical and electronics industries.

The rank order of the branches of industry by share of privatized R&D units is as follows: 1) mechanical engineering – 38 per cent of all units; 2) electrical and electronics industry – 34 per cent; 3) chemical and oil industry – 22.5 per cent.

The distribution of units by the volume of private shares is 6 units (19 per cent) <25 per cent; 11 units (34 per cent) 25–67 per cent; 15 units (47 per cent) >90 per cent.

By profile of their main activity the units are distributed as follows: research activity – 31.3 per cent; development and engineering – 28.1 per cent; technological and production activity – 40.6 per cent. It is important to note that one third of the units rely on their own research results; the other two thirds derive their knowledge externally (Simeonova, 1998a).

The impact of privatization has become a major and controversial issue in the restructuring of the R&D system. It has been noted that privatization is likely to result in a shift in units' activities towards enterprise at the expense of research.

The changes in the institutional system are related to the appearance of *private firms* in the R&D sector. R&D firms have been set up by researchers who have left state research organizations or continue to be employed by them on a part-time basis only (Tchalakov, 2001). According to a 1996 survey (Slavova, 1996), there are 1 375 such firms and they comprise 0.3 per cent of the country's registered companies. Personnel engaged in private R&D account for 1.5

per cent of all private-sector employment in the country. Two thirds of R&D firms are concentrated in Sofia, followed by Plovdive and Varna. They comprise between 0.1 and 0.4 per cent of the private firms in these regions. Some of the small firms operate in technological centers, with modern facilities and large building complexes. These were established by some big branch institutes, like the Central Research Institute of Complex Automation in Sofia (giving 70 small technological firms access to its infrastructure). In 2001 the latter was privatized like many other state branch institutions.

More than half of all entrepreneurs in R&D firms registered in the sector in 1994 had moved from state enterprises, which explains why their branch and product structure has been duplicated. The market share of the small and medium-sized enterprises (SMEs) is limited and unstable with respect to both demand and supply. The interest in branch diversification is still very weak. The main advantage of SMEs is their ability to react quickly to market demand and shortages of particular goods. This was considered the key factor in their embracing of technological transfer. A large share of SMEs have modern equipment; more than 50 per cent of the entrepreneurs surveyed pointed out that the average age of their equipment is less than 5 years. SMEs active in R&D suffer mostly from the lack of venture capital in the country and the substantial decline in production; these are the crucial factors that limit the demand for their services from other small firms as well as from state-owned enterprises. Over 55 per cent of SMEs have not introduced any innovations or significant improvements to their products since their inception.

The new law on SMEs adopted in 1999 aroused certain expectations. The legislation created important instruments for financial support, program development, staff training and infrastructure improvements. However, in the list of priorities that accompanies the legislation, SMEs engaged in R&D are in last (11th) place. Top priority is given to manufacturing firms, which in 1999 accounted for only 8–12 per cent of all firms. The important criterion for the granting of state support to these SMEs is that they should be involved in the transfer of advanced technologies. Unfortunately, according to expert evaluations, only 0.1 per cent of all SMEs fell into this category in the second half of the 1990s. Even if all firms engaged in software transfer are included, the share will still not exceed 0.4 per cent (Pachev, 2000). Analysts also point to the fact that no provision has been made for a fixed percentage of the state subsidy to be devoted to developing SMEs, which makes the future completion of the government programs doubtful.

Non-profit organizations (various foundations) began to appear in 1992. Out of a total of 136 such organizations, 30 per cent include R&D activity within the scope of their sponsorship; nine of them are involved in innovation. The number of those involved in innovation declined to five in 1994 and four in 1995. However, in 1999 their number increased again to ten, with a total staff of 81 people including 11 scientists. In 1999 the issue of establishing high-tech parks came to the fore again and the decision was taken to set up two regional high-tech parks. In the year 2000 one was set up in Plovdive, the country's second city, with a well developed R&D infrastructure and industrial base. The priority areas of this 'Hebros' high-tech park are optical and laser technologies, new materials for metal construction, computer systems and technologies. It is planned to locate the second park, which is still under consideration, in the north-west of the country, which, in contrast, is one of Bulgaria's less developed regions. This decision is a core element of the regional strategy adopted by the government in 2000. The expected success of the parks is linked to the favorable impact of negotiations on accession to the EU.

3. SITUATION IN 2000 AND OUTLOOK

The reforms introduced prior to 1997 were implemented without any firm resolve or strong consensus as to their aims and objectives. Unpopular restructuring measures were postponed and the economic policy changes were introduced slowly, which resulted in shrinkage in the market and in demand for R&D results. Privately owned manufacturing firms accounted for about 5 per cent of GDP in 1995, and privately owned service firms for about 10 per cent. The situation worsened dramatically in 1996 and brought about the political changes that occurred in 1997. In consequence, the environment for a long time was not conducive to the re-orientation and restructuring of R&D institutions. There were no new actors and agencies involved either in setting new research priorities or stimulating demand for R&D results.

Although positive moves towards restructuring have taken place since 1997, scientific organizations and scientists are still following a survival strategy, coping with job insecurity, a deteriorating infrastructure, loss of social prestige and an uncertain future. Instead of developing common strategies, the scientific community is becoming more and more fragmented as organizations and scientists vie with each other for diminishing state subsidies and grants. Through their participation in international projects organized as part of EU programs such as PHARE, COPERNICUS, TEMPUS and the Fifth Framework, some of the institutes have improved their information facilities, extended their collaboration and increased the number of joint publications in international journals (BAS, 2001).

As Bulgarian science opened up to international networks, so researchers began to move, either abroad or to other jobs within the country. By 1996, approximately 600 had left the country, while about 5 400 people had migrated to other sectors of the economy. This 'domestic' migration is particularly prevalent among the younger, promising scientists, who are leaving science to find better jobs and income. Few young people are embarking on careers in the research institutes. This is in stark contrast to the socialist past, when the very best graduate students entered scientific careers. However, between 1997 and 2000, the number of PhD students increased by 34 per cent, reaching a total of 3 069.

The legislative process had stopped at the point of giving greater autonomy to the various R&D institutions. It recently got under way again with new measures being introduced, such as a better definition of the role of state R&D institutions, the establishment of new institutional structures for the transfer of R&D into industry and the restructuring of some big institutions in order to adapt them more closely to the country's social and economic needs.

State bodies such as the MES allocate funds through competitive financing, with a well-developed system of 'ex ante' and 'ex-post' evaluations. However, the share of costs covered by these funds is very small. Direct institutional funding is regulated by the Budget Act and is left to internal institutional control.

The need for stronger budget control is linked to the establishment of the Currency Board in 1997. In response to the requirements of the International Monetary Fund, the number of institutions directly subsidized from the budget was drastically reduced in 1998.

The attempt to develop a 'Strategy for the development of science' failed, although in 1999 such a draft document was submitted to the National Council for Science and Technology. It had been drawn up by the MES and most of the key issues were formulated and fixed as a part of the country's preparation for entering into negotiations on EU membership. Science and education were among the first group of issues discussed in the negotiations, which began in March 2000, and the relevant chapter was successfully closed after four months of negotiations. Unfortunately, the requirements that had to be fulfilled for this purpose do not include the main problems of the sector, in particular the inadequate funding of research. However, the general context of enlargement still remains the decisive factor in accelerating institutional reform and improving S&T policy in the country.

The above-mentioned strategy document has not been adopted, due to the political changes in the country. However, the observation and analysis of the most acute problems relating to science are still valid. These include insufficient impact on socio-economic development, outdated infrastructure coupled with inefficient utilization of existing expensive facilities, ageing of research personnel and lack of favorable conditions for the reproduction of the science labor force, inefficient use of resources, low quality of results, outdated legislation and regulation of R&D and continuing outflow of scientists.

The R&D priorities set out in the draft strategy document are a matter of consensus among the research community and politicians. They are:

1. to strengthen the role of R&D as a factor in accelerating the restructuring of Bulgarian society and the integration of the country into Euro-Atlantic structures;
2. to reorient R&D activity towards improving the competitiveness of Bulgarian industry and renewal of its technology base;
3. to mobilize R&D resources for the improvement of the basic infrastructure and environment in the priority sectors of the economy – such as telecommunications, energy, transportation, etc.;
4. to consider a thriving science and education system as the main precondition for the successful adaptation of human resources to the requirements of the market economy and to efforts to improve the quality of life.

Consideration was also given to mechanisms for reaching these goals, which deserve to be seriously considered by the government.

In conclusion, two factors of importance for the development in the last ten years can be singled out. First, the reform of R&D in Bulgaria is lagging behind those in the other post-socialist countries. This is due to the general slowdown of economic reforms and the country's political instability.

Second, by the year 2000 the political stability enjoyed by the country since the 1997 election had led to a considerable quickening of the pace of reform in all spheres of social life. Since Bulgaria was invited to join the other post-communist countries in negotiations on EU membership, the situation has changed for the better. One promising and important factor in this respect was the consensus reached by the different political parties represented in parliament on the major objective of EU accession.

The situation in the country in mid-2001 and its possible implications for the institutional system of R&D require stability and continuity of policy. The parliamentary elections of June 2001 again broke this continuity on the political level. A new political movement, the 'National Movement Simeon the Second' (NMSS) set up by the ex-Bulgarian tsar, came into power with radical ideas about reforms in the country. It has said it will pursue the same priorities in international policy, but implement far-reaching changes in economic and social legislation in order to speed up the reforms at lower cost to Bulgarians. Of the short programmatic statements on education and science to be found in the NMSS program, the most promising point is the plan to increase GERD annually by a fixed percentage. The expectation is that the positive developments that have emerged in recent years will continue and that much progress will be achieved in R&D policy. Membership of the EU is one of the declared priorities of the new government elected on July 24, 2001 and this is an important guarantee that the enlargement process will continue to have a positive impact on Bulgarian science policy.

REFERENCES

BAS (1991) *Otcheten doklad na Bulgarskata Akademia na naukite za 1990 godina* (Annual Report of the Bulgarian Academy of Sciences for 1990). Sofia: Publishing House of the BAS.

BAS (2001) *Otcheten doklad na Bulgarskata Akademia na naukite za 2000 godina* (Annual Report of the Bulgarian Academy of Sciences for 2000). Sofia: Publishing House of the BAS.

Eurostat (2000) *R&D and Innovation statistics in candidate countries and the Russian Federation in 1996–1999.* Ed. Eurostat. Luxembourg: Office for Official Publications of the European Communities.

Laafia, I. (2000) *R&D expenditure and personnel in candidate countries and the Russian Federation in 1998.* Luxembourg: Office for Official Publications of the European Communities.

MES (Ministry of Education and Science) (1992) *Biala kniga za bulgarskoto obrazovanie i nauka* (White paper on Bulgarian education and science). Sofia: Publishing House of Sofia University 'Svety Kliment Ochridski'.

NSI (1990–1995) *Statistical Yearbook.* Sofia: Nazionalen Statisticheski Institut (NSI).

NSI (1994; 1995) *Nauchno-technicheska i razvoina deinost* (Research and development activity). Sofia: NSI.

NSI (2001) *Statisticheski spravochnik* (Statistical Handbook). Sofia: NSI.

NSI Data base (1989–2000). Sofia: NSI.

OECD (1994) *Proposed standard practice for surveys of research and experimental development – Frascati Manual 1993.* Paris: OECD.

Pachev, P. (2000) "Zakonat za malkite I srednite predpriatia – zakasnyal opit za normativno urezdane" (The Law of SME – a delayed trial for normative regulation). *Ikonomicheska misal* (Economic Thought) no. 2: 48–62.

Simeonova, K. (1995) "Radical and Defensive Strategies in the Democratization of the Bulgarian Academy of Sciences." *Social Studies of Science* 25, no. 4: 755–775.

Simeonova, K. (1998) "The Two-Edged Sword of Autonomy: Changes in the Academy-Institute Relations." *East European Academies in Transition.* Ed. R. Mayntz, et al. Dordrecht: Kluwer Academic Publisher, 125–139.

Simeonova, K. (1998a) "The Struggle for the Survival of Industrial R&D in Bulgaria." *Transforming Science and Technology Systems – the Endless Transition? NATO Science Series 4: Science and Technology Policy – Vol. 23.* Ed. W. Meske, et al. Amsterdam: IOS Press, 253–265.

Simeonova, K., M. Ivanova, S. Grivekova, and S. Roshkov (1995) "Kontextbedingungen der Transformation des Wissenschaftsystems in Bulgarien." *Transforma-*

tion mittel- und osteuropäischer Wissenschaftsysteme – Länderberichte. Ed. R. Mayntz, et al. Opladen: Leske+Budrich, 1044–1124.

Slavova, M. (1996) "Darzavnata podkrepa za inovacionata deinost ha malkia I srednia biznes" (The state support for innovative activities in small and medium businesses). *Proceedings of the conference 'Nauchna politika i ikonomichesko razvitie' (Science policy and economic development) May 27, 1996 in Sofia.* Sofia: University publishing house 'Stopanstvo', 148–167.

Stoynov, Z. (2001) "Perspectivi za naukata I texhologichnoto razvitie v Bulgaria" (Perspectives of Science and Technology Development in Bulgaria). *Nauka* (Science – Review of the Union of Scientists in Bulgaria) 11, no. 5: 23–27.

Tchalakov, I. (2001) "Innovating in Bulgaria – two cases in the life of a laboratory before and after 1989." *Research Policy* 30: 391–402.

Tzonev, M., E. Lazarova, N. Savov, D. Minchev, and N. Gergova (2001) "Nauchno-obrasovatelnata deinost na BAN" (The education activity of the BAS). *Bulgarskata akademia na naukite po patya na reformite 1989–2000 (The Bulgarian Academy of Sciences on the road to reform 1989–2000).* Ed. K. Simeonova, et al. Sofia: Center for Science Studies and History of Science at the BAS, 257–280.

Yuhnovski, I. (2001) "The interview with the editor of the review 'Nauka'". *Nauka* 11, no. 5: 7–10.

16. FEDERAL REPUBLIC OF YUGOSLAVIA: RESTRUCTURING THE S&T SYSTEM – INDICATORS OF TRANSFORMATION

Duro KUTLACA

An understanding of the economic situation in Yugoslavia over the past decade is obviously crucial to any investigation of the country's S&T system. Four separate periods can be identified.

1. In the period before 1991, specific patterns of technical change could be observed in the self-management system. Obviously, this period was marked by an absence of economic growth as the country effected the transition to a market economy. The self-management system had developed mechanisms for technological development based mainly on imported technologies, a relatively large public R&D system (measured in terms of researchers per capita) and weak industrial R&D.

2. In the period between 1991 and 1999, the economic system remained largely unchanged, but international isolation and war affected the extent and nature of technical change. This was a period characterized by a strong downturn in GDP and all industrial activities, as the country was faced with a brain drain and only limited communication with the outside world. As far as the S&T system was concerned, one of the country's main achievements during this period was the allocation of about 1 per cent of GDP to the development of S&T in 1994. Between 1996 and 1998, the economy began to recover, international links were re-established and the pace of privatization accelerated, but all progress came to a halt on March 24, 1999, as a consequence of NATO aggression against Yugoslavia.

3. The third period, 1999/2000, was marked by NATO aggression against Yugoslavia, destruction of the country's infrastructure and devastation of its environment, with industry and the economy as a whole being seriously affected. This aggression lasted 78 days. The period that followed saw the rebuilding of destroyed infrastructure, industrial plants, homes and schools. According to the country's background report prepared for UNDP in January 2001, the direct costs resulting from UN sanctions are estimated at 36 billion USD and the losses inflicted by NATO bombing at tens of billions of USD (UNDP, 2001).

4. A fourth period began with democratic change at the end of 2000, followed by the re-integration of the country into the international community and its main financial institutions and the provision of emergency aid to the exhausted economy and society. Political changes at the end of 2000 brought in a new democratic government, which is now adapting the legal system to economic transition and restructuring. New democratic governments (at the federal level and in the Republic of Serbia) began to function at the beginning of 2001; however, the country is still beset by serious problems. The average real wage is about 50 USD per month and pensions are even lower. Although family budgets are augmented by work in the informal economy, two thirds of households in Serbia (without Kosovo and Metohia) live on less than 80 USD per month per family. In Central Serbia and Vojvodina, 2.8 million people were below the poverty line in 1999 and the unemployment rate was 26 per cent. The country's foreign debt is in excess of 13 billion USD. Inflation in the year 2000 was close to 50 per cent. Per capita GDP decreased from 3 000 USD in 1989 to 1 400 USD in 1999 (1 220 at 1994 prices) (UNDP, 2001). Substantial international aid for the stabilization of the economy was expected in 2001 (some of which was forthcoming at the end of 2000). New regulations to govern the massive restructuring process are also expected to come into effect in 2001, while a re-evaluation of the privatization program leading to the introduction of new regulations and an acceleration and extension of the program are expected in the 2001–2004 period. National S&T policy is going to be modified to deal with the necessary restructuring of the R&D system. Because of the bad economic situation, it is unrealistic to expect new R&D projects to be financed from the state budget to any large extent. R&D organizations should probably try to be involved in international (primarily EU) R&D programs and to find alternative sources of funding, including foreign ones, at least for the near future.

Tables 16.1 and 16.2 present the basic macroeconomic data for the Socialist Republics of Serbia and Montenegro in the former SFRY until 1992 and for the Republics of Serbia and Montenegro in the FR of Yugoslavia since 1992.

1. S&T POLICY

The S&T system in the FR of Yugoslavia is governed both at the federal and republic level. Until the end of the year 2000, policy at federal level was drawn up by the Federal Ministry of Development, Science and the Environment, whose power was substantially reduced in 2001 when it was reorganized to the Federal Secretariat for Science and Development. The policy-making body in the Republic of Serbia was the Ministry of Science and Technology, which

Table 16.1 FRY: selected macroeconomic indicators, 1987–1991

Indicator	Period I				
	1987	1988	1989	1990	1991
Export (mill. USD)	4 063	4 298	4 461	5 815	4 704
Import (mill. USD)	4 851	4 915	5 383	7 460	5 548
Trade balance (mill. USD)	-788	-617	-922	-1 645	-844
GDP (mill. USD)(1994 prices)	27 585	27 948	26 637	28 139	24 568
GDP per capita (USD) (1994 prices)	2 667	2 684	2 544	2 672	2 360
DPI* (mill. USD) (1994 prices)	11 550	12 010	11 490	11 628	9 845
Private-sector GDP (mill. USD) (1994 prices)	3 036	3 040	2 896	3 556	4 078
Consumer price indices (previous year = 100)	220	294	1365	680	222
Exchange rate (1 USD = new YU dinars)	1.81	1.76	1.87	1.63	1.65
Population (thousands, June 30th)	10 342	10 411	10 471	10 529	10 408
Employment (thousands)	2 762	2 784	2 790	2 707	2 625
Employment in industry (thousands)	1 073	1 089	1 090	1 067	992
Employment in private sector (thousands)	48	52	58	66	187
Unemployment (thousands)	558	578	607	663	714

*DPI = domestic product of industry.
Source: EI, 1995–1998; FSO, 1998; MAP, 1995–2000.

Table 16.2 FRY: selected macroeconomic indicators, 1992–2000

	Period II							Period III			2000(1998)/ 1990 in %
	1992	1993	1994	1995	1996	1997	1998	1999	2000		
Export (mill. USD)	2 539				1 842	2 677	2 858	1 498	1 727		30
Import (mill. USD)	3 859				4 102	4 826	4 849	3 296	3 698		50
Trade balance (mill. USD)	-1 320				-2 260	-2 149	-1 991	-1 798	-1 971		120
GDP (mill. USD) (1994 prices)	18 497	12 264	12 809	12 510	14 295	15 364	15 758	12 969	13 883		49
GDP per capita (USD) (1994 prices)	1 770	1 170	1 218	1 186	1 354	1 449	1 484	1 220	1 306		49
DPI* (mill. USD) (1994 prices)	7 931	4 700	4 846	4 651		5 236	6 098	4 280			(37)
GDP-private sector (mill. USD: 1994 prices)	4 065	3 415	3 854	4 006		5 819	5 849				(164)
Consumer price indices (previous year = 100)	9 026	222×10^9	103	179	192	122	130	145	186		
Exchange rate (1 USD = new YU dinars)	1.58	1.65	1.62	1.76	4.97	5.72	9.22	–	–		(18)
Population (thousands, June 30)	10 448	10 482	10 516	10 547	10 557	10 600	10 617	10 629	10 634		101
Employment (in 1000)	2 536	2 464	2 413	2 379	2 367	2 332	2 289	2 092	2 035		75
– in industry	940	915	894	870	852	820	796	715	676		63
– in private sector	207	221	236	265	288	318	327	306	321		486
Unemployment(1000)	747	738	726	775		819	814	811	806		122

*DPI = domestic product of industry.

Source: EI, 1995–1998; FSO, 1998; MAP, 1995–2000.

became the Ministry of Science, Technology and Development in 2001, and in the Republic of Montenegro it was (and still is) the Ministry of Education and Science. In the Republic of Serbia, the Council for Scientific and Technological Policy was established in the 1990s as a high-level advisory body to the Minister of S&T.

1.1. FEDERAL LEVEL

Important changes in the infrastructure have followed from changes in the country's legal environment. New legislation was enacted during the 1990s in the FR of Yugoslavia in the following areas.

1. *Firms' adjustment to market economic conditions* (privatization, management, working conditions, financial questions, etc.).

   In the R&D sector, the introduction of these laws will affect independent institutes and those that belong to the state sector, despite the government's readiness to sacrifice them to market competition (or willingness to sell them). This process is still in the preparatory phase and proposals concerning its practical implementation are awaited.

   Some independent institutes were partly privatized under the legislation in force in the former Yugoslavia, but the present privatization authority in the Republic of Serbia has effectively annulled these transactions on the grounds that the value of firms was not accurately calculated because of errors in the inflation rate used. A new privatization law in the Republic of Serbia was adopted in 2001. Under the terms of this new legislation, the transition from social ownership to private (public/state) ownership will have to be completed within three years. A small number of companies as well as some R&D institutions are included in a special privatization program, but there is still no specific government policy on restructuring and privatization in the R&D sector in the Republic of Serbia. In the Republic of Montenegro almost all industrial firms are privatized, but all R&D activities take place in the public sector.

2. *Industrial and intellectual property rights*.

   A new patent law adopted by the Federal Assembly in 1995 significantly altered national practice in respect of invention protection by preventing employees from applying for patent rights without their firm's permission. They were entitled to do so under earlier legislation, and as a result, only a small number of patents are registered as company patents, while the greater share is registered as individually owned (Kutlaca, 1998).

   Further patent laws have adopted the recommendations of the World Intellectual Property Organization (WIPO) and of the European Patent Office (EPO) concerning the protection of pharmaceutical products, which had

been the subject of negotiations between the former Yugoslavia and these organizations.
3. *Environmentally sound and sustainable development.*
The legislation in this area deals with various aspects of environmental protection. It impacts on R&D in so far as it defines the obligations of R&D organizations to prevent, control and measure the direct and indirect effects of the use and/or processing of dangerous materials, products, processes, etc.
4. *Quality standards.*
Activities were organized at the Republic level. International cooperation in the field of science and technology is a federal responsibility. One of the top priorities for the Federal Secretariat for Science and Development following the country's re-admission to the international community in 2001 is the re-establishment of cooperation with countries of particular interest for S&T and industrial development in Yugoslavia.

1.2. REPUBLIC LEVEL: REPUBLIC OF SERBIA

Since 1991, all R&D activities in the Republic of Serbia have been subject to a planning, monitoring and evaluation process, which is regulated by the 'Law on Scientific and Research Activities', the 'Policy on Scientific and Technological Development' and the 'Regulation on the Financing of Programs and Projects.' As a result of action pursuant to this regulation, just one year after adoption of the new scheme for Technological Development Projects – strategic, innovation and pre-competitive projects – the third group of projects was canceled in 1993 as a result of an evaluation of their effectiveness and their contribution to the country's technological development. Nevertheless, the planning, monitoring and evaluation of R&D activities all require improvement, especially with respect to:

1. the contribution of independent evaluators, if possible from abroad;
2. the development of appropriate methods and tools;
3. increasing public awareness and public evaluation of R&D activities.

In order to implement the Program for Quality Assurance adopted in 1992 and to develop policy in this area, the Ministry of Science and Technology (MST) established the Quality Council, which was to act as an expert and professional advisory body to the minister. Between 1993 and 1996, following a competitive tendering process, more than 300 industrial firms and more than 20 R&D institutes were given financial support to implement the program. According to the Chamber of Commerce, over 1 000 firms are already actively working on the ISO 9000 program.

In 1993, the MST adopted a program to support the development of young scientists. For the three years of the program's existence, more than 1 000 young research fellows worked on projects in the R&D organizations. Their work was supervised by mentors, who included eminent scientists, researchers and professors.

In 1995, the MST announced a call for tenders for basic research projects seeking full funding from the Republic's budget between 1996 and 2000. The three main criteria in selecting project proposals were the following:

1. their likely contribution to the resolution of the country's most pressing development problems;
2. their contribution to rational use of the Republic's natural and human resources and its productive potential;
3. their contribution to global knowledge and the extent to which they might facilitate the integration of Serbian science into world science.

Between 1996 and 2000, 250 basic research projects were funded. The next program, which was to provide funding for 70 strategic technological development projects between 1998 and 2000, was adopted by the MST at the end of 1997.

Research programs in the field of technological development adopted in 1992 were implemented through either strategic or innovation projects. The strategic R&D projects focused on the creation of technologies crucial to the country's development and lasted up to three years. Funding was provided for projects in six technological fields: information technologies, energy technologies, new materials, high-quality food production, technology upgrading and quality enhancements, and management of technological development. Three cycles of strategic R&D projects, each lasting three years, were financed, with about 150 projects running between 1992 and 1994, about 90 between 1995 and 1997 and about 70 between 1998 and 2000. More than 500 innovation projects (of 6 months' to 1 year's duration) were financed between 1992 and 1998.

Application to production as well as the commercialization of R&D results were continuously monitored by the MST and the Chamber of Commerce of the Republic of Serbia. These project are funded in part by the MST (70 per cent for strategic and 30–50 per cent for innovation projects). Each project must have end-users in industry and other sectors. Firms could make (human, material or financial) resources available, but their main contribution to this program was their readiness to apply the results produced. The Regulation on the Financing of Programs and Projects included provisions governing the intellectual and industrial rights of all the partners, that is R&D organizations, firms and the Ministry. Should the results not be applied by contracting firms, they must be offered to other possible users on special terms. Each autumn, the MST supports R&D organizations in presenting their results, particularly those originating from strategic and innovation projects, at the 'NOVOTEH' Fair of New Technologies. The objective of this fair is to attract the attention of industrial and other potential users of research results. Similar presentations are also organized in other industrial/regional centers in the Republic.

The S&T information network links all the major Serbian cities, and a connection has also been established with Podgorica, the capital of the Republic of Montenegro. The network provides all basic Internet services (e-mail, Telnet, ftp, talk, finger, WWW). During 2001, a SinYu (Scientific Information Network Yugoslavia) project proposal was drawn up by the Max Planck Institute (2.5 gigabyte capacity network) and there are ongoing negotiations between Serbia and the EU for financial support for the network.

In 2001, a new S&T development policy was drawn up by the new government based on the following principles (outlined by Dragan Domazet, Minister for Science, Technology and Development of the Republic of Serbia in a document on July 28th, 2001):

- R&D activities should have a more significant impact on economy and society;
- higher priority should be given to the quality of R&D projects by applying rigorous selection criteria to project proposals;
- international collaboration to be encouraged for all R&D programs;
- brain drain to be reduced by launching a special program to improve the R&D environment;
- knowledge and technology transfer from local and foreign sources to be directed to industry as much as possible;
- a new model for the rapid economic growth of Serbia to be developed, based on technology development and knowledge.

The new S&T development strategy has two phases:

Phase 1 – revitalization of the S&T sector (2001 – 2002);

Phase 2 – building the innovative infrastructure needed for the rapid and sustainable development of Serbia (2003 – 2008).

Phase 1 got under way in 2001 with the following programs:

- basic research program, which provides grants for R&D projects in physics, chemistry, biology, mathematics and mechanics, geo-sciences, medicine, social sciences and humanities;
- technology development program, which provides grants in information technologies, electronics and electrical engineering, mechanical engineering, construction industry and civil engineering, biotechnology etc.;
- national programs intended to solve important national problems:

+ energy efficiency;
+ biotechnology and agro-industry;
+ water management and treatment;
+ business efficiency;
- grants program for innovation projects.

All preparing procedures relating to the programs outlined above were to be completed by February 2002. Project proposals for the basic research program were to be evaluated in Italy and Germany, while all other proposals were to be evaluated nationally.

Serbia's R&D budget was increased to 41 million EURO in 2002, more than three times the 2000 level of 13 million EURO. Nevertheless, these figures are very moderate and insufficient to bring about any noticeable improvement in the R&D sector. The next step in Phase 1, the restructuring of the R&D system, was to be completed during 2002.

2. THE S&T SYSTEM IN THE FRY

S&T systems in the two republics of the FRY are organized in very similar ways. They comprise:

1. the higher education sector (HE);
2. R&D or independent institutes (RDIs);
3. R&D units (RDUs) in industry;
4. S&T infrastructure.

There are further organizations or offices for intellectual property rights, standardization, measurements and precious metals, which are registered as federal institutions.

There is an enormous imbalance in the system between the two republics, with more than 95 per cent of it concentrated in the Republic of Serbia. The Republic of Montenegro's contribution increased in 1999 as scientists migrated from Serbia to Montenegro because of heavy political pressure on universities and institutes in Serbia, especially in 1998–1999 (cf. Table 16.3).

The decline in the number of R&D organizations in both republics is the result of smaller organizations being merged with larger ones. The decline in the total number of employees and researchers working in R&D in Serbia is a result of migration ('brain drain') and (normal and forced) retirement. The decline in the number of researchers (measured in terms of full-time equivalents) is a result of researchers moving from institutes to university faculties, where the main activity is teaching and only one third of working time on average is allocated to R&D activities.

In Montenegro, the growth in the total number of employees resulted from the establishment of one new (medical) faculty. The decline in the number of

Table 16.3 FRY: structure of the S&T system by republic, organization and personnel

Type of organization		Organizations			Employees			Researchers		
		Number		1999/1990 %	Persons		1999/1990 %	FTE		1999/1990 %
		1990	1999		1990	1999		1990	1999	
FRY	HEO	6/140	7/92	65.7	16 729	14 122	84.4	2 729	3 108	113.9
	RDI	152	80	52.6	12 913	9 010	69.8	4 610	3 129	67.9
	RDU	49	52	106.1	2 646	2 194	82.9	752	730	97.1
Total S&T system		341	224	65.7	32 288	25 326	78.4	8 091	6 967	86.1
Serbia	HEO	5/128	6/79	61.7	16 214	13 319	82.2	2 578	2 877	111.6
	RDI	140	74	52.9	12 539	8 708	69.5	4 486	3 046	67.9
	RDU	46	50	108.7	2 453	2 171	88.5	678	724	106.8
Total S&T system		314	203	64.7	31 206	24 198	77.5	7 742	6 647	85.9
% of FRY		92.1	90.6		96.6	95.5		95.7	95.4	
Montenegro	HEO	1/12	1/13	108.3	515	803	155.9	151	231	152.9
	RDI	12	6	50.0	374	302	80.8	124	83	66.9
	RDU	3	2	66.7	193	23	11.9	74	6	8.1
Total S&T system		27	21	77.8	1 082	1 128	104.3	349	320	91.7
% of FRY		7.9	9.4		3.4	4.5		4.3	4.6	

Note: FTE—author's calculations (1 FTE researcher at universities, involved in research and teaching = 3 employees); Source: FSO, 1980–1999.

researchers resulted from migration, albeit on a smaller scale than in Serbia because of the somewhat better economic and political situation in Montenegro, which encouraged scientists to move from Serbia to Montenegro.

Table 16.4 presents data on the changes between 1990 and 1999 in the S&T system in the FR of Yugoslavia broken down by scientific fields:

Natural and mathematical sciences: growth in the total number of employees, mostly attributable to an increase in university staff accompanied by a decline in the number of researchers (FTE);

Technical and multidisciplinary sciences: the decline in absolute numbers and FTEs is the result of migration, while the growth in the HE sector is a consequence of the shift from institutes to universities.

Medical sciences: organizational changes resulted in the abandonment of R&D activities in some hospitals and the concentration of human resources in better-equipped organizations. This field of science suffers not only from a considerable brain drain but also from the emigration of support staff.

Agricultural sciences are a traditional field of R&D in the country and some degree of brain drain is being offset by new recruitment. The decline in the FTE number is a consequence of a shift from institutes to universities.

Social sciences show a decline in all three sectors, having become less attractive for employees. The bad economic situation has forced many of them to find new careers.

Humanities: the substantial growth in university staff is the result, in part, of a shift from institutes to universities as well as of increasing interest among young people. There is a general trend in the country towards a decline in the number of students in technical sciences and increased interest in the humanities.

The *number of R&D organizations* in almost all sectors and all sciences declined as a result of mergers, with just a few being closed down or re-organized.

3. S&T RESOURCES

3.1. FINANCING

The volume and sources of financing for R&D activities and the share of R&D expenditure in GDP are shown in Table 16.5.

Figure 16.1 shows the share of income from R&D in the total income of Yugoslav R&D organizations (RDOs) and the changes over time in the R&D system following the transition to a market economy.

1. From 1980 to 1989, R&D work was the basic activity and the main source of income for RDIs – 70 per cent on average in the R&D system as a whole, and as much as 90 per cent for RDUs. It is interesting to note that HE establishments earned up to 65 per cent of their income from R&D work.

Table 16.4 FRY: structure of the S&T system by scientific field

Field of science	Type of organization	Organizations			Employees				Researchers				
		Number		1999/1990 %	Persons		1999/1990 %	Employees 1999	FTE		1999/1990 %	Researchers		1999/1990 %
		1990	1999		1990	1999			1990	1999		1990	1999	
Natural and mathematical sciences	HEO	21	11	52.38	1 483	1 655	111.60		311	360	115.76			
	RDI	30	10	33.33	1 576	1 557	98.79		843	754	89.44			
	RDU	4	4	100.00	81	207	255.56		57	74	129.82			
	Subtotal	55	25	45.45	3 140	3 419	108.89		1 211	1 188	98.10			
Technical and multidiscipl. sciences	HEO	40+0	23+4	67.50	3 835+0	3 656+231	101.36		715+0	847+43	124.48			
	RDI	50+0	22+2	48.00	7 792+0	4 718+73	61.49		2 326+0	1 351+39	59.72			
	RDU	24+0	25+1	108.33	2 195+0	1 408+9	64.56		498+0	455+1	91.57			
	Subtotal	114+0	70+7	67.54	13 822+0	9 782+313	73.04		3 539+0	2 653+83	77.31			
Medical sciences	HEO	24	7	29.17	6 453	2 561	39.69		708	577	81.50			
	RDI	5	2	40.00	246	59	23.98		82	39	47.56			
	RDU	4	5	125.00	56	209	373.21		38	51	134.21			
	Subtotal	33	14	42.42	6 755	2 829	41.88		828	667	80.56			
Agricultural sciences	HEO	11	5	45.45	1 509	1 738	115.18		288	320	111.11			
	RDI	23	20	86.96	2 053	1 837	89.48		596	392	65.77			
	RDU	10	10	100.00	169	282	166.86		62	105	169.35			
	Subtotal	44	35	79.55	3 731	3 857	103.38		946	817	86.36			
Social sciences	HEO	31	25	80.65	2 580	2 035	78.88		533	481	90.24			
	RDI	24	13	54.17	646	503	77.86		391	351	89.77			
	RDU	3	3	100.00	46	30	65.22		28	11	39.29			
	Subtotal	58	41	70.69	3 272	2 568	78.48		952	843	88.55			
Humanities	HEO	13	18	138.46	869	2 246	258.46		174	480	275.86			
	RDI	20	10	50.00	600	263	43.83		372	203	54.57			
	RDU	4	4	100.00	99	49	49.49		69	33	47.83			
	Subtotal	37	32	86.49	1 568	2 558	163.14		615	716	116.42			
All sciences	HEO	140	92	65.71	16 729	14 122	84.42		2 729	3 108	113.89			
	RDI	152	80	52.63	12 913	9 010	69.77		4 610	3 129	67.87			
	RDU	49	52	106.12	2 646	2 194	82.92		752	730	97.07			
Total S&T system		341	224	65.69	32 288	25 326	78.44		8 091	6 967	86.11			

Note: FTE—author's calculations (1 FTE researcher in universities, involved in research and teaching = 3 employees).

Source: FSO, 1980–1999.

Table 16.5 FRY: R&D financing

Funding for R&D activities	1989	1990	1991	1992	1993	1994	1995	1996	1997	1998	1999	1999/1991 in %
Mill. USD; 1994 prices												
Public funds	78.8	116.5	109.3	56.4	–	75.9	75.1	87.5	110.8	95.6	100.2	92
Industry	113.7	148.4	139.3	136.3	–	56.8	39.2	60.5	46.8	32.1	29.8	21
Other national sources	13.7	32.8	25.2	29.1	–	16.5	23.3	29.2	35.3	63.9	49.8	198
From abroad	17.7	14.7	8.8	0.2	–	0.6	1.1	2.9	5.4	3.9	1.8	20
Total funds	223.8	312.3	282.5	222.0	–	149.9	138.7	180.1	198.2	195.4	181.6	64
Share of sources (%)												
Public funds	35.2	37.3	38.7	25.4	–	50.7	54.1	48.6	55.9	48.9	55.2	
Industry	50.8	47.5	49.3	61.5	–	37.9	28.2	33.6	23.6	16.4	16.4	
Other national sources	6.1	10.5	8.9	13.1	–	11.0	16.8	16.2	17.8	32.7	27.4	
From abroad	7.9	4.7	3.1	0.1	–	0.4	0.8	1.6	2.7	2.0	1.0	
GERD/GDP (%)	0.84	1.11	1.15	1.20	0.86	1.17	1.11	1.26	1.29	1.24	1.4	122

Source: FSO, 1980–1999.

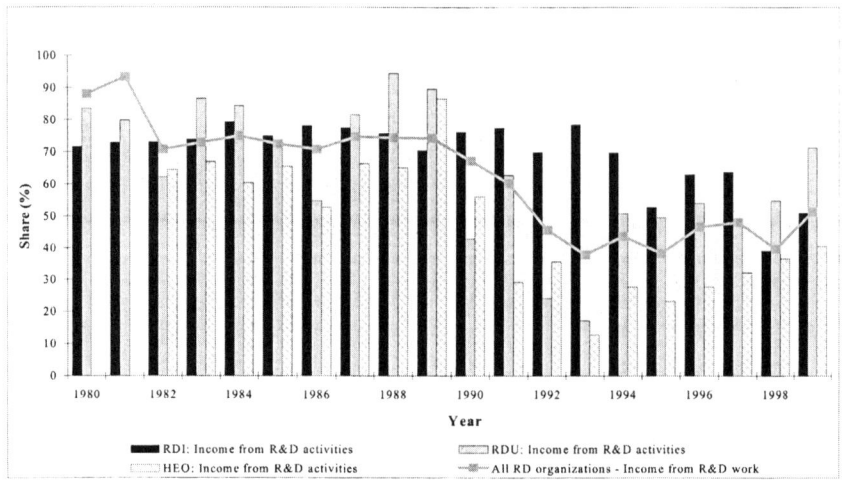

Fig. 16.1 FRY: share of income from R&D in the total income of R&D organizations, 1980 – 1999 (%)
Source: FSO, 1980 – 1999.

As RDIs were also engaged in design, small-series manufacture and other non-research activities, income from R&D work accounted for 70 per cent.

2. From 1989 to 1995, all organizations within Yugoslavia's R&D system were looking for additional funding sources. The share of income from R&D was decreasing constantly and amounted to no more than one third of the RDIs' total income in 1993 and 1995. More precisely, HE establishments relied almost exclusively on educational activities and earned only some 10 – 30 per cent of their total income from R&D, while RDUs depended on R&D for 25 – 60 per cent of their total income; RDIs were the least diversified organizations, earning on average 70 per cent of their income from R&D.

3. From 1996 to 1999, there was a notable increase in the share of income from R&D work within the total income of HE establishments (up to 40 per cent) and RDUs (some 71 per cent), and a decrease in RDIs to about 51 per cent. For the R&D system as a whole, there was an increase to 51 per cent in 1999 due to a growing demand for domestic technologies as substitutes for inaccessible foreign ones.

3.2. Personnel

The following methodological comments are necessary prior to any analysis. Figures for employees do not necessarily (and in the FRY S&T system definitely do not) equate to FTEs as defined in the OECD's Frascati Manual (OECD, 1994). This is largely because they cover all researchers in the HE sector, some of whom are engaged in R&D full-time and others only part-time (teaching staff are not included in these figures). I have calculated that 1 FTE researcher in HE involved in both research and teaching is equivalent to 3 employees.

The basis for this calculation is the price of a research-year in HE; funding for one third of the year is provided by the Ministry of Science and Technology, while two thirds are being paid by the Ministry of Education. This is the case in the Republic of Serbia, but similar calculations were made for the S&T system in the Republic of Montenegro.

The country's statistical system classifies the sciences as follows:
1. natural sciences and mathematical sciences,
2. technical sciences,
3. medical sciences,
4. agricultural sciences,
5. social sciences,
6. humanities.

Staff are classified by type of work (researchers, technicians, administrative staff and others) and by working time (full-time/part-time).

In terms of human resources, the evolution of the Yugoslav R&D system between 1980 and 1999 can be characterized as follows.

1. The total number of persons employed in the R&D system had grown since 1980, reaching its peak in 1988. From 1988 onwards, employment levels began to fall. The sharp decline in 1991 was the result of a brain drain caused by the dissolution of the former SFRY. In 1999, 30 per cent fewer people were employed than in 1988 (cf. Table 16.6).
The number of researchers increased from 9 522 to 13 874 between 1980 and 1988, then decreased between 1988 and 1991 by 17 per cent. It then started to grow again, reaching 13 220 in 1997. This change is a direct consequence of a timely intervention by the relevant ministries, which established programs for the training and employment of young researchers. The worsening of the economic situation and the status of the R&D system in the country resulted in yet another decrease in the number of researchers as well as in the total workforce in this sector; by 1999, the number of researchers was down to 12 740.
The *trend* in the FTE number of researchers, however, is different from that in the absolute numbers over the 1990–1994 period and is the first *indicator of structural changes within the R&D system* (Figure 16.2). The number of FTEs decreased from 1988 to 1992, was relatively constant from 1993 to 1994 and since 1994 has followed the trend of the absolute number of researchers.

Table 16.6 FRY: selected R&D indicators

Indicator	Period I							Period II							Period III
	1987	1988	1989	1990	1991	1992	1993	1994	1995	1996	1997	1998			1999
Employees in RDO	34 290	35 973	35 453	32 288	30 554	31 246	30 445	29 369	30 196	26 763	27 194	24 350			25 326
Researchers in RDO	13 073	13 874	13 761	12 786	11 522	11 944	12 041	12 520	13 263	13 110	13 220	12 151			12 740
Researchers in RDO (FTE)	8 725	9 137	9 010	8 091	8 013	6 954	6 970	6 995	7 576	7 423	7 476	6 887			6 967
Researchers in industry (FTE)	366	394	677	752	931	807	706	575	878	958	952	822			730
Brain drain from FRY: Researchers (only R&D sector; official data)	53	67	83	126	180	190	223	201	89	84	58	72			103

Source: EI 1995 – 1998; FSO, 1998; MAP 1995 – 2000; Matejic et al., 1996.

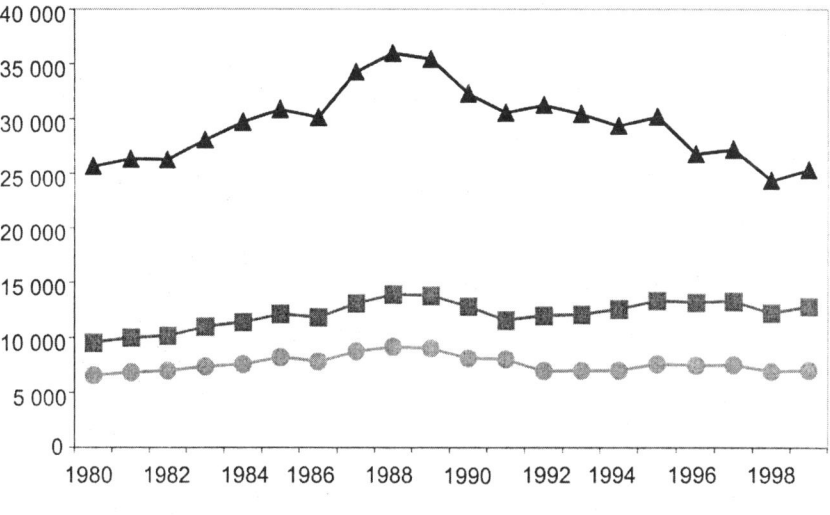

Fig. 16.2 FRY: R&D personnel
Note: FTE – author's calculations (1 FTE researcher in universities, involved in research and teaching = 3 employees). Source: FSO, 1980–1999.

2. The next indicator of structural changes in R&D is a *change in the allocation of human resources* in all three sectors under consideration (Figure 16.3). In 1980, of the total number of employees in R&D, 56.7 per cent were in RDIs (50.4 per cent of researchers), 5.6 per cent in RDUs (2.8 per cent of researchers) and 37.7 per cent in HE (46.8 per cent of researchers); in 1999, 35.6 per cent were in RDIs (24.6 per cent of researchers), 8.6 per cent were in RDUs (5.7 per cent of researchers) and 55.8 per cent were in HE (69.7 per cent of researchers).

More detailed analysis shows a considerable increase in the absolute number of researchers in HE (21 per cent more in 1999 than in 1988, i.e. twice the 1980 figure). At the same time, the absolute number of researchers in RDIs decreased and in 1999 stood at only 51 per cent of the 1988 figure. Fluctuations in the number of researchers in RDUs are not a reliable indicator (because of the small total number working in this sector). It is, however, significant that the number of researchers in this sector grew constantly until 1996, when it stood at 958 (compared to 269 in 1980). The number of researchers in industry (RDUs) decreased to 730 between 1996 and 1999.

The change in the structure of the workforce resulted in a considerably higher concentration of R&D resources in the HE sector than the average in OECD countries and, conversely, a very low concentration of R&D resources

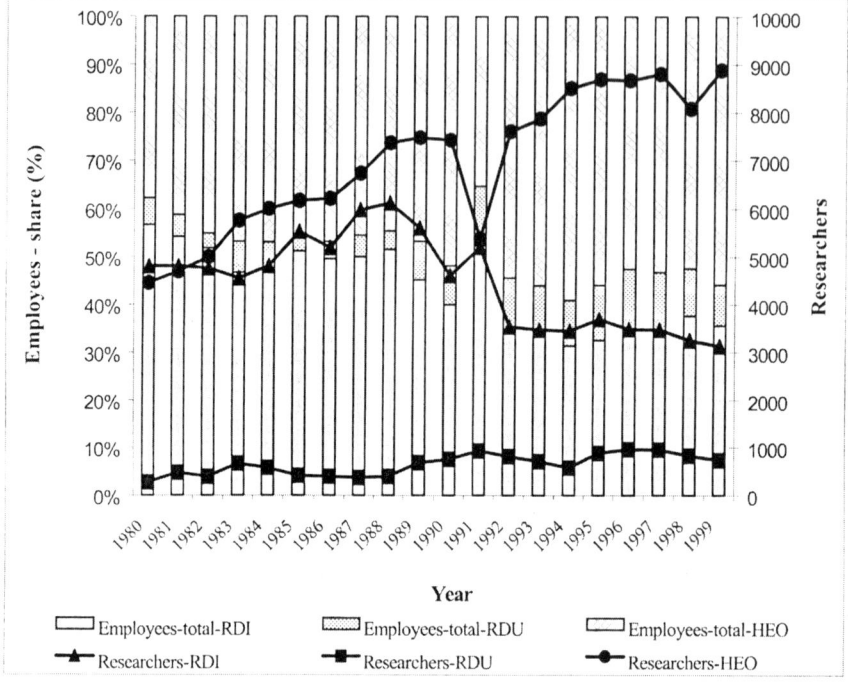

Fig. 16.3 FRY: allocation of R&D personnel
Source: FSO, 1980–1999.

in the industrial sector (20 – 25 per cent of the average in OECD countries). At the end of the period in question, the share of the RDI sector in the national R&D system, which corresponds to the governmental sector in the OECD classification, was similar to the average in OECD countries.

Between 1992 and 1995, many researchers moved from independent institutes and R&D units in industry into the universities, for the following reasons.

1. The transition to a market economy led to R&D being concentrated mainly in universities.
2. R&D staff were seeking greater job security: university staff are moderately paid, though without major delays, unlike in other sectors during that period. However, the universities are under government control and there are limits on the numbers of researchers that can be accommodated, especially due to the decline in the number of students financed by the Ministry of Education. Additionally, the universities would need more equipment for R&D if the number of research and non-teaching staff were to be significantly increased.
3. There was a *change in the functional structure of the workforce* in RDIs, RDUs and HE between 1988 and 1999 (Figure 16.4). In all three sectors,

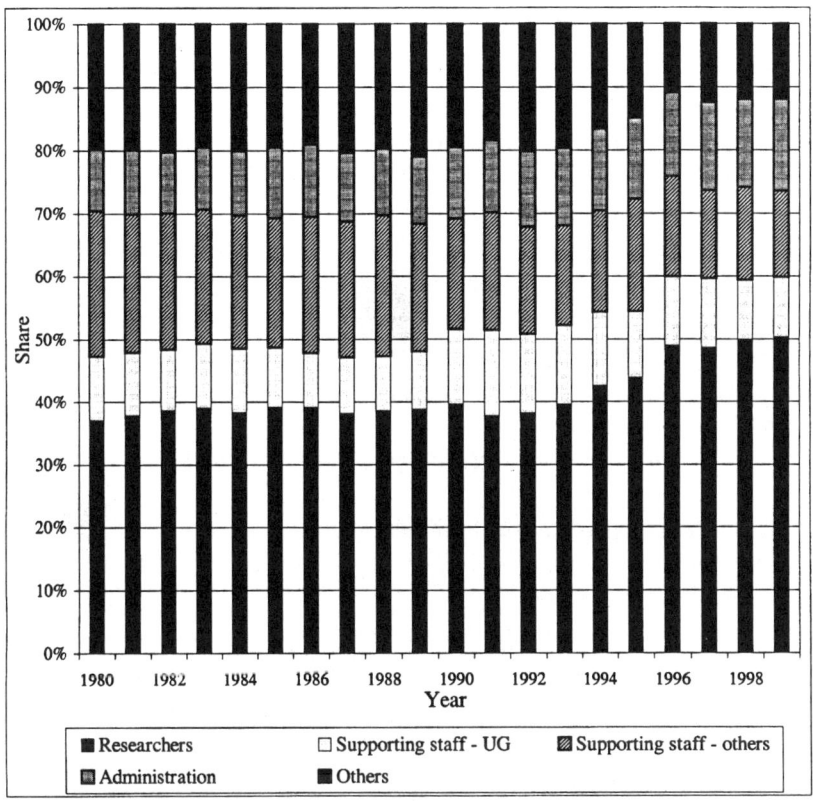

Fig. 16.4 FRY: structure of the R&D workforce by function
Note: UG = University graduates (supporting staff) Source: FSO, 1980–1999.

there was an increase in the share of researchers and graduate professional associates and a decrease in the share of non-graduate professional associates and other staff within the total workforce. However, the share of administrative workers in all three sectors remained more or less the same during the whole period. Since 1997, the structure has remained constant.

3.3. SUMMARY OF CHANGES IN S&T RESOURCES

Analysis of the evolution of the funding and resource structure of Yugoslavia's RDOs indicates that structural changes to the R&D system started in 1989–1990 as a result of:
– individual researchers transferring on their own initiative to HE;
– a slight increase in R&D work in industry;
– a greater share of non-research activities in the overall activities of RDOs.

However, all these changes resulted from the country's unstable political and economic situation. The transition to the market economy did not get under way until the change in the political situation at the end of 1999. This heralded a period of renewed structural change and more pain, as a still exhausted R&D system undergoes further restructuring, particularly in the Republic of Serbia.

The updated and longer-term analysis presented in this report provides us an opportunity to present some additional findings:

Structural changes to the R&D systems in both republics started in the late 1980s (1989–1990). Five indicators of these changes can be identified:

Strong movement of researchers from RDIs and RDUs to HE

This change might be positive (in accordance with the Western model of R&D), but the HE sector in Yugoslavia is not equipped to act as the lead player in R&D.

An increasing share of researchers in the total number of employees in RDO

This is to be regarded as a positive change.

A decrease in the share of researchers in the technical, natural, agricultural and medical sciences

This contrasts with an increase in researchers in the social sciences and humanities – a negative trend from the point of view of technological development.

A decrease in the share of R&D in total RDO activity

This could be a sign of structural change at the level of the individual organization (phase 2 in Meske's three-phase model – see Meske, 2000), which means that a number of RDOs will become manufacturing or service organizations.

A slight increase in the number of researchers in the industrial sector

This is unlikely to be sufficient for RDUs to make a more significant contribution to technological development. The heavy reduction of funding from industry for R&D activities (BERD/GDP $\leq$ 0.3 per cent) points to the incompetence of this sector in organizing R&D work. Despite all these negative trends, innovative activities in industrial firms (and especially in those with their own RDU) are still important for the country, since they help to prevent brain drain and to maintain firms' technological capabilities (cf. the findings of the second Innovation Survey, STPRC, 1997; see also paragraph 4.2 in this chapter).

The S&T system in the Republic of Montenegro shows how difficult it is to deal with the case of a small country: because of the small number of individuals involved, just a few people can be responsible for more or less major indices of change. Indeed, in small S&T systems, such change may indicate nothing less than the establishment or the disappearance of an entire scientific discipline. There are similar cases in the Republic of Serbia as well. The international support offered to the country as a display of goodwill in the wake of the Republic of Montenegro's confrontation with Serbian and federal authorities, particularly from 1998 to 2000, benefited the R&D system in Montenegro by providing financial support, access to the international S&T community, better living standards for researchers, etc. The university in Podgorica became better-equipped, a number of researchers and some teaching staff from Serbia moved to Montenegro and a new medical faculty was established.

4. INDUSTRIAL R&D AND INNOVATION ACTIVITIES

Two innovation surveys were carried out by the Science and Technology Policy Research Center of the Mihajlo Pupin Institute (STPRC). They analyzed innovation activities in industrial firms during two periods, 1987–1991 and 1992–1995.

4.1. THE FIRST INNOVATION SURVEY

This survey (STPRC, 1992) covered the 1987–1991 period in metal processing and the chemical and textile industries in the Republics of Serbia and Montenegro. The sample comprised 25 per cent of employees in those industries. The analysis is based on OECD methodological documents (OECD, 1992).

I would like to emphasize several topics which could lead to a general conclusion about the innovative capabilities of the industries analyzed on the basis of this innovation survey.

1. Innovation activities were mostly financed and realized by the firms themselves (innovation projects may be carried out or canceled if not successful) and focused on the development of new products or processes likely to improve the firm's market position. Most of the enterprises carried out in-house R&D and/or financed external R&D (Table 16.7).
2. In looking for ideas, firms drew on the sources typically used in such circumstances, including clients, customers, conferences, fairs, exhibitions, universities and so on (Table 16.8).
3. Inter-sectoral analysis produces a picture similar to that in the developed countries. The metal-processing industry developed products and processes not only for itself, but also for a number of other industries and sectors; there was a similar pattern in the chemical industry, while the textile industry differed by being 'closed', with firms performing R&D on their own and with their own resources.
4. The effects of innovation activities were visible in all areas analyzed (profits, production costs, marketing expenses, salaries, business risk, etc.). Due to the undeveloped market economy and import restrictions, domestic firms

were not sufficiently interested in marketing matters and bore little business risk.

Table 16.7 FRY: R&D in industrial firms (1987–1991)

	SR of Serbia and SR of Montenegro			
R&D activities:	Total	Metal	Chemical	Textile
intramural and external	n = 40	n = 25	n = 12	n = 3
Basic research	2.5	2.9	1.7	1.7
Development of new products	38.6	37.6	44.6	23.3
Improvement of existing products	27.2	24.1	34.2	25.0
Development of new process technologies	10.3	10.7	7.9	16.7
Improvement of existing process technologies	12.4	12.4	8.3	28.3
Development of new technological services	3.8	5.2	1.1	3.3
Improvement of existing technological services	4.2	5.4	2.2	1.7
Others	1.0	1.7	0	0
Total (%)	100	100	100	100

Source: STPRC, 1992.

4.2. THE SECOND INNOVATION SURVEY

The second innovation survey (STPRC, 1997) covered the same branches, but only firms from the Republic of Serbia. Also, the number of firms surveyed was different, although the sample in both surveys is virtually the same when measured in terms of the number and share of occupational groups in the FR of Yugoslavia (see Table 16.9). More than two thirds of the firms surveyed in the second survey were included in the first survey. The analysis is based on EC and OECD methodological documents (EC, 1993; EC, 1994; OECD, 1992).

The main innovation activities in the firms analyzed were product innovation with and without changes in process technologies. Process innovation was limited in scope because of serious obstacles to the acquisition of new technologies abroad (see Table 16.10). In the metal-processing and chemical industries, internal R&D and pilot production were the main items of innovation expenditure. In the textile industry, experimental development was the most significant item. Low expenditure on external R&D and market analysis reflected these firms' self-sufficiency and a lack of market competition in the country due to its international isolation. The main R&D activity in the firms

Table 16.8 FRY: sources of ideas/information for innovation activities (1987–1991)

Sources of important ideas/information	Number of firms			
	SR of Serbia and SR of Montenegro			
	Total	Metal	Chemical	Textile
	n = 46	n = 29	n = 13	n = 4
Internal:				
Administration	8	5	3	0
R&D department	26	21	5	0
Marketing	26	17	7	2
Production	15	9	5	1
Other internal sources	1	1	0	0
External:				
Suppliers	5	3	1	1
Clients or customers	33	22	9	2
Other firms	11	9	2	0
Subsidiaries	8	8	0	0
Competitors	19	11	7	1
Professional conferences	27	17	8	2
Fairs/exhibitions	35	24	9	2
Patent office/documents	11	7	3	1
Technical journals	36	22	11	3
Software firms	5	4	1	0
Consulting firms	4	4	0	0
R&D institutes	21	16	5	0
Universities	20	17	2	1
Regulative/standards (documents like ISO, DIN..)	3	3	0	0
Agencies for technology transfer	5	5	0	0
Other external sources	10	8	2	0

Source: STPRC, 1992

analyzed was the development of new products, followed by the improvement of existing products. No data on research activities in the textile industry were obtained.

The main source for the acquisition of technologies necessary for innovation activities in industrial firms was contracts for technology-based services. The purchase of capital equipment was the second most important source, mainly because of the country's international isolation (Table 16.11).

Table 16.9 FRY: sample of firms – Innovation Survey II (innovation activities in industrial firms: 1992–1996)

Industry	Number of firms	Number of employees	Share of total industry
Metal processing industry	18	67 663	25.4
Chemical industry	10	12 777	24.9
Textile industry	5	20 237	20.4
Total	33	100 677	24.2

Source: STPRC, 1997

Table 16.10 FRY: expenditure on innovation activities (1992–1996; %)

Innovation activities	Total n = 25	Metal n = 16	Chemical n = 8	Textile n = 1
R&D				
Internal	34.26	23.53	57.68	11.00
External	6.64	5.94	8.88	9.00
Others				
Technology transfer	1.10	0.45	2.56	–
Experimental develop.	18.34	20.93	9.06	50.00
Pilot production	30.45	38.41	15.19	25.00
Market analysis	7.27	8.54	5.00	5.00
Training	1.73	1.88	1.63	–
Other	0.21	0.32	–	–
Total	100.00	100.00	100.00	100.0

Source: STPRC, 1997

Compared with the first survey, there were much closer relationships in all the industries in the sample with domestic R&D institutes and universities (even in the textile industry, which had been more self-sufficient in the previous period) and sporadic liaisons with foreign companies. However, there were no formal contracts with foreign R&D institutions.

Internal factors were of major importance for the firms in the sample. Assessment of the importance of the various internal and external factors influencing firms' innovation activities revealed that the pattern varied from industry to industry. A firm's financial situation and its management's vision of future development with skilled personnel were crucial in all three industries. R&D capabilities were another important internal factor in the metal-processing industry. In the chemical and metal-processing industries, the im-

Table 16.11 FRY: innovation activities and acquisition of technology (1992–1996)

Channels of technology acquisition for innovation activities	Number of contracts			
	Republic of Serbia			
	Total	Metal	Chemical	Textile
	n = 25	n = 19	n = 4	n = 2
R&D contract	5	2	1	2
R&D cooperation	6	1	1	4
License for:				
– Patents	2	–	2	–
– Models	–	–	–	–
– Design	–	–	–	–
– Others	1	–	1	–
Information system with new technologies	2	1	–	1
Capital equipment	8	1	3	4
Technology-based services	12	5	2	5
New process technologies	1	–	1	–
Parts and materials with new technologies	1	1	–	–

Source: STPRC, 1997

portance of market information was stressed. In none of the three industries analyzed were external factors a great concern, although some importance was attached to the external financing of R&D and pilot projects in the metal-processing and textile industries and to support for capital equipment in the chemical industry.

Products and processes not subject to innovation made a bigger contribution to firms' sales and exports in this period compared to the previous one. New products and processes contributed more to firms' sales and exports than products and processes with incremental improvements. That means that technological change was limited and concentrated in a few new products. The textile industry is an exception, with new products making a 60 and 80 per cent contribution to total sales and exports respectively, as a result of a highly specialized production system geared to manufacturing a variety of products with technologies acquired at the end of the previous period.

The main conclusions on innovation in the metal-processing, chemical and textile industries between 1992 and 1996 to be drawn from the second survey are as follow:

– the main objective of innovation activities was import substitution;

- because of the country's isolation and a drastic decline in output, internal innovation activities were in effect the main (and in some cases the only) activities going on in the firms investigated;
- the government's technology development program was the main source of financial, moral and psychological support for innovation in industry;
- the firms in the sample drew mainly on domestic sources of ideas and information for their innovation activities (clients, customers, conferences, fairs, exhibitions, universities, etc.) because communication with the outside world was severely curtailed;
- innovation activities were a factor in preserving the firms' technological capabilities and in staunching the brain drain.

4.3. CHANGES IN INNOVATION ACTIVITIES IN THE PERIODS UNDER INVESTIGATION

The main differences in the patterns of innovation activities in the two periods in question are as follows:

1. In the first survey, innovation activities were identified as important activities in all the firms that were surveyed. In the second survey, internal innovation activities were the main (in some firms almost the only) innovation activities that the firms engaged in.
2. In the first survey period, innovation activities might or might not lead to the development of new technologies, since technologies, whether domestic or foreign, were accessible through the market. In the second survey period, innovation activities were the only source of the technologies firms required.
3. In the first period, R&D departments were not very important parts of their firms' organizational structures. During the second period, however, these departments became a core element in most firms, maintaining human resources and preventing brain drain.
4. Communication with the outside world became a regular activity in the first period. In the second period, communication was limited to domestic institutions and organizations. A lack of information and of foreign experience, along with limited access to new technologies, led the industries analyzed into technological regression.

The second period under analysis (1992–1996) saw the beginning of decline (because of the war and dissolution of the SFRY) and some degree of disruption (suspension, but not lifting of the sanctions), all of which merely foreshadowed the country's subsequent troubles. This period was very difficult and it was a serious test of the national innovation system's capabilities, but industry passed it successfully by supporting R&D and other innovation activities. Support from public funds, mainly from the MST, was crucial, both financially and organizationally.

The end of this period was one of a recovery for the Yugoslav economy. The firms analyzed were also being privatized and reintegrated into the world economy. The innovative capacities built up over the past 10 years made it seem possible that these processes could be completed successfully.

Soon after, in 1998, the second major decline set in; it worsened with the bombing campaign at the beginning of 1999 and continued for the next 18 months. Industry had no time to recover or to utilize the technological and human resources it had preserved. Reconstruction of the country employed some of the capacities, but real investment in technological development and also FDI were lacking. Even worse was the absence of public support (with industry being left behind universities and independent institutes in the struggle for the ever declining amount of money available for R&D in the republic's budgets) and the lack of understanding of the importance of this part of the national innovation system for the country's future development. In this period, industry suffered some brain drain, though not as much as the other sectors of the R&D system.

Thus only part of the innovation capabilities developed earlier remain intact. The question is: who is counting on them? Is it possible strategic partners, government authorities, company managers, employees or engineers? This is probably the fundamental question that the country is now facing as it effects the transition to a market economy, because it will be impossible in the near future to compensate for human resources lost as a result of inappropriate transitional reforms, restructuring and other changes.

5. SUMMARY AND OUTLOOK

The existing R&D programs, developed in 1999–2000, a period of international isolation for the country, are now being reformulated and reoriented away from the substitution and independence in production and services that led to a closed economy. These programs have to be redefined in line with global tendencies in technological development so that the Yugoslav economy can be re-integrated into the international environment.

There was a setback in 1999/2000. *The funding of R&D activities was irregular*, with major delays and without the increases necessary to cover inflation. In the absence of official data, some estimates indicate a decrease in GERD to 0.3 per cent of GDP. This will cause considerable disruption to the R&D system. *R&D activities in industry* prospered in 1999, with expenditure on R&D accounting for 1.1 per cent of industrial output. This was due mostly to industrial firms' well-conceived policy of preventing brain drain from their

R&D laboratories, despite a sharp decline in DPI (30 per cent less than in 1998).

However, industrial activities were halted during the war. A number of large firms of great importance to the country were destroyed. It is very difficult to make any forecast about their recovery. It is realistic to expect scientists to begin emigrating again because of a serious decline in GDP and economic recession.

In 1999–2000, immediately prior to the democratic changes, the R&D system in the Republic of Serbia was in the so-called 'waiting' phase, characterized by strong involvement on the part of leading political parties in all aspects of the R&D system.

The monitoring, evaluation and planning of R&D was the province of a small group of interested people, selected on political rather than scientific criteria. The main results of this political interference were a 'Strategy for S&T Development in the Republic of Serbia until 2010' and a new 'Law on R&D Activities in the Republic of Serbia'. These two documents have never been approved by the Republic Assembly, mostly because of disagreement within this special interest group, which includes scientists belonging to the same leading political parties but working in different fields of R&D who expect their own field to be accorded the greatest priority. The never-ending consultations and negotiations dragged on for almost 15 months. During this period the Ministry of S&T practically froze all R&D activity in the country.

The main priorities of the new democratic leadership since the political changes of September-October 2000 have been focused on stabilizing the financial system, integrating the FRY into Europe and the rest of the world and solving problems with the supply of energy, food and drugs, poverty, etc.

The activities and R&D programs defined by the new government in 2001 (cf. paragraph 1.2) meet the needs of S&T in the FRY. They are similar to the 'urgent actions for the preservation and transformation of the S&T system in Serbia' proposed by a group of former and present members (including the author of this chapter) of the Science and Technology Policy Research Center of the Mihajlo Pupin Institute in Belgrade in 2001. These proposals prioritized four main areas of activity:

1. *Preservation and development of human resources in the R&D system in Serbia;*
2. *Technological and industrial development policies for the transition period;*
3. *Developing an S&T policy for the Republic of Serbia up to the year 2010;*
4. Establishing *programs and schemes for training and education in public administration* (staff in ministries, government agencies, etc., including ministers and other leading figures in public administration) with the aim of

supporting the emergence, implementation and adaptation of policies intended to stabilize the country and to facilitate the process of transition and strategic development.

The Serbian R&D system has entered the transition period ten years later than the other CEECs. Their wide-ranging experience of the transition of S&T systems is now available to the FR Yugoslavia. There is every reason to suppose, therefore, that good practice will be adopted and that mishaps and faulty decision-making can be avoided.

REFERENCES:

EC (1993) *EC Harmonized Innovation Surveys 1992–1993 – Final Questionnaire.* Brussels: European Commission (EC), DG XIII.
EC (1994) *The Community innovation survey – status and perspectives.* Brussels: EC, DG XIII.
EI (1995–1998) *Konjukturni barometar* (Economic trends). Belgrade: Economic Institute (EI).
FSO (1998) *Statistical Yearbook of Yugoslavia.* Belgrade: Federal Statistical Office (FSO).
FSO (1980–1999) *Institutions of Scientific-Technological Development – Statistical bulletin.* Belgrade: FSO.
Kutlaca, D. (1998) "Patent-Related Activities in Serbia from 1921 to 1995." *Scientometrics* 42, no. 2: 171–193.
MAP (1995–2000) *MAP – Monthly Analysis and Prognoses.* Belgrade: EI.
Matejic, V., D. Kutlaca, V. Grecic, and O. Mikic (1996) *Brain-drain from Yugoslavia.* Research project granted by the Federal Ministry for Development, Science and Environment. Belgrade: Science and Technology Policy Research Center of the Mihajlo Pupin Institute (STPRC), Institute of International Politics and Economics.
Meske, W. (2000) "Changes in the innovation system in economies in transition: basic patterns, sectoral and national particularities." *Science and Public Policy* 27, no. 4: 253–264.
OECD (1992) *OECD Proposed Guidelines for Collecting and Interpreting Technological Innovation Data – Oslo Manual.* Paris: OECD/Eurostat.
OECD (1994) *Proposed standard practice for surveys of research and experimental development – Frascati Manual 1993.* Paris: OECD.
STPRC (1992) *Science and technology in Europe and their implications on technological, economic and social development in Yugoslavia.* First phase – Project granted by the Federal R&D Fund and the STPRC 1988–1992. Belgrade: STPRC.
STPRC (1997) *Serbia's Innovation System Research.* A strategic R&D project sponsored by the Ministry of Science and Technology of the Republic of Serbia and the STPRC 1994–1997. Belgrade: STPRC.

UNDP (2001) *Suspended Transition: Vulnerability Trends and Perceptions. Background report.* Eds. United Nations Development Program (UNDP), Federal Republic of Yugoslavia/Republic of Serbia, European Movement in Serbia. Belgrade: UNDP.

17. SLOVENIA: TRANSFORMATION OF THE S&T SYSTEM

Peter STANOVNIK

Like other Central and Eastern European countries (CEECs), Slovenia has traditionally had a relatively highly developed R&D system and relatively high levels of educational capital. R&D and knowledge capital is defined here as an aggregation of scientists, engineers, other R&D personnel, laboratories and associated equipment and related expenditures (Mansfield, 1982). Educational capital denotes the average number of years of schooling in the adult population of a given country (Bevc, 1995; Bevc and Stanovnik, 1996). It can be further argued on the basis of research studies that R&D and production capabilities constituted the inherited comparative advantages of CEECs in transition (Radosevic, 1995). After the introduction of radical institutional and economic reforms in the late 1980s, it became obvious that they should also include the reorganization and restructuring of the science and technology system (STS) so that the acquired R&D capabilities could be put to use in order to foster innovation, improve market outcomes and increase the international competitiveness of CEEC economies.

Slovenia has undergone radical political and socio-economic changes since the introduction of market reforms in the late 1980s. Compared to other CEECs, Slovenia inherited some of the advantages of the former self-management system, including decentralized research institutions independent of the Academy of Sciences and government bodies, public institutes open to cooperation with the business enterprise sector, absence of rigid branch research institutes and contacts with neighboring Western research and educational institutions.

After Slovenia gained its independence in 1991, the transformation process got under way. At the same time, the country had to adjust to its new economic status: it was now a national economy rather than part of a regional economy. The reforms that accompanied the transition included privatization, the elimination of monopolies and the introduction of new institutional and legal frameworks impacting on all spheres (business sector, public administration and, to some extent, education and the STS as well).

This chapter takes as its starting point the thesis that in any new market economy emerging in a post-communist country, a restructured STS is a prerequisite for sustainable economic growth, further structural adjustment, increased international competitiveness and successful integration into the EU (Meske, 1997). It begins with an analysis of the organizational structure of the research and technology development (RTD) system, which did not undergo radical change or active restructuring during the transition process in the period 1993 – 2000. Secondly, the changes in S&T policy during the transition are analyzed. This period also saw the increased inclusion of Slovenia's research and business organizations into various European RTD programs, particularly the EU 5th Framework Program, which led to international networking and improved research quality. Thirdly, the process of building up the technological capabilities of manufacturing and service industries is discussed, with particular emphasis on the development of small and medium-sized companies. By way of conclusion, some views are expressed on human capital formation in R&D and the mobility of researchers in Slovenia. Proposals are also put forward for new innovation and technology policy measures intended to create the required linkages between the public RTD sector, the private business sector and intermediate institutions with a view to building a modern national innovation system.

1. THE ORGANIZATION AND PERFORMANCE OF THE RTD SYSTEM BETWEEN 1993 AND 2000

Unlike other, comparable CEECs (the Czech Republic, Hungary, Poland, Slovakia, the Baltic states), gross expenditure on R&D (GERD) in Slovenia did not fall drastically in the transition period from 1993 onwards. In 1993, GERD stood at 1.6 per cent of GDP. In 1994, it increased to 1.8 per cent and between 1996 and the year 2000 remained stable at 1.4 to 1.5 per cent of GDP. This is a level comparable with some of the less developed OECD countries but still somewhat below the OECD average, which was 2.2 per cent in 1998.

The sources of funding for R&D in Slovenia in 1999 were: business 56.9 per cent, government 36.8 per cent, HE funds 0.6 per cent, funds from abroad 5.6 per cent (SURS, 2002). Approximately 55 per cent of all R&D is performed by public institutions (universities, research institutes), which carry out mainly basic and, to a much lesser degree, applied research, but practically no industrial experimental development. This research is oriented mainly towards investment in human capital and in basic knowledge, not towards innovation and market acceptability.

A survey of the structure of R&D expenditure between 1993 and 1999 (SURS, 2002) shows a decline in the funds allocated to R&D by the Slovenian government. The government's share of total R&D funding dropped from 48.3 per cent in 1993 to 37.1 per cent in 1997 and to 36.8 per cent in 1999. On the other hand, there was a notable increase in the share of funding coming from the business enterprise sector, which rose from 38 per cent in 1993 to 52.6 per cent in 1997 and 56.9 per cent in 1999. The main disadvantage of the R&D system is the very low level of cooperation between public R&D organizations and the private business enterprise sector.

The number of researchers decreased slightly, but is still relatively high. In 1995, approximately 12 500 people were working in R&D, of whom 6 094 were full or part-time researchers (4 897 in FTE).[1]

In the four years between 1995 and 1999, the number of full-time or part-time researchers increased to 6 721. The small decrease in total personnel can be attributed to the dismissal of administration and technical personnel. The number of researchers per 10 000 of the labor force (51) is slightly under the average OECD figure (56) (OECD, 1999).

There are five types of *research organizations* in Slovenia: the Academy of Science and Arts (AoS), universities, the so-called independent government institutes, research units in the business enterprise sector and those in the private non-profit sector. Branch research institutes never existed in Slovenia. The AoS research center consists of 14 institutes with 139 researchers working primarily on basic research projects in the humanities and natural sciences. The two universities (Ljubljana and Maribor) had 36 research organizations operating under their auspices in 1999. University research was financed by the Ministry for Science and Technology until the year 2000; since then, the funding body has been the newly established Ministry of Education, Science and Sport (MESS). Most of the research carried out at the universities is of a basic nature.

Surprisingly, the number of state-owned research institutes increased by 30 per cent between 1993 and 1999. In 1999, there were 60 public institutes with 3 230 R&D personnel (on average 32 researchers/institute) (SURS, 2000a; 2002). These institutes have no direct link to the universities, but some can be categorized as national institutes, receiving direct institutional (program) and indirect (project) funding from the government. The legal provisions in this field are changing, as is the legal status of independent research institutes. The 1992 law on institutions in research, education and some other social services had given ownership of the institutes to the original founders.

At the beginning of the transition there was a lack of intermediary institutions and mechanisms linking domestic basic and applied research with the

---

[1] Since 1993 the Slovenian statistical office has used Frascati methodology to measure R&D activities. Because of some doubts about the accuracy of the data, a special survey was conducted to check the R&D data for the business enterprise sector. The survey confirmed relatively high levels of in-house R&D expenditure in manufacturing industry (Stanovnik et al., 1998).

mainstream of social and economic development and with international R&D networks. Some of the accumulated knowledge in R&D and production engineering disappeared in the process of opening the economy and restructuring the oversized, underutilized and technologically unspecialized manufacturing enterprises, especially following the break-up of several vertically integrated companies and previously strong R&D units (for example Iskra holding, Litostroj, Metalna, Slovenijales, Smelt etc.). The growth of service industries did not compensate for this outflow of researchers from the manufacturing sector. The majority of larger Slovenian manufacturing firms did not succeed in diminishing the relative gap in R&D input (and technological level) between themselves and their counterparts in the developed market economies. The lag in high and intermediate-tech manufacturing reflected in relative R&D spending (2–4 times lower) is also expressed in the low share of exports in these sectors. High levels of in-house R&D activity are found only in the electrical engineering, transport equipment, pharmaceuticals and rubber industries. In other industries, R&D intensity has declined due to an 'internal brain drain' from large enterprises. A lot of highly skilled engineers and scientists left R&D units in public enterprises and founded new private firms. The process of building new in-house R&D capacities in the private business sector is still ongoing and should be more actively supported by government policies. The transition to the use of intangible production factors will demand skills and know-how hitherto unknown to Slovenian enterprises and research institutions. It is evident that a substantial part of this know-how should be obtained from foreign business and academic partners.

The trends in 1999–2000 show that in-house R&D and other innovation activities have recovered somewhat in the business enterprise sector, but not sufficiently to compensate for the internal brain drain of about 3 000 researchers in the early 1990s. The highest growth rates for researchers were achieved in the higher education sector (see Table 17.1).

Many R&D units in industrial firms which, in 1995, accounted for 67.6 per cent of all research organizations but only 25.5 per cent of total employment in the Slovenian R&D sector, have been absorbed into the everyday operation of running and maintaining technological equipment and, with the exception of a few medium-sized companies, have suffered serious financial and personnel cuts. The situation had changed by 1999, when 273 R&D business units employed 26.4 per cent of all researchers. A private non-profit sector has emerged in the last few years but plays no significant role (accounting for only 0.8 per cent of R&D personnel in 1999) (SURS, 2000a; 2002).

In 1995, the structure of *R&D personnel* by scientific discipline was as follows: engineering sciences 46 per cent, medical sciences 17 per cent, natural sciences 13 per cent, social sciences 10 per cent, bio-technical sciences 7 per

Table 17.1 Slovenia: R&D personnel

	R&D Personnel (full- and part-time)							
	1993		1995		1999		2000	
	persons	%	persons	%	persons	%	persons	%
Total	10 323	100	12 416	100	12 286	100	12 220	100
a) by function:								
– Researchers	4 864	47	6 094	49	6 721	55	6 562	54
– Expert Personnel	1 541	15	1 799	14	1 598	13	1 708	13
– Technicians	2 337	23	2 744	22	2 547	21	2 512	20
– Managerial Personnel	227	2	295	2	378	3	354	3
– Other Personnel	1 354	13	1 484	12	1 042	8	1 084	9
b) by sector								
+ Total R&D personnel								
– Business Enterprise Sector	3 370	32.6	4 484	36.1	4 939	40.2	4 824	39.5
– Government Institutes	3 067	29.7	3 045	24.5	3 230	26.3	3 132	25.6
– Private Non-Profit Sector	46	0.5	140	1.1	91	0.7	122	1.0
– Higher Education Sector	3 840	37.2	4 746	38.2	4 026	32.8	4 142	33.9
+ Researchers								
– Business Enterprise Sector	1 145	23.5	1 506	24.7	1 772	26.4	1 585	24.2
– Government Institutes	1 735	35.7	1 760	28.9	1 963	29.2	1 919	29.2
– Private Non-Profit Sector	34	0.7	128	2.1	68	1.0	104	1.6
– Higher Education Sector	1 950	40.1	2 651	43.5	2 918	43.4	2 954	45.0

Source: SURS, 2002.

cent and humanities 6 per cent. In the next 4 years, the structure shifted in favor of engineering sciences and natural sciences. The trends in the quality of R&D personnel were positive. The number of personnel with PhDs and MAs increased by almost 50 per cent in the last decade, with their share in total personnel in the R&D sector reaching 36.8 per cent (see Table 17.2). The only exception is the business enterprise sector, where only 10.6 per cent of those employed in business research units held the highest degrees. The age structure improved as a result of two simultaneous processes: the retirement of older researchers and the effects of a national program called 'Young Researchers'. The average age of researchers, at between 38 and 40, is now lower than it was in the 1980s, when it was between 44 and 46.

2. THE EVOLUTION OF SCIENCE AND TECHNOLOGY POLICY DURING THE TRANSITION

The science and technology system in Slovenia did not undergo radical changes during the transition period. The existing research and educational institutions and coordination mechanisms (predominance of basic research, technology-push system supported by government funding, fostering of investment in research equipment) remained in place. The Ministry for Science and Technology (MST) was responsible for science and technology policy. For the entire period between 1993 and 1999, the largest share of the budget was allocated to research (i.e. the national research program), followed by investments in infrastructure, the young researchers program and technology support schemes.

The highest share of public funding was allocated to the *national research program* run by the former MST. In 1995, about five billion Slovenian tolars (SIT) were spent on 613 basic and 255 applied projects; the figure for 1998 was about 7 billion SIT out of a total MST budget of 21.5 billion SIT and for 1999 it was 8 billion SIT out of the total MST budget of 22.6 billion SIT. Basic research projects were financed exclusively by the MST, whereas the applied and 'targeted' projects were co-financed, with 25–50 per cent of the funds coming from other ministries or business enterprises.

A new financial system was put in place in 1999, which introduced program financing for basic and applied research (and to a great extent for other research projects as well). This system gave public R&D institutions greater financial stability. On the other hand, however, it made those institutions less flexible and reduced the competition between them.

An expert system has been put in place for the selection of R&D programs, with peer reviews being used for ex-ante and ex-post evaluation. The criteria applied are overwhelmingly scientific in nature (SCI index, other sci-

Table 17.2 Slovenia: human capital in R&D sectors (persons in paid employment by educational attainment; in 1999)

	TOTAL	Structure by professional attainment (in %, total = 100)					
		Postgraduate level - level 7 by ISCED*		HE–university degree	HE–non-university degree	Secondary education	Less than secondary education
		Doctors (PhD)	MA, MSc**	level 6	level 5		
		%	%	%	%	%	%
Slovenia – total	12 286	21.4	15.4	29.0	9.1	20.8	4.2
– Researchers	6 721	37.0	24.7	35.3	2.2	0.8	
– Expert personnel	1 598	4.4	8.0	50.3	22.0	14.9	0.4
– Technicians	2 547		0.3	5.6	18.2	72.0	3.9
– Managerial personnel	378	17.7	20.9	51.3	7.4	0.3	
– Other personnel	1 042		1.3	4.9	12.3	41.1	40.4
Business enterprise sector							
– total	4 939	2.9	7.7	38.4	13.7	32.2	5.1
– researchers	1 772	6.3	16.1	67.4	7.8	2.4	
Government institutes							
– total	3 230	25.4	21.4	24.3	6.6	17.2	5.1
– researchers	1 963	38.9	32.2	28.6	0.3		
Private non profit sector							
– total	91	11.0	29.7	47.3	3.3	6.6	2.1
– researchers	68	13.2	35.3	44.1	2.9	2.9	1.6
Higher education sector							
– total	4 026	41.2	19.7	20.8	5.6	10.1	2.6
– researchers	2 918	55.0	24.5	20.1	0.1	0.2	

* International Standard Classification of Education.
** Including the specialization after attainment of graduate level.

Source: SURS, 2000a; 2002.

entific excellence), whereas the criteria relevant to the country's economic and technological development have been to a large extent neglected.

The research and information *infrastructure* (partially developed and co-financed by the Ministry for Education, Science and Sport (MESS), the Ministry for Information Society, and the Ministry for Economic Affairs) also includes the development of a physical framework for the integration of Slovenian R&D institutions into the EuropaNet, Ebonet and Internet information systems. In the past decade, the purchase of hardware for numerous research organizations was co-financed by the appropriate ministries. Between 1996 and 1999, the Slovenian library information system, called COBISS, was developed.

The prevailing conceptual approach favored academic research, although since 1991 the Slovenian government has introduced several *schemes for promoting technological development*. These schemes have been oriented toward supporting the experimental phase of R&D projects, with subsidies available for pre-competitive research and researchers' salaries (PhDs and MAs) employed in the business sector, for technological parks, incubators and information centers, plus preferential interest rates for investment in R&D. Despite the declared objective of increasing state support for S&T, the severe budget constraints adversely affected R&D expenditures. Moreover, between 1993 and 1999, the funds allocated by the state to R&D activities dropped from 48.3 per cent of national gross domestic expenditure on R&D in 1993 to 37.2 per cent in 1997 and 36.8 per cent in 1999. Consequently, the funds for technological support schemes were diminishing from year to year, so that in 1998 they represented only 8 per cent of all funds spent by the ministry in charge. In consequence, the so-called 'targeted research programs', which were introduced in 1994 and geared towards different applied topics (natural resources, ecology, education, land, strategy for international economic relations, tourism, national identity, etc.) and co-financed by different ministries, did not reach the planned level (ten per cent of public R&D expenditures) (MESS, 2001).

In 1999, a new "Act on state support to enterprises developing new technologies and support to industrial R&D departments 2000–2003" was adopted. It promised more incentives for innovative enterprises and for innovation in general. Due to the delay in selling state property, the legislation had not been implemented by the middle of 2001.

A special *program for young researchers* was introduced in the mid 1980s to provide funding for postgraduate and doctoral studies. This comprehensive program also offers financial support for public or private organizations to employ young researchers for a limited period until they finish their studies (two and a half years for MAs and an additional two years for the PhDs). In the past decade, the overwhelming majority of participants in this program took jobs in universities and research institutes, where employment and working conditions were more attractive than in the business sector. In 1998, about 900 students were participating in the program, which accounted for 18 per cent of the Ministry for Science and Technology's total budget. The increasing attractiveness of the private business sector will help to increase the employment of young researchers in industry (Stare and Bucar, 1997). In the year 2000, this scheme was open to 250 new applicants from all scientific disciplines.

Between 1993 and the year 2000, the government, with EU support, established several new institutions in order to foster human capital formation, the establishment of new technology-based companies and the development, application and diffusion of new products and technologies. Examples included the Technology Development Fund, which merged in 1997 with the Slovenian Development Agency, the Slovenian Science Foundation, two technology parks and numerous information centers, a network for the promotion of innovation in small and medium-sized enterprises, the National Regional Agency and regional agencies and the Innovation Relay Center. The primary role of the first venture capital fund was to increase investment in innovative enterprises, to stimulate the marketing of domestic innovations and to provide assistance in launching new products and services based on technological advances. The initial phase of the establishment of the Technology Development Fund was financed by PHARE, while later operations were financed through privatization schemes. The seed capital was provided by revenue from privatization. Innovation activity and new technology-based enterprises are concentrated in central Slovenia.

The government's declared S&T policy objectives over the mid term period 1997–2000 (MST, 1996) were as follows:
- to maintain R&D at an internationally competitive level;
- to increase the involvement of R&D institutions in the technological upgrading of the Slovenian economy;
- to foster a demand for innovation and promote the dissemination and transfer of knowledge;
- to enhance R&D capacities both qualitatively and quantitatively by linking academic research activities with the needs of the business sector.

In order to achieve these objectives, various programs and measures were drawn up. These included an annual national research program, special incentives for technological development, a young researchers program, investment in the information and research infrastructure, incentives for international activities in science and technology and tangible and intangible investments in the STS. These objectives were only partly fulfilled. For this reason, a new national research program was to be launched at the beginning of 2002.

3. THE GROWING PARTICIPATION IN EUROPEAN R&D PROGRAMS

Slovenia's participation in EU research programs has grown in the last decade.

The TEMPUS program, which is primarily intended for the development and renewal of higher education, has benefited the Slovenian education system through the introduction of several new curricula in both universities. Between 1991 and 1998, Slovenian higher education institutions participated in 73 TEMPUS projects (MESS, 2001).

Through the ACE (the Action for Cooperation in Economics) program, itself part of PHARE, the largest single assistance program to operate in CEECs, support has been forthcoming for research studies in economics and finance and grants have been made available for fellowships and scholarships and for those attending or organizing conferences. ACE has also enabled Slovenian researchers to put together numerous economic research networks with partners from the EU and from other CEECs. The main topics of investigation between 1996 and 2000 were European integration and preparation for accession to the EU, private sector development, the restructuring of agriculture, the reform of institutions and public administration, the reform of health and social services, educational reform, training and science, the development of the transport and telecommunications infrastructure and environmental protection.

From 1992 onwards, the PECO program fostered cooperation on technology and development among CEE institutions participating in European programs. It came to an end in the mid-1990s, thus enabling Slovenian and other CEE organizations to enter the 4th and 5th Framework Programs. Slovenian institutions participated in 58 PECO projects.

INCO-COPERNICUS, launched in 1994 with the aim of co-financing joint industrial research projects involving EU and CEE organizations, has proved to be one of the most successful programs for Slovenian partners.

A total of 107 Slovenian organizations took part in the 4th Framework Program. They were involved in 49 projects in 1995/96 and in 28 in 1997/98. Slovenian research institutes have been actively participating in the COST program, which is one of the oldest EU research programs. They are currently active partners in more than two thirds of on-going COST projects (MESS, 2001).

As the Slovenian economy had been lagging behind in technological development, and especially in high and intermediate-tech industries, it was appropriate that the country should become a full member of EUREKA in 1994. In the next 6 years, Slovenia was represented in a total of 30 projects, but with insufficient involvement on the part of manufacturing companies.

In the middle of 1999, research and business organizations applied to the call for proposals under the 5th Framework Program. In all, 432 Slovenian organizations put forward 331 project proposals, with 75 being accepted for funding. Preliminary analyses of projects completed in 1999 and 2000 showed that Slovenian organizations performed quite successfully (MESS, 2001).

Generally speaking, the increasing participation of Slovenian research and other organizations in these EU programs has produced good results: increased international networking and researcher mobility, increased quality of scientific output, international exchange of expertise and know-how etc. On the other hand, several deficiencies can be identified. R&D teams are small compared to those in EU countries and there are insufficient international joint-ventures in R&D. There has also been insufficient impact on technology transfer and technology diffusion, and R&D results have not been adequately commercialized.

In order to improve its innovation capabilities and technology transfer, Slovenia, with financial assistance from the EU, established its Innovation Relay Center – IRC Slovenia – by forming a consortium of 5 domestic partners in 2000. The main task of the IRC is to help local industry by conducting technological audits, assisting with negotiations between the providers and receivers of new technologies and supporting the implementation of technology transfer. The twinning partner of the Slovenian IRC is the Finnish Innovation Relay Center.

4. THE BUILDING OF TECHNOLOGICAL AND INNOVATION CAPABILITIES

The transformation of science and technology in Slovenia has been examined in various studies and assessed by several different methodologies (EC, 1999; Gliha, 1998; Gopa consultants, 1994; IER, 2000; SURS, 1997–2002). Various indicators have been used to measure technological development:
– the number of product and process innovations;
– the number of years needed to bring new products to market and to renew process technology;
– the structure of manufacturing industry by high, intermediate and low-tech branches;
– the level of IT implementation;
– patents granted to residents;
– the number of patents secured abroad by Slovenian residents;
– the number of patents in force;
– the availability of qualified engineers and IT skills;
– the level of cooperation between universities and business companies;

- the development and application of new technology;
- the relocation of R&D facilities;
- the adequacy of the education system in producing the required skills.

Despite the fact that some of the indicators show the country is performing relatively well in international terms (overall R&D expenditure, the relative number of researchers, R&D infrastructure, the level of scientific infrastructure), the rest of them indicate there is a substantial technological lag compared with EU member states. The result is 3 times lower value added per employee in Slovenian manufacturing and service industries compared to the EU average and an above-average share of traditional industries in the economic structure (IMD, 1999–2001).

As far as technological development and exports more generally are concerned, the analysis of innovation processes in manufacturing industries is of particular importance (EC, 1995). Only 10 per cent of manufacturing enterprises have their own R&D department. These departments are usually very small, employing on average only 10 engineers and spending relatively little on product and process innovations. The exceptions are several enterprises in the chemical, pharmaceutical, rubber and electro-engineering industries that have larger R&D departments. The modest level of innovation activity in manufacturing industry is reflected in the length of time it takes to technologically renew products and production processes (see Table 17.3). Such lagging behind the European competitors is especially critical in the following industries: leather and textile products, the pulp and paper industry, fabricated metal production and mineral products. These industries were very slow in making technological improvements to their products between 1994 and 1998 (SURS, 1997–2002). The most innovative industries seem to be rubber and plastics, electrical engineering, electronics and to some extent also wood processing.

Like other developed European market economies, in-house R&D is the most significant item of expenditure in innovative industries and companies. Forty-three per cent of all intramural R&D expenditure goes on product research, 34 per cent on general research and the rest on process research (SURS, 2000a). Eighty per cent of all in-house R&D expenditure is accounted for by the following manufacturing industries: chemical products, pharmaceuticals, mechanical engineering, manufacture of telecommunication and audio-visual products, electrical machinery, rubber and plastics and medical and optical products. Expenditure on external R&D is low and concentrated in two industries, namely chemical products and pharmaceuticals. The majority of Slovenian manufacturing companies are not involved in innovation processes based on strong links with other companies and research institutions and the diffusion of information technologies. Manufacturing industries (excluding chemi-

Table 17.3 Slovenia: product and process innovations in manufacturing industries (1997 – 1998)

	Product innovations (in %) (total products = 100)			Process innovations (in %) (total products = 100)	
	new	improved	unchanged	innovative processes	unchanged processes
TOTAL MANUFACTURING	5	33	62	16	84
Food and beverages	8	10	82	16	84
Textiles	5	11	84	7	93
Garment, fur		54	46	57	43
Leather, leather products	8	16	75	15	85
Wood processing	11	26	63	11	89
Pulp and Paper	1	7	93		100
Printing	13	20	67	17	83
Chemical products	6	4	89	9	91
Rubber, plastics	1	87	12	16	84
Mineral products	11	21	68	14	86
-Metals	6	6	88	28	72
Metal products (without machinery)	5	14	82	5	95
Machinery	8	29	63	21	79
Computers, bureau machines	3	5	92		100
Electro engineering industry	3	4	93	32	68
Telecommunications	8	15	77	10	90
Optical products	8	8	84	26	74
Motor vehicles	18	10	72	4	96
Other transport means	6		94	5	95
Furniture	3	31	66	25	75
Recycling industry				100	

Source: SURS, 2000a; 2002.

cals, electrical engineering, rubber, plastics and pharmaceuticals) account for 64 per cent of physical output but for only 20 per cent of total R&D expenditure and an even lower share of R&D personnel (engineers and scientists). These industries are making little use of their innovation potential because of the need to restructure both enterprises and product ranges. The prospects for improving international competitiveness are therefore poor in capital and technology-intensive industries with low innovation potential and huge marketing problems (Stanovnik et al., 1998).

Since the introduction of market reforms, 2 technology parks have been established in Slovenia: the Ljubljana Technology Park and the Steir Technology Park in Maribor. Both parks provide premises and basic infrastructure together with promotional activities, information services, cooperation with universities, advice on potential financial sources, assistance with protecting industrial property rights and management of joint R&D projects. Ljubljana Technology Park also offers the use of laboratories and test equipment. Currently there are about 50 technology-based companies operating within both technology parks. They employ around 350 people, mostly highly skilled personnel.

The concept of technology parks and incubators certainly meets some of the development needs of the area. Although they are constrained by a lack of financial resources, both for the further development of the parks' operations and for the member companies, these first steps do establish a basis for the further development of technology-based companies. Both parks need further support from central government and local communities to help them to grow into fully operational technology parks.

The Slovenian Science Foundation plays an indirect role in the functioning of Slovenian STS by organizing conferences and workshops, co-funding applied R&D projects and financing post-docs studying abroad.

The Slovenian Business Innovation Network (SBIN) was established in 1992 by the Center for the Promotion of Small Businesses. The SBIN provides consultancy services to innovators (intellectual property rights, preparation of business plans, etc.). Its activities are financed primarily by the state. This financial support is limited and dependent upon the bargaining positions of the various ministries in the annual negotiations on the distribution of budget funds. The SBIN acts as an intermediary for innovators (predominantly individuals and independent researchers), for whom it is very difficult to struggle on their own through all phases of the innovation process.

The transition to a new development paradigm based on knowledge and the acquisition of competences at firm level needs greater emphasis on cooperation between S&T institutions and business enterprises and the diffusion of new technologies. Slovenian industrial firms need to pay more attention to global marketing, financing, system integration at product level, cost reduction through outsourcing and downsizing and international networking. The OECD economic survey on Slovenia confirms that the country's competitive advantages seem to lie more in quality or product differentiation than in prices. Science and technology policy should alleviate the search for comparative advantages, which appear to be concentrated in light industries with a relatively high R&D content (OECD, 1997, p. 14). Economic restructuring could be enhanced by increasing the participation of strategic partners and by fostering international networking.

Different channels of technology transfer should be used to improve the process of technological learning. These include foreign direct investments, joint ventures, subcontracting, exporting, strategic alliances and licensing. Radosevic (1999) argues that the product development gap could be most effectively closed through FDI, joint ventures and international clustering.

5. CONCLUSIONS

The process of change in the Slovenian STS has been gradual and passive. The starting point at the beginning of the transition period was quite favorable in respect of gross expenditure on R&D, the structure of R&D personnel in the business enterprise sector and public research institutions and competitiveness of the manufacturing industry.

The organization of the public RTD system in Slovenia remained unchanged during the transition period, while R&D units in the privatized business sector suffered extensive cuts at the beginning of the privatization process, losing some 3 000 researchers in all (Kos, 1996; 1999). The process of revitalizing R&D units in the business sector started in the late 1990s.

The following deficiencies in the Slovenian STS were identified during the transition period:
- despite a strong national R&D effort to which about 1.5 per cent of GDP was devoted in the 1990s, R&D is focused more on academic and research interests and much less on the problems of production and manufacturing;
- there is an emphasis on basic research at the expense of applied technological research of relevance to industry;
- the current RTD system is characterized by rigid institutional structures (state-owned national research institutes, universities), the role of private institutions is insignificant;
- the relevance and contribution of the RTD system to the Slovenian economy is weak due to the low level of cooperation between research institutions and business;
- government research priorities are not clearly defined, especially as regards technological specialization;

– government financial support for RTD has been falling since 1993. The new national research program for the period 2003–2006 is underway (MESS, 2001).

Closer examination of industrial R&D expenditure and the technological content of Slovenian products reveals substantial gaps between Slovenia and the developed EU economies. The country's goods and services embody three times less value added than those of developed EU countries. Financial constraints and inappropriate corporate governance practices (high risks, lack of appropriate management, high costs and long pay-back periods) constitute the greatest barriers to innovation.

The system lacks flexible linkages and coordination mechanisms between public R&D institutions and the business enterprise sector, especially intermediary organizations such as innovation agencies, consultants, venture capital funds and banks. International networking is developing as a result of the increased participation of Slovenian RTD organizations in EU research programs.

Special attention should be paid to the creation and transfer of knowledge. Greater mobility of R&D personnel between research organizations and the business sector as well as between SMEs, together with informal contacts within innovation networks, should be encouraged.

The Slovenian government needs to play a new role in managing knowledge and making technology and innovation policy an efficient tool for enhancing the competitiveness of S&T and of the whole economy. Technology and innovation policy in Slovenia should focus on the following key objectives:

– building a culture of innovation and entrepreneurship;
– improving technology diffusion;
– promoting networking; experiences from the 5th Framework Program should be used in the forthcoming 6th Framework Program;
– stimulating market-driven research.

The priority objectives for the planned RTD system in the period 2001–2006 can be summarized as follows:

– increasing the quality of the RTD system;
– inter-ministerial support for target/market-driven research in Slovenia;
– enhancing internal mobility and collaboration of university and business enterprise sector researchers;
– increasing the scope and quality of international R&D collaboration;
– modernization of RTD infrastructure;
– fostering innovation and new technology diffusion in the private business sector.

Following the examples of the more developed EU countries Slovenia will use technology forecasting to identify the most promising areas of technology. Representatives of public and private research institutions, civil servants and NGOs will take part in this exercise. Priority areas should be harmonized with EU directives and take into account the specific characteristics of Slovenian society. Targeted research programs are built on interdisciplinarity and collaboration between several domestic and foreign institutions. These programs are oriented towards internationally competitive RTD, the reform of public administration, enhancement of human resources, more balanced regional and spatial development, sustainability in the agro-food sector, national security and identity and the building of an information society.

Due to weak internal mobility among researchers and an imbalance between universities, public institutes and the private business sector, it is planned to provide additional funding for the young researchers program, thereby enabling 300 new researchers to undertake postgraduate studies each year. These measures will encourage the employment of young researchers in R&D units in the private business sector.

International R&D collaboration will not be concentrated only on the 6th Framework Program, but also include bilateral research projects, especially with neighboring countries. A special program of infrastructure modernization is planned (centers and networks of excellence) in compliance with the EU policy 'Towards a European Research Area'.

REFERENCES

Bevc, M. (1995) *Education*. Ljubljana: Imad.
Bevc, M., and P. Stanovnik (1996) *Education and R&D as Levers for Innovation*. Ljubljana: Institute for Economic Research (IER).
EC (1995) *Green Book on Innovation*. Brussels: European Commission (EC).
EC (Ed.) (1999) *Impact of the enlargement of the European Union towards the associated Central and Eastern European countries on RTD-innovation and structural policies*. Luxembourg: Office for Official Publications of the European Communities, European Commission (EC).
Gliha, M. (1998) *Tehnološka raven slovenske industrije* (Technological level of Slovenian manufacturing). Ljubljana: IER.
Gopa consultants (1994) *A Science and Technology Strategy for Slovenia*. Ljubljana: Ministry of Science and Technology (MST).
IER (2000) *Predlog Programa o podpori gospodarskim dru bam pri razvoju novih tehnologij in vzpostavljanju in delovanju njihovih razvojnih enot od leta 2000 do 2003* (Basis for the elaboration of a support program for business companies in fostering new technologies and establishing R&D units between 2000 and 2003). Ljubljana: IER.

IMD (1999–2001) *The World Competitiveness Yearbook.* Lausanne: International Management Development (IMD).
Kos, M. (1996) *Strategic Technological Orientations of Slovenian Manufacturing.* Ljubljana: MST.
Kos, M. (1999) *Inovacije in inovativno poslovanje* (Innovation and innovative business). Ljubljana: IER.
Mansfield, E. (1982) "How Economists View R&D." *Research Management* 15, no. 4: 23–29.
Meske, W. (1997): "R&D-Systems in Transition – Lessons for Eastern Europe from the East German Experience." Unpublished Paper presented to Workshop IV (Inter-regional cooperation & research and technological development) of the 6th European STRIDE Conference, March 3–4 in Bremen, Germany.
MESS (2001) "Informacije o RR kazalcih" (Information on Research and Development Indicators). Mimeo. Ljubljana: Ministry for Education, Science and Sport (MESS).
MST (1996) *Znanstvena in tehnološka politika za leto* (Science and Technology Policy). Ljubljana: MST.
OECD (1997) *Economic Surveys – Slovenia.* Paris: OECD.
OECD (1999) *Managing National Innovation Systems.* Paris: OECD.
Radosevic, S. (1995) *Technology transfer and restructuring of technology capability in global competition: The case of economies in transition.* Brighton: SPRU/University of Sussex.
Radosevic, S. (1999) *International Technology Transfer and Catch-up in Economic Development.* Cheltenham: Edward Elgar.
Stanovnik, P., V. Lavrac, and V.K. Prevolnik (1998) *Kvaliteta statisticnih podatkov o vlaganjih v RR dejavnost in o RR kadrih v poslovnem sektorju v letu 1996* (The quality of statistical data on R&D expenditures and R&D personnel in the business enterprise sector in 1996). Ljubljana: IER.
Stare, M., and M. Bucar (1997) *Diffusion of Innovations and Technology Transfers in the present R&D System in Eastern Europe.* Ljubljana: Center for International Relations, Faculty of Social Sciences.
SURS (1997–2002) *Raziskvoanje in razvoj, znanost in tehnologija – Statisticne informacije* (R&D, Science and Technology – Statistical information – Rapid reports). Ljubljana: Statisticni urad Republike Slovenije (SURS – Statistical Office of the Republic of Slovenia).
SURS (2000a) *Raziskovanje in razvoj, znanost in tehnologija, 1997; Inovacijska dejavnost, 1998* (R&D, Science and Technology, 1997; Innovation Activity, 1998). Ljubljana: SURS.

Part III

Common features, particularities and results: a comparative analysis of S&T transformation by country

18. THE REORGANIZATION OF S&T SYSTEMS IN CEECS DURING THE 1990S

Werner MESKE

In Part III, the most significant empirical and statistical findings on the course and results of the changes in S&T systems in the individual CEECs during the 1990s (Part II) are summarized, compared and evaluated. The primary objective is to demonstrate the common features and particularities in the course and outcomes of the transformation in the individual countries as well as to determine those factors that played a fundamental role in the process. Underpinning the following presentation is the recognition that, by the turn to the 1990s, the socialist S&T system was beginning to disintegrate, both internationally and in the individual CEECs. As part of this process, considerable changes in the volume and structure of resources made available for S&T, including scientific personnel, the most important factor in S&T activities, had already occurred or were predicted (cf. Meske and Nadiraschwili, 1994). These quantitative changes were caused, accompanied or followed by fundamental institutional changes in S&T.

1. DISSOLUTION OF THE SOCIALIST WORLD SYSTEM

Until the late 1980s, the supranational features of S&T systems (STS) in the socialist bloc were ultimately the decisive determinants, rather than the particular interests and features of individual countries. This began to change with the remarkable political events of 1989/1990. Stimulated by Gorbachev's 'perestroika' policy of the late 1980s, which was actually designed to maintain Soviet rule by way of internal reforms, radical political changes were taking place by late 1989 and early 1990 in Poland, Hungary, the GDR, Czechoslovakia, Romania and Bulgaria. These changes called into question not only the socialist control system but also the Soviet Union's claims to political leadership and the economic integration of these countries within the socialist bloc. As early as July 1, 1990, the GDR merged economically with the FRG and on October 3, 1990 the political transition took place. With this transition came entry into the EU and NATO. This in turn encouraged other countries in their striving for independence and at the same time increased the internal tensions in the USSR. As a result, by January 1991, the CMEA had disintegrated in all but name. On August 19, 1991 there was a coup in which Gorbachev was overthrown, and following this the USSR disintegrated. In December 1991, the Commonwealth of the Independent States (CIS) was established, comprising all the former republics of the USSR, except the three Baltic states and Georgia. The USSR had ceased to exist. In the same year, the break-up of the SFR Yugoslavia occurred, followed by war between Serbia and Croatia and (civil) war in Bosnia-Herzegovina. In Czechoslovakia, tensions increased between the two parts of the country and on January 1, 1993 they officially split into the Czech Republic and the Slovak Republic.

Thus in just three years the former 'socialist bloc' around the USSR, with Russia as its heart, had turned into a conglomerate of several independent, mostly small states (cf. Lavigne, 1995, p. 96–129). Only Russia strived to take up the legacy of the USSR as a world power and to retain as much influence as possible over the former Soviet republics at least. However, like all the other former socialist countries, these latter were concerned to rid themselves of this influence. To this end, most CEECs were even prepared to join the EU and NATO, especially since most of the new political actors considered the systems of multi-party democracy and market economy practiced there to be the best, if not the only alternative to a socialist system. As well as political and military protection against Russia, they hoped that they would also gain economic advantages; relatively small states, which is what most CEECs are, are unavoidably dependent on close cooperation with larger national economies (such as Russia or Germany) or with blocs like the EU.

The first characteristic of the transformation was the collapse of the socialist bloc. The existing systems and the actors and rules upon which they rested (decision-making bodies, regulations, operational organizations) collapsed as a result of the dissolution of international treaties, such as the Warsaw Pact, the Council for Mutual Economic Assistance (CMEA), Interkosmos and so on, the collapse of individual states, such as the GDR, the USSR, SFR Yugoslavia and the CSSR, and the removal of hierarchical leadership systems in politics, the economy and science. These were essentially top-down processes leading to the dissolution of the institutional structure of S&T in the socialist world, which was the heart of its innovation system. Consequently, each of the former socialist countries strove, or was forced, to reposition itself nationally and internationally and to find its own way in doing so. Cuba, for example, has tried to survive as a socialist country without essential changes (cf. Becker, 2001), while China has relied on its socialist political system to introduce the market economy in a gradual and controlled way and also to adjust its S&T system accordingly (cf. Gu, 1999; Yang, 1998). Vietnam has also gone down this path, although even here our own analyses have revealed specific transitional and development problems (cf. Meske and Dang Duy Thinh, 2000).

One result of this first wave of societal transformation in the CEECs was the extensive fragmentation of the former STS and, in most cases, the survival of only a few of the individual elements of the former system. This process was particularly pronounced in the successor states of the USSR: most of them had no 'traditional' STS of their own with the corresponding decision-making bodies and regulations and had inherited not only organizations specific to the particular republic but also parts of former 'all-union' institutes of the former USSR.

Analysis of the transformation processes in the various CEECs confirms the initial hypothesis that each of these countries tended to be in a specific situation at the end of the socialist era. The differences between them at that time meant, of course, that their starting conditions for the transformation of the STS were also different, which definitely had a strong influence on the subsequent processes. The decisive factors were:
- each country's position and role in the global socialist S&T system;
- the degree of institutional deviation from the Soviet model of STS and
- the basic societal conditions in each country when the socialist system collapsed.

Examination of these factors reveals differences in the initial conditions in the individual CEECs. These differences were strongly influenced by each country's position within socialism (at the center, C, or in circles 1 – 3; cf. Figure 1.2), but were no longer wholly determined by it. The Baltic states (circle 1) in particular moved away from their close ties with Russia (C) and were some of the first countries to align themselves strongly with the EU. Generally speaking, it would appear that orientation towards the EU was more strongly influenced by each CEEC's pre-socialist traditions (not only with regard to S&T) and by their geographic proximity to EU nations than by their degree of integration into the socialist world system. 'Pre-socialist' traditions in S&T in particular and their preservation (supported by intensive international contacts), especially in the Central Eastern European and Baltic countries, provided scientists with a solid foundation from which to launch early initiatives to gain autonomy and democratize scientific bodies. Thus these traditions played a major role in determining the subsequent course of the transformation process.

2. THE FRAGMENTATION OF NATIONAL S&T SYSTEMS AND THE TASTE OF THEIR REORGANIZATION

While the first steps toward *state independence* can be seen as the 'dissolution' of the former global socialist S&T system, the subsequent political and economic *changes in the individual countries* generally went beyond this and led to a further 'top-down fragmentation' of their former S&T systems (cf. Figure 18.1).

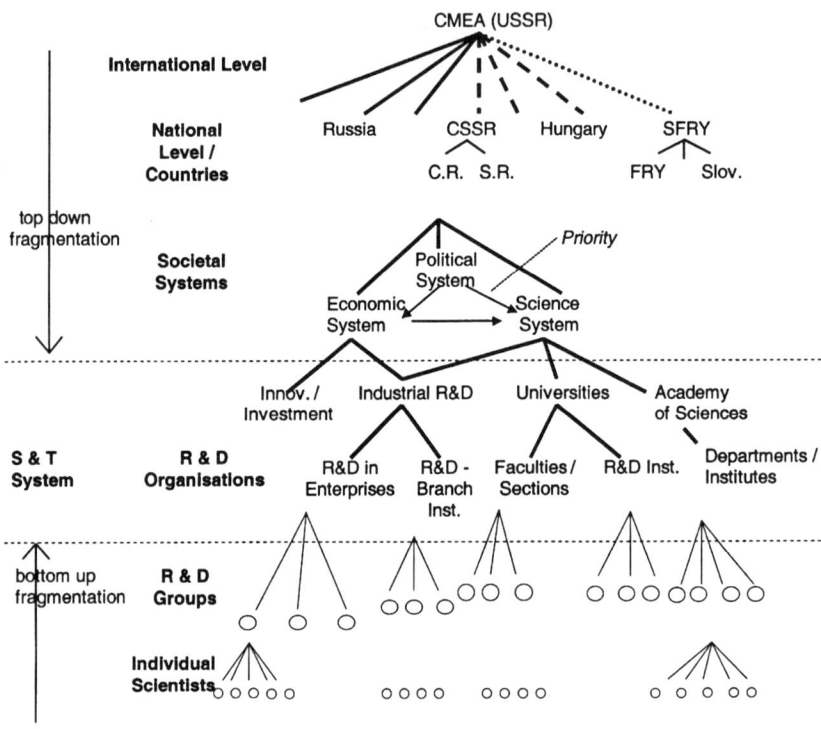

Fig. 18.1 Dissolution and fragmentation of the former socialist S&T system

Although they differed from country to country, the main factors were as follows:
- the political vacuum left behind after the dissolution of former union states and the emergence of new independent states, such as the Baltic States;
- state withdrawal from responsibility for science and technology through the abolition of state planning, the dissolution of ministries and of other bodies formerly responsible for S&T and the granting of autonomy to universities and to the Academy of Sciences (AoS). In all cases this was associated with a substantial reduction in state funding. In most countries, these reductions far exceeded the general level of economic downturn, in reaction to the earlier overestimation of science. In some cases this reduction was less severe

because the status of science tended to be enhanced as part of the process of developing 'national heritage and pride' (in the Ukraine and Poland, for example). Either directly or indirectly (via financial cutbacks), this led to the dissolution of scientific organizations, especially those directly subordinate to the ministries and those with political-ideological and other functions linked to the 'socialist' system, including some units within the universities and AoS;
- state withdrawal from responsibility for the business enterprise sector. The introduction of market economy mechanisms, the devolvement of responsibility to enterprises and, in particular, their privatization all led to the collapse of the old system of state-managed industrial R&D and innovation. With the dissolution of industrial branch ministries, the branch R&D institutes lost not only their management and funding basis but also the most important coordinating body linking them to the enterprises. And since the enterprises themselves were struggling for survival in the market economy, they were seldom interested in maintaining or, even less, financing contracts with external R&D institutes, many of which had been imposed upon them. On the contrary, they were usually eager to cut in-house R&D capacities in order to reduce costs. As part of the rush to privatize, many of the industrial R&D institutes and departments were liquidated or converted into independent firms;
- the S&T management and coordination structures and mechanisms that were dismantled in this way left in their wake a fragmented multitude of relatively autonomous but more or less 'isolated' S&T organizations, such as universities, Academy institutes, R&D institutes and new, R&D-intensive enterprises. In many cases, their status was initially not absolutely clear and established in law; most of them had no formal links either with each other or with their non-scientific environments.

These top-down processes leading to the dissolution of the former S&T system, driven or organized primarily by forces outside science, were supplemented and aggravated by bottom-up processes within the S&T system itself. The crucial processes involved here were, and continue to be:
- mobility in the scientific community: emigration of leading and up-and-coming scientists; migration of scientists between countries for ethnic reasons (cf. Toren, 1994); withdrawal of younger scientists from scientific activity or failure to enter the profession in the first place; migration between scientific institutes and fields;
- institutional spin-offs or foundation of new institutes by scientists and engineers: establishment of private universities and research institutes; foundation of R&D service providers, research enterprises, transfer institutions;

transformation of former technology departments into independent entities, partly through their conversion into high-tech businesses;
- establishment of self-governing bodies within scientific institutions (science councils, and so on) and for the whole S&T system (various types of research support organizations and also decision-making bodies).

This meant that even the remaining scientific organizations and their subunits experienced substantial changes and loss of significance. Their position was often unclear, since the failure to rescind employment contracts and scientists' departure for foreign countries with the option to return (among other factors) frequently led to the collapse of formal affiliations to organizations and even of entire occupations.

For the most part, these top-down and bottom-up processes came into being spontaneously at the beginning of the transformation process. At the same time, however, it was precisely these processes that made the formation of new types of organizations and networks in S&T possible and therefore marked the often scarcely perceptible beginning of the restructuring of STS.

Following the dissolution of the socialist bloc and the extensive fragmentation of the S&T systems in the now independent CEECs, these countries generally faced the task of restructuring their R&D organizations and adapting their S&T institutions to the new conditions arising out of the transition to a market economy and multi-party democracy. All thinking and proposals in this area were based on the (at least implicit) assumption that the institutional framework of S&T should be adapted to, or at least be made to converge towards, that found in the leading OECD countries. This was in line with the approach to 'transformation' as the implementation of democratic systems and market economy frameworks by neo-liberal economists (cf. Sachs, 1993) and 'theoreticians of modernization'. Thus Zapf, for example, characterized the transformation as 'remedial modernization', "in which the goal is known" (Zapf, 1994, p. 138, own translation).

However, opinions as to how this convergence might be achieved often diverged considerably. Thus recommendations by OECD experts for a complete reorganization of Czechoslovakia's R&D system met with skepticism from the Czechs and Slovaks "as to the viability or the feasibility of implementing the institutional and structural changes proposed by the (OECD)Examiners. Whether they were skeptical as to the short-term effectiveness of the changes or of the viability of the more long-term perception of the evolution of the S&T sector remained unclear. A second area of disagreement between the Examiners and their Czech and Slovak colleagues appeared to be the latter's belief that *existing* institutions could be modified or provided with better incentives, and that, given these, they could fulfill new functions, achieve higher levels of efficiency and make effective contributions to a market economy. Finally, there seemed to be a view among the Czechoslovak participants that weak S&T institutions – the universities and the private industry sector – would somehow be strengthened and improved *without* the elimination or radical downsizing of existing S&T institutions" (OECD, 1992, p. 186).

We can start from the assumption that the political transformation in the CEECs offered opportunities for the institutional reorganization of S&T. Largely because of the economic transformation also taking place, these opportunities had to be realized in an environment characterized, on the one hand, by considerable constraints caused by a shortage of resources and, on the other, heightened expectations of the contribution research policy could make to the re-establishment of economic and political stability (cf. Schimank, 1995, p. 10–11). It was, therefore, presumed (also in the light of experiences in East Germany; cf. Meske, 1993 and 1994) that the transformation of the S&T system would be a long-term process involving various types of changes on various levels. Although the situation at the outset varied in the individual CEECs, the tasks facing them were very similar.

As far as scientific organizations were concerned, the main objective was to introduce greater autonomy at all levels and to strengthen the role of higher education organizations in the national R&D system. With regard to the public research institutes, the main debate revolved around whether to retain or abolish the Academy of Sciences (AoS). The general view was that the academies would have to change in character at least, with a greater separation between scholarly societies and research institutes. At the same time, it was expected that the often very pronounced combinative nature of AoS research would be eliminated by greater differentiation of institutional profiles towards either basic or applied research and the abolition or out-sourcing of non-research (service and production) units. These changes were to be realized and encouraged by evaluating institutions and personnel, restructuring and reprofiling institutes and making project funding much more dependent on competitive bidding. In the manufacturing sector, finally, it was expected that the hitherto dominant branch institutes would be largely abolished, mainly in favor of the development and/or strengthening of in-house R&D capacities in privatized or newly established companies. All these changes required corresponding changes in the political and economic spheres or, conversely, were to be instigated and supported by such changes. For this reason, attention was initially focused on putting in place a legislative framework applicable to research and to making changes in both the institutional structures of research and the (financial and human) resources available for research (Mayntz et al., 1995, in particular the preface).

These were assumptions and expectations on the part of Western experts, at least, but also of wide circles of East European scientists and politicians (for example, in Russia, cf. Gokhberg, 1996, p. 2)

In general, it was hoped or expected that the changes in the political, economic and scientific spheres would promote and strengthen democracy and thereby lead to economic revival. The precondition for and driving force behind this renewed economic growth was to be innovation and the upstream scientific activities, which were to be harnessed much more efficiently than under the previous socialist regimes. It was expected that there would be a shift from a science push to a technology pull model, the driving force being the considerably more active role in innovation being played by autonomous private or state-owned firms operating according to market principles. This shift was necessary because "enterprises in the Western sense did not exist in socialism" (Radosevic, 1997, p. 376) and, therefore, "centrally planned systems have proved to be relatively weak in applying and diffusing the output of R&D and other innovative activities" (Hanson and Pavitt, 1987, p. 86).

3. UNCOMPLETED CHANGES IN THE FIRST HALF OF THE 1990s

A record and comparison of the most important institutional changes that took place in the CEECs in the first half of the 1990s and up to 1996/97 was compiled as part of the above-mentioned TSER project (cf. Meske, 1998; SPRU, 1999; cf. also Chapter 3 in this volume). This analysis drew on the available documents and reports in order to examine the changes in the sectors undertaking S&T as well as in their political and economic environments; the same indicators were used for all CEECs, although they were not always totally and unconditionally determinable, quantifiable or comparable. We then evaluated them according to a rough three-tiered scale (the criterion was the situation in the CEECs, which was often not comparable with that in the OECD countries, particularly as far as industrial R&D was concerned). In the mid-1990s, that is at a time when the greatest changes had already taken place or were prepared in most of the CEECs, the indicators for each segment of the STS crucial to the transformation varied remarkably from country to country.

In the case of *organizations carrying out R&D* (cf. Table 18.1), the most substantial advances were the granting of autonomy to HE institutions and non-university research facilities and their reorganization, which was based largely on the evaluation of units and individuals. However, realizing the other two objectives, namely strengthening research in the HE sector and introducing competitive financing for public research, proved to be more problematic. While advances in institutional restructuring tended to dominate in the higher education sector and were in progress in the Academies of Sciences and other public R&D institutes, the situation in industrial R&D must, in contrast, be assessed as unsatisfactory. Here, restructuring was largely unresolved or still in its initial stages. Thus the reorganization of the former branch R&D institutes was far from being completed – with the exceptions of successor states of the SFR Yugoslavia, which never had such institutes, and the Czech Republic and the Baltic states, where they were rigorously dissolved or privatized. Often there was even a lack of proposals as to what should be done with these institutes. The aim of strengthening in-house R&D (personnel, funding) and building it up as a core area of the new R&D and innovation system remained largely unfulfilled. Exceptions to this failure to build up new bases for innovation included the emergence of new small enterprises and new types of transfer organizations (Webster, 1996) and, in some cases, the establishment of R&D units in subsidiaries by multinationals, for example in Hungary (cf. Chapter 13). By and large, however, standards in industrial R&D were even lower than they had been in the socialist era.[1]

In the sphere of *S&T policies* (cf. Table 18.2), the main advances lay generally in the creation of a new state administration and the introduction of new legislation rather than in the formulation or, in particular, the implementation of new policies or the introduction of competitive forms of financing and increased budget funding for S&T. Nearly all of the 17 countries analyzed had, within a relatively short space of time, restructured or established state authorities for S&T with (more or less) clearly defined competencies. The exceptions were Bulgaria and Slovakia where, due to political instability and restrictive science policies, new state authorities and legal systems governing S&T were not actually established in the first half of the 1990s (cf. Chapters 12 and 15). In the European CIS countries, state authorities were indeed set up, but there were delays in passing and, particularly, implementing new legislation and in reaching agreement on a clearly defined policy for S&T. As far as the state funding of R&D was concerned, efforts were made in virtually all the countries, it is true, to introduce and/or extend competitive project funding as a supplement to basic institutional funding. In all the countries, however, those efforts were hampered by the sharp reduction in the public resources devoted to S&T. For this reason, diversity of funding did not become a reality until the mid-1990s, and even then only in the small number of countries that had already begun to experience renewed economic growth and were able at least to stabilize their budgets and state funding for S&T.

---

[1] Further information, in addition to that presented in Part II of the present volume, on the situation in industrial R&D and on the changes introduced in branch R&D institutes in several countries will be found in Bouché, 1998; Couderc, 1996; Mosoni-Fried, 1998; Schneider, 1998 and 1998a; and Tichonova, 1998.

Table 18.1 Institutional changes in S&T sectors in individual CEECs in the first half of the 1990s

Country	S&T sector						Rank
	Higher Education		AoS/Public Sector		Industrial R&D		
	Autonomy/ Diversification	Role of research	Autonomy/ evaluation	Competitive funding	Transformation of branch/ state institutes	Role of in-house R&D	
	1	2	3	4	5	6	7
Russia	0	0	0	0	0	–	3
Ukraine	0	0	–	0	0	–	3
Belarus	–	–	–	–	–	–	3
Moldova	–	–	–	–	–	–	3
Latvia	+	+	+	+	+	–	1
Estonia	+	+	+	+	+	–	1
Lithuania	+	0	+	–	+	–	2
Poland	+	+	+	0	0	+	1
Czech Republic	+	0	+	+	+	+	1
Slovakia	0	0	+	+	0	+	2
Hungary	+	+	+	+	0	+	1
Romania	+	0	+	0	0	0	2
Bulgaria	0	0	+	+	0	–	2
Slovenia	+	+	+	+	+	+	1
Croatia	0	+	+	0	+	+	1
Serbia	0	+	0	0	0	0	2
Montenegro	0	+	0	0	+	0	2
Total +	8	8	11	7	7	6	(1) 7
0	7	7	3	7	8	3	(2) 6
–	2	2	3	3	2	8	(3) 4

Key:
1: + realized
    0 partially realized
    – not realized
2: + relatively high
    0 low
    – only marginal
3: + realized
    0 partially realized
    – not realized
4: + relatively high
    0 low
    – only marginal
5: + complete conversion into private or departmental R&D institutions
    0 only partly converted, still in existence
    – largely unchanged situation
6: + (relatively) developed/retained or newly created
    0 only in initial stages
    – non-existent or marginal
7: 1 predominantly + in the preceding columns
    2 predominantly 0 and + in the preceding columns
    3 predominantly – and 0 in the preceding columns

Table 18.2 Institutional changes in S&T policies in individual CEECs in the first half of the 1990s

Country	Area of S&T policy					Rank
	State authorities and competencies	Legal system	State S&T policy	State funding		
				Mode	Trend (1995/96)	
	1	2	3	4	5	6
Russia	+	0	–	0	–	3
Belarus	+	–	–	0	–	3
Ukraine	+	0	–	–	–	3
Moldova	+	–	–	0	–	3
Latvia	+	+	+	+	+	1
Estonia	+	+	+	+	+	1
Lithuania	+	+	+	–	+	1
Poland	+	+	+	+	+	1
Czech Republic	+	+	0	+	+	1
Slovakia	0	0	–	0	+	2
Hungary	+	+	+	+	+	1
Romania	+	+	+	+	+	1
Bulgaria	0	0		0	–	3
Slovenia	+	+	+	+	–	1
Croatia	+	+	+	+	0	1
Serbia	+	+	+	+	+	1
Montenegro	+	+	+	+	+	1
Total +	15	11	10	10	10	(1)11
0	2	4	1	5	1	(2) 1
–	–	2	6	2	6	(3) 5

Key:
1:    +   in place or clearly regulated
     0   only partly in place or regulated
     –   no stable or clear regulations
2:    +   mostly new and complete system
     0   new system only partially in place
     –   new system only in initial stages
3:    +   priority setting in S&T and innovation policy
     0   science policy formulated only in general terms
     –   lack of clear S&T policy/laissez-faire policy
4:    +   relatively high proportion of competitive financing
     0   competitive financing not widespread or in early stages
     –   almost wholly institutional financing
5:    +   growth
     0   stagnation/stabilization
     –   further drop
6:    1   predominantly + in the preceding columns
     2   predominantly 0 in the preceding columns
     3   predominantly 0 and – (particularly regarding funding levels) in the preceding columns

Table 18.3 Changes in the economy in individual CEECs in the first half of the 1990s

Country	Changes in the economy (based on Stern 1997, Tables 2 and 3 and Büschenfeld 1997[a])				Rank
	Evolution of GDP (1995/96)	Enterprises			
		Restructuring	Privatization		
			Large-scale	Small-scale	
	1	2	3	4	5
Russia	−	0	0	+	3
Belarus	−	0	−	−	3
Ukraine	−	0	−	−	3
Moldova	0	0	0	0	2
Latvia	0	0	−	+	2
Estonia	+	+	+	+	1
Lithuania	+	0	0	+	2
Poland	+	+	0	+	1
Czech Republic	+	+	+	+	1
Slovakia	+	+	0	+	1
Hungary	+	+	+	+	1
Romania	+	0	−	0	2
Bulgaria	−	0	−	0	3
Slovenia	+	+	0	+	1
Croatia	+	0	0	+	2
Serbia[a]	0	−	−	+	3
Montenegro[a]	0	0	0	+	2
Total +	9	6	3	12	(1) 6
0	4	10	8	3	(2) 6
−	4	1	6	2	(3) 5

Key:
1:    +   growth
      0   stagnation/stabilization
      −   further drop
2:    +   significant and sustained actions to harden budget constraints and   (Stern's category 3)
           effectively promote corporate governance (e.g. through privatization
           combined with strict credit and subsidy policies and/or enforcement
           of bankruptcy legislation)
   0   moderately strict credit and subsidy policy but weak enforcement of   (Stern's cat. 2)
           bankruptcy legislation and little action taken to break up dominant
           firms
   −   soft budget constraints (lax credit and subsidy policies weakening   (Stern's cat. 1)
           financial discipline at the enterprise level); few other reforms to pro-
           mote corporate governance
3:    +   more than 50% of state-owned enterprise assets privatized in a   (Stern's cat. 4 and 4*)
           scheme that has generated substantial outsider ownership; standards
           and performance close to those of advanced industrial economies

	0	more than 25% of large-scale state-owned enterprise assets privatized or in the process of being sold, but possibly with major unresolved issues regarding corporate governance	(Stern's cat. 3)
	−	comprehensive scheme almost ready for implementation; some sales completed	(Stern's cat. 2)
4:	+	complete privatization of small companies with tradable ownership rights; standards and performance close to those of advanced industrial economies	(Stern's cat. 4 and 4*)
	0	comprehensive program nearly implemented, but design or lack of government supervision leaves important issues unresolved (e. g. lack of tradability of ownership rights)	(Stern's cat. 3)
	−	substantial share privatized	(Stern's cat. 2)
5:	1	predominantly + in the preceding columns	
	2	predominantly 0 and + in the preceding columns	
	3	predominantly − (in particular regarding changes in GDP) and 0 in the preceding columns	

The situation in the *economic sphere* (cf. Table 18.3) was very different in the CEECs in the mid-1990s. Up to 1994/1995, all the CEECs had had to accept a considerable decline in their economies, in some cases to less than 50 per cent of the 1989 figures for GDP (cf. Figure 18.2), and, with the exception of Ukraine, even greater losses in industrial output (cf. Figure 2.2). By 1995, none of these countries had yet reached pre-1989 levels of GDP or real industrial output and only about half of them had even been able to halt the economic decline and start growing again. The country analyses conducted at the time showed very clearly that stabilization and growth of GDP and the restructuring of enterprises were of much greater importance for the transformation of the STS than the completion of privatization. For this reason, these first indicators were given a relatively high weighting in the ranking of the countries.

It is clear from the comparative analysis and summary of the findings for the situation in the middle of the 1990s given in Tables 18.1–18.3 that the countries can be divided into three clearly defined groups according to the advances made in the transformation of their STS (cf. Table 18.4).

Group I comprises five countries, each with a total rank estimate of 1, that clearly demonstrated the greatest advances in their institutional changes. Group III comprises a further five countries (Bulgaria and the four European CIS countries) that lagged considerably behind the first group with regard to the reorganization of their STS. The other countries, which make up Group II, tend to lie somewhere between these two groups; they had made some advances but lagged behind in various other areas.

Allocation to Group I and Group III was determined primarily by the stage of transformation achieved in both the economy as a whole and in the various S&T sectors in particular. Both these groups of indicators revealed clear con-

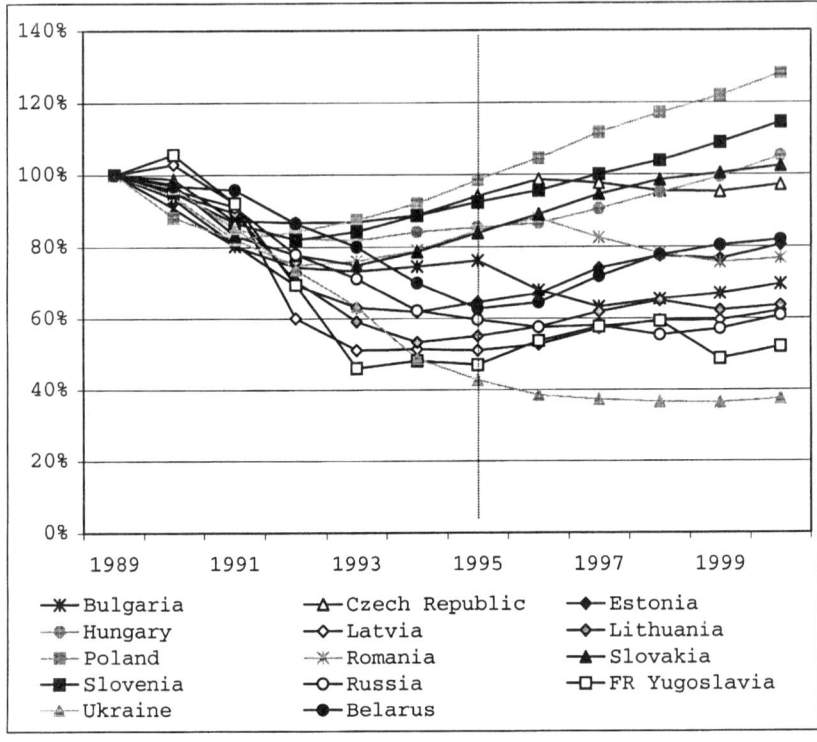

Fig. 18.2 CEECs: real GDP levels since 1989 (1989 = 100%)
Source: Own compilation based on EBRD, 2000 (all data at constant prices); FRY: cf. Tab.16.1 and 16.2 (all data at 1994 prices).

gruence within each group of countries as well as equally clear differences between the two country groups. Group II countries, on the other hand, had made substantial advances in the area of S&T policies that were often comparable to those made in the Group I countries. Without a stable economic basis, however, these countries had been unable as yet to follow up these policy changes with radical changes in the organizations actually carrying out S&T. This confirmed Nesvetailov's statement, made with regard to the CIS countries, that "a law on S&T policy is just a piece of paper as there is no money to implement it" (Nesvetailov, 1998, p. 10).

A study of the situation in research and technological development (RTD) in 10 CEECs (Bulgaria, Czech Republic, Estonia, Hungary, Latvia, Lithuania, Poland, Romania, Slovak Republic and Slovenia) that were candidates for entry to the EU was carried out by a consulting firm in 1997; it is more or less contemporaneously with our TSER project but unknown to us at the time. Its findings, which were published in 1999 (EC, 1999), largely concur with our

Table 18.4 Classification of individual CEECs by degree of institutional transformation in the national S&T system in the first half of the 1990s

Country	Transformation of fundamental elements of the STS (accord. to Tables 18.1 – 18.3)		
	S&T sectors	S&T policies	Economy
I. Countries furthest advanced in the transformation of their STS (total rank: 1)			
Estonia	1	1	1
Poland	1	1	1
Czech Republic	1	1	1
Hungary	1	1	1
Slovenia	1	1	1
II. Countries in intermediate position (total rank: 2)			
Latvia	1	1	2
Croatia	1	1	2
Slovakia	2	2	1
Lithuania	2	1	2
Romania	2	1	2
Serbia	2	1	3
Montenegro	2	1	2
III. Countries lagging considerably behind (total rank: 3)			
Moldova	3	3	2
Bulgaria	2	3	3
Russia	3	3	3
Belarus	3	3	3
Ukraine	3	3	3
Total: 1	7	11	6
2	6	1	6
3	4	5	5

Key:
1 S&T sectors   1 = Considerable changes in all 3 sectors realized, in particular diversification (and strengthening of research) in the HE sector, democratization, evaluation and increasing competitive funding in the AoS or public sector, changes in the former branch institutes and retention or establishment of (still small) in-house capacities
              2 = Only partial or incomplete realization of the changes as per 1
              3 = Changes usually only just beginning, without fundamental transformation of the structure and mode of operation of the individual sectors
2 S&T policies 1 = New institutional framework in S&T politics and funding in place; state S&T funding increasing
              2 = New framework not fully in place, problems with state funding
              3 = Changes in institutional framework beginning but not yet actually realized; continuing decrease in S&T budget
3 Economy   1= GDP growing; enterprise restructuring and privatization largely completed
              2= GDP stabilizing; restructuring and privatization of enterprises not yet completed
              3= GDP falling; restructuring and privatization still in initial phase

Table 18.5 CEECs: economic situation and resources devoted to R&D in the second half of the 1990s

Country	Economic situation: GDP in 2000 in relation to			Resources devoted to R&D			R&D personnel
	1995 (growth; %)	1989/90 (level; %)	EU average (level; %)	GERD/GDP (%)		year with lowest level	2000 (1999) to 1995 (%)
				1995	2000 (1999)		
1	2	3	4	5	6	7	8
Poland	30	128	39	0.74	0.58	2000	95.3
Estonia	25	80	38	0.63	0.70	1996/97	90.2
Slovenia	22	115	72	1.71	(1.50)	1997	98.4
Hungary	20	105	52	0.75	0.82	1996	120.0
Belarus	20	82	.	0.89	0.82	1998+2000	83.8
Slovakia	19	102	48	0.98	0.69	1999	94.1
Latvia	12	62	29	0.52	0.48	1999	120.6
Lithuania	8	64	29	0.48	0.60	1995	107.0
FR Yugoslavia	6	52	.	1.11	(1.40)	1995	(83.9)
Czech Republic	3	97	60	1.04	1.37	1995	106.7
Russia	1	61	.	0.79	1.09	1995	83.9
Ukraine	-5	38	.	0.60	0.30	2000	65.3
Bulgaria	-7	70	24	0.62	0.55	1996/97	55.0
Romania	-8	77	27	0.80	0.37	2000	59.5

Source: Own compilation based on EBRD, 2000; Table 19.1 and Fig. 19.1 in this volume.

finding that none of the countries had completed the transformation of its S&T system by the mid-1990s.

According to the EC report, "none of the countries has a reasonably coherent RTD system in which all the actors work together as an RTD or innovation system [...] All of the ten CEECs have policies in place for various parts of RTD, but not always coherent with each other and not always with a complete coverage over the whole field, and in most cases the responsibilities for RTD policies are divided over several ministries." (EC, 1999, p. 22 and 24). The considerable differences between the countries in the actual process of transformation are also confirmed by the EC report. For example, "the transition from a centrally planned RTD system towards a market-oriented system has been slower, and less progressive, in Bulgaria than in most transition economies"(EC, 1999, p. 73). At the same time, the report draws attention to the fact that, by 1997, new problems had emerged but that new measures had been introduced in some of these countries. As noted at the outset (cf. Chapter 3), this vigorous dynamic and the variable course of the transformation in the individual CEECs were also the main reasons for extending and updating our analyses of S&T transformation, which now cover the second half of the 1990s and on up to the years 2000/2001 (cf. Chapters 4 – 17 in Part II).

4. STABILIZATION IN THE SECOND HALF OF THE 1990S

Our analysis of the first half of the 1990s had showed that domestic R&D had virtually no influence on economic development during that period but that, conversely, the S&T system clearly could not be successfully transformed without the stability and renewed growth in the economy required to give policy-makers a genuine capacity to take action.

For this reason, we begin our analysis of the changes that took place in the last years of the 1990s by looking at the development of the economy.

Between 1996 and the year 2000, the downward economic spiral was halted in most of the CEECs and some countries even achieved stable growth (cf. Figure 18.2). In Poland, Estonia, Slovenia, Hungary, Belarus, Slovakia and Latvia GDP grew by between 12 and 30 per cent between 1995 and the year 2000 (an annual average of between 2 and 5 per cent); as a result, GDP in Poland, Slovenia, Hungary and Slovakia reached or considerably exceeded its pre-transformation level (cf. Table 18.5).

Another four countries (Lithuania, FR Yugoslavia, Czech Republic and Russia) achieved only a slight increase in GDP of between 1 and 8 per cent (that is in most cases an average of less than 1 per cent per year) over the five-year period. However, the main feature of these countries' economic development was a series of temporary setbacks, such as the Russian financial crisis of

1998, the NATO bombing of Serbia, which had negative effects on neighboring countries as well, and the recession in the Czech Republic. By 1999/2000, GDP in all four countries had started to grow again. In Ukraine, Bulgaria and Romania, it is true, GDP declined further over the period to the year 2000; even in these countries, however, a slight increase in GDP was recorded again or, in Ukraine, for the first time in the latter part of the period, in Bulgaria from 1998 onwards and in Romania and Ukraine in the year 2000. Nevertheless, all the countries in these last two groups are clearly still a long way away from their earlier level of GDP (except for the Czech Republic, which has recovered to 97 per cent of its former level).

In contrast to the first half of the 1990s, the second half of the decade, and particularly the situation in the years 2000 and 2001, was characterized in virtually all the CEECs by stabilization in both the political and economic spheres. Support for this view is to be found in the EBRD's prognosis of continued growth in this region in the year 2001, despite the deterioration in the international situation (EBRD, 2001). The main reason for these generally positive developments is the changes that have taken place in the economic sphere, which have involved extensive, though in some cases incomplete, liberalization, privatization and enterprise restructuring. The process of change has been subject to delays and setbacks, particularly in the three CIS countries and the FR Yugoslavia, while in Romania, following the elections of the year 2001, fresh attempts are being made to rapidly privatize and restructure the economy (cf. EBRD, 2001, Country assessments).

In most countries, the improvements in the economic situation were accompanied by the stabilization and/or establishment of democracy in the political sphere. As a result, even Bulgaria and Slovakia were able to make good the deficit in S&T legislation that was still evident in the mid-1990s, while the other countries were able to make further progress in the establishment of state authorities, the drafting and implementation of legislation and, above all, in the development of economic policy priorities and strategies. Consequently, the process of institutional adaptation to OECD norms is far advanced in the area of S&T policy. The European Commission, in its 'Regular Reports 2001' published in November 2001 (EC, 2001), also confirms further progress and/or a relatively high level of alignment with the 'acquis communautaire' in the area of science and technology (chapters 17 in the country reports) in virtually all the ten candidate countries studied here; the only reservations expressed concern Bulgaria, Romania and Latvia. To judge from the information presented in the country chapters in Part II of the present volume, these reservations regarding the completion of the process of institutional restructuring in this area must be extended to the CIS countries and, as a result of the recent political changes, to the FR Yugoslavia as well, and in particular to the Republic of Serbia.

Account also has to be taken of the fact that the significantly more favorable economic situation in the second half of the 1990s gave S&T policy-makers more room for maneuver. There is space for further optimism, because in the year 2000, for example, GDP increased for the first time in *all* the 14 CEECs analyzed here! However, the opportunities offered by the more favorable economic situation have been utilized to very different degrees, as is shown particularly by the provision of state funds for S&T, the evolution of research intensity (GERD/GDP) and R&D personnel in this period (cf. Tables 18.5 and 19.1).

It can be concluded from this survey of the changes that took place in the S&T systems of the CEECS in the 1990s that considerable progress was made in creating new actors in politics, industry and science and in introducing and enforcing new rules governing their behavior. This process is virtually complete in all the countries, although some countries still have some catching up to do, particularly as regards industrial R&D.

The most pressing of these problems concerns the former branch R&D institutes, particularly in Romania and the CIS countries, but also in Poland; they have either failed to find a new status, or at least not one that is sustainable over the long term, or are fighting for their survival as independent contract research institutes competing for a share of an already inadequate demand. However, the problems also concern companies themselves, which in this respect can be divided into at least three groups. In most countries, there is a relatively strong group of foreign subsidiaries which, it is true, have at their disposal modern products and technologies. To date, however, they have acquired them solely from their parent companies and will continue to give priority to these imports of technology in the future. It is in Hungary only that preliminary and partial attempts have been made to use and develop domestic scientific capacities by this group of companies. The second group includes the remaining large-scale enterprises, now restructured and often already privatized, which for reasons of cost seldom have their own R&D capacities or are in a position to give contracts to external R&D organizations. Because of a lack of resources for investment, these firms' innovation activities are usually tightly constrained and seldom geared to the strategic use of domestic R&D. The third group comprises the newly formed innovative companies, which are usually too small and too weak financially to engage in R&D to any considerable extent as a strategic exercise, whether independently or in cooperation with external organizations.

For these reasons, it is not yet possible, even in those countries in which market economic structures and new actors dominate the manufacturing sector, to assume that industrial R&D capacities and innovation activities have been consolidated on a lasting basis. The investigations that have now been conducted into the founding and slow growth processes of these firms in East Germany provide further evidence of the uncertain situation faced by the reorganized R&D institutes and domestic R&D-intensive SMEs (cf. Kohn, 2001; Legler et al., 2002, p. 3; Pleschak et al., 2000; Pleschak et al., 2002) (cf. Figures 18.3 and 18.4). The situation is even more difficult in those countries in which there has not yet been any real economic recovery but at best stabilization at a low level, as in the FR Yugoslavia, Bulgaria and Romania.

Although the situation now differs considerably from country to country, this general weakness in the manufacturing sector in respect of R&D and innovation means that the processes of linking the actors and creating new regional and national S&T systems based on multiple linkages is at best in its very early stages in all the CEECs.

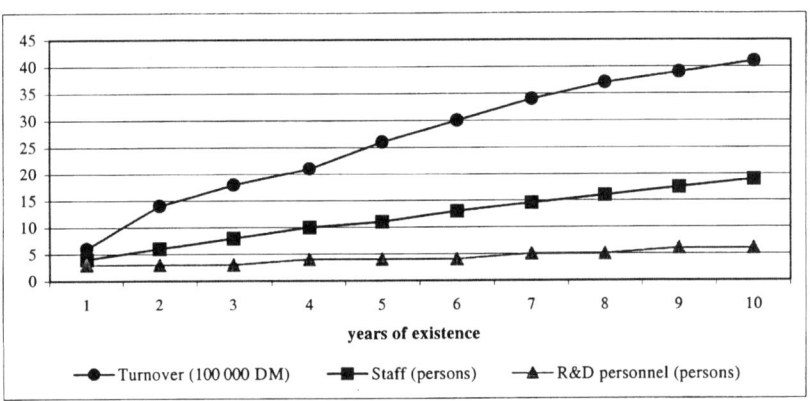

Fig. 18.3 East Germany: development of newly founded R&D-intensive enterprises
Source: Pleschak et al., 2002, p.59/60 (n = 390)

For a long time, science and economic policymakers in most countries did little or nothing at all to foster integrative processes within R&D. In some cases, this lack of intervention was a result of deliberate, neo-liberal policy, such as the laissez-faire S&T policy initially adopted by the Czech Republic; in others, such as Hungary for example, there was a bias in favor of foreign firms as providers of know-how. In most cases, however, policy-makers were for a long time simply unable to cope with this task because they themselves had been slow to consolidate, had little competence in this area and, above all, had insufficient financial resources at their disposal. "The CEE countries have only recently started restructuring policies for industrial sectors, [...] lacking political will to come to grips with the large social and regional unemployment

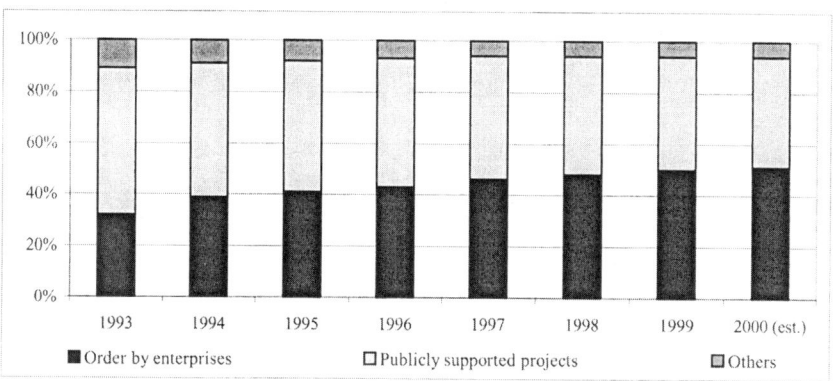

*Fig. 18.4 East Germany: structure of turnover in R&D-oriented innovative SMEs (n = 77)
Source: Kohn, 2001, p. 103/Table 26.*

problems of troubled sectors, together with limited administrative capacities of governments, led to the postponement of industrial policy measures for troubled industries. [...] Only some CEE countries have passed documents on industrial (for example Hungary) or innovation policies (for example Poland). However, we should not exaggerate the impact of these explicit policies. The effects on industry and on innovation have so far been much stronger through other policies than through innovation and industrial policy, which attempt to explicitly address these issues. After 10 years of pursuing the transition policy agenda, CEECs are now searching for alternative policy solutions that will also address the problem of their technological competitiveness. Given the current role of the state in these countries, it is unlikely that we will see the implementation of highly selective structural (industrial and technological) policies aimed at a strengthening of inter-firm and inter-sectoral technological linkages" (Radosevic, 1999, p. 3).

It was not until the end of the 1990s that renewed attempts were made to put in place an integrated science and technology policy combining support for science and education with measures to stimulate investments, exports and regional development, with the aim of restructuring domestic industries and creating SMEs. Examples are the governmental resolution on "National Policy on Research and Development of the Czech Republic" from January 5, 2000 and the "Science and Technology Policy 2000" initiative prepared by the S&T Policy College of the Hungarian Government. In the applicant countries, the main factor behind this more active approach has been the negotiations on accession to the EU.

To summarize, it can be said that considerable progress has been made in the CEECs with the establishment and consolidation of new S&T actors and that they have been partially integrated into national and international networks. In none of the countries, however, has a permanent, dynamically evolving S&T *system* able to meet the challenges of the new millennium yet emerged.

REFERENCES

Becker, J.M. (2001) 'Dieses Land haben sie noch nicht erobert... ' Kuba – Modell für Entwicklungsländer oder Fossil? *Utopie kreativ* no. 125: 253–267.

Bouché, P. (1998) "Alternative approaches to industrial R&D institutes in Hungary and Russia." *Transforming Science and Technology Systems – The Endless Transition? NATO Science Series 4: Science and Technology Policy – Vol. 23.* Ed. W. Meske, et al. Amsterdam: IOS Press, 183–197.

Couderc, M.L. (1996) *Enterprisation: Adaptation of Some Former Research Units to the New Economic Environment in Russia.* Research Paper 96-B-2. Paris: CERNA.

EBRD (2000) *Transition Report 2000.* London: European Bank for Reconstruction and Development (EBRD).

EBRD (2001) *Transition Report update – April 2001.* London: EBRD.

EC (Ed.) (1999) *Impact of the enlargement of the European Union towards the associated Central and Eastern European countries on RTD, innovation and structural policies.* Luxembourg: Office for Official Publications of the European Communities, European Commission (EC).

EC (2001) *Regular Reports on Progress towards Accession.* European Commission (EC). Internet: europa.eu.int/comm/enlargement/report2001.

Gokhberg, L. (1996) *Transformation of the Soviet R&D system.* Economic Transition and Integration Project. Laxenberg: International Institute for Applied Systems Analysis.

Gu, S. (1999) *China's Industrial Technology. Market reform and organizational change.* London and New York: The United Nations University Press, Institute for New Technologies.

Hanson, P., and K. Pavitt (1987) *The Comparative Economics of Research Development and Innovation in East and West: A Survey.* Chur: Harwood Academic Publishers.

Kohn, H. (2001) *Externe Industrieforschung im Wettbewerb.* Schriftenreihe Verband Innovativer Unternehmen e.V. (VIU) no. 5. Dresden: VIU.

Lavigne, M. (1995) *The Economics of Transition. From Socialist Economy to Market Economy.* New York: St. Martin's Press.

Legler, H., H. Belitz, B. Gehrke, C. Grenzmann, and R. Marquardt (2002) *Industrieforschung in Deutschland – Positionen im internationalen Vergleich. Materialien zur Wissenschaftsstatistik Vol. 12.* Essen: Wissenschaftsstatistik gGmbH im Stifterverband für die deutsche Wissenschaft.

Mayntz, R., U. Schimank, and P. Weingart (Eds.) (1995) *Transformation mittel- und osteuropäischer Wissenschaftssysteme. Länderberichte.* Opladen: Leske+Budrich.

Meske, W. (1993) "The restructuring of the East German research system – a provisional appraisal." *Science and Public Policy* 20, no. 5: 298–312.
Meske, W. (1994) *Veränderungen in den Verbindungen zwischen Wissenschaft und Produktion in Ostdeutschland*. WZB-Paper 94–402. Berlin: Wissenschaftszentrum Berlin für Sozialforschung (WZB).
Meske, W. (1998) *Institutional Transformation of S&T Systems in the European Economies in Transition – Comparative Analysis*. WZB-Paper P 98–403. Berlin: WZB.
Meske, W., and Dang Duy Thinh (ed.) (2000) *Vietnam's Research & Development System in the 1990's – Structural and Functional Change*. Research Report P 00–401. Berlin: WZB.
Meske, W., and A. Nadiraschwili (1994) "Umbruch der Wissenschaft in Mittel- und Osteuropa." *Institutionenvergleich und Institutionendynamik. WZB-Jahrbuch 1994*. Ed. W. Zapf und M. Dierkes. Berlin: Edition Sigma, 349–376.
Mosoni-Fried, J. (1998) "Structural Changes in Industrial R&D in Hungary: Losers and Winners." *Transforming Science and Technology Systems – The Endless Transition? NATO Science Series 4: Science and Technology Policy – Vol. 23*. Ed. W. Meske, et al. Amsterdam: IOS Press, 171–182.
Nesvetailov, G.A. (1998) "Changes in STS of Russia, Ukraine, Belarus, and Moldova – Comprehensive overview." Unpublished paper. Berlin: WZB.
OECD (Ed.) (1992) *Reviews of National Science and Technology Policy: Czech and Slovak Federal Republic*. Paris: OECD, Center for Cooperation with the European Economies in Transition.
Pleschak, F., M. Fritsch, and F. Stummer (2000) *Industrieforschung in den neuen Bundesländern. ISI-Schriftenreihe Technik, Wirtschaft und Politik – Vol. 42*. Freiberg: Fraunhofer Institut für Systemtechnik und Innovationsforschung (ISI).
Pleschak, F., H. Berteit, B. Ossenkopf, and F. Stummer (2002) *Gründung und Wachstum FuE-intensiver Unternehmen – Untersuchungen in Ostdeutschland. ISI-Schriftenreihe Technik, Wirtschaft und Politik – Vol. 47*. Freiberg: ISI.
Radosevic, S. (1997) "Systems of Innovation: From Socialism to Post-Socialism." *Systems of Innovation. Technologies, Institutions and Organizations*. Ed. C. Edquist. London and Washington: Pinter Publishers, 371–389.
Radosevic, S. (1999) "After 10 years of transformation of S&T in Central and Eastern Europe. Policy lessons." Paper presented at the TSER Project final conference 'Restructuring and reintegration of S&T systems in countries of central and eastern Europe: restructuring patterns and policy experiences', held at Hove (SPRU, University of Sussex, Brighton).
Sachs, J. (1993) *Poland's Jump to the Market Economy*. Cambridge (Mass.)/London: MIT Press.
Schimank, U. (1995) "Die Transformation der Forschungssysteme der mittel- und osteuropäischen Länder: Gemeinsamkeiten von Problemlagen und Problembearbeitung." *Transformation mittel- und osteuropäischer Wissenschaftssysteme – Länderberichte*. Ed. R. Mayntz, et al. Opladen: Leske+Budrich, 10–39.
Schneider, C. (1998) "Institutional Transformation in the Industrial R&D Sector – Changes in Organizational Structures, Functions, and Interrelations: Analysis by Country: Poland (Country Report)." Unpublished Paper. Berlin: WZB.

Schneider, C. (1998a) "Institutional Transformation in the Industrial R&D Sector – Changes in Organizational Structures, Functions, and Interrelations: Analysis by Country: Czech Republic (Country Report)." Unpublished Paper. Berlin: WZB.

SPRU (1999) *Science, Technology and Growth: Issues for Central and Eastern Europe*. Brighton: Science and Technology Policy Research Unit (SPRU).

Tichonova, M. (1998) "How to Ensure a Future for industrial R&D Institutes in Russia?" *Transforming Science and Technology Systems – The Endless Transition? NATO Science Series 4: Science and Technology Policy – Vol. 23*. Ed. W. Meske, et al. Amsterdam: IOS Press, 198–210.

Toren, N. (1994) "Professional-Support and Intellectual-Influence Networks of Russian Immigrant Scientists in Israel." *Social Studies of Science* 25: 725–743.

Webster, A. (Ed.) (1996) *Building New Bases for Innovation. The Transformation of the R&D System in Post-Socialist States*. Cambridge: Anglia Polytechnic University.

Yang, Q. (1998) "The Structural Reform of the R&D System in China." *Transforming Science and Technology Systems – The Endless Transition? NATO Science Series 4: Science and Technology Policy – Vol. 23*. Ed. W. Meske, et al. Amsterdam: IOS Press, 153–159.

Zapf, W. (1994) *Modernisierung, Wohlfahrtsentwicklung und Transformation: soziologische Aufsätze 1987 bis 1994*. Berlin: Edition Sigma.

19. THE REDUCTION IN SCIENTIFIC RESOURCES DURING THE 1990S

Werner MESKE

A preliminary overview of quantitative changes within S&T in the transformation countries of Central and Eastern Europe reveals the same basic pattern, namely a rapid and pronounced reduction, both relative and absolute, in scientific resources, especially those allocated to R&D, during the first half of the 1990s. Since then there has been rather stagnation than a new increase of resources devoted to R&D in most of the countries.

1. THE DISMANTLING OF R&D RESOURCES AS THE DOMINANT TREND

The extent of the fragmentation of the S&T systems, which was a general feature of the early stages of transformation in all CEECs, is evident from the substantial reductions in the financial resources allocated to S&T and also, and more particularly, in R&D personnel. Given the low levels of innovative activities and technological change arising out of improvements to existing products and technologies in state-owned enterprises and the relative scarcity of other forms of technology transfer (through investment in imported machinery, for example), R&D personnel were the main actors in S&T during the socialist era. For this reason, this category is used here as the main indicator of quantitative change in S&T systems.

Over and above the organizational changes, changes in the financing of S&T during the time of economic transition were influenced by different and mostly very high inflation rates, changes in wage levels and payment methods, changes to funding channels, methodological problems with statistics, taxes and fees, the frequent failure to distinguish between current and investment expenditure and so on. In consequence, the data on science expenditure in most countries is hardly comparable on a long-term basis; for this reason, changes in R&D intensity, that is the ratio of gross expenditure on R&D (GERD) to gross domestic product (GDP) will be used here to provide a preliminary overview (cf. Table 19.1). From this perspective, R&D intensity has fallen to a minimum of between 10 per cent (Ukraine) and 58 per cent (Poland) of its pre-transformation level. The exceptions are Slovenia, which maintained a relatively high level of 75 per cent and the FR Yugoslavia, which actually increased R&D intensity to 167 per cent. GDP itself had also declined significantly in all of these countries and with only a few exceptions (Poland, Slovenia, Hungary and Slovakia) has not yet recovered its initial level (EBRD, 2000); taking these changes in GDP into account, the changes in R&D intensity can be used to calculate approximate values for the level of GERD in the year 2000 relative to the position in 1989/1990. Calculated in this way, GERD in the Ukraine in the year 2000 was less than 5 per cent of the 1989 level; in most of the other countries it was less than 50 per cent and even the 'league leader' Slovenia had still not reached the 1990 level (cf. Table 19.1, last column). Even in the FR Yugoslavia R&D expenditure fell to about 81 per cent; consequently, the rise in R&D intensity there is attributable solely to the even greater decline in GDP.

In the following, we focus our quantitative analysis of S&T resources on the changes in the number and structure of R&D personnel. Of course the personnel statistics are also problematic and underwent considerable methodological changes as well, but our experience with the analysis of S&T transformation in East Germany has shown that this is the most reliable indicator available for characterizing the fundamental changes in the size and structure of the R&D potential. The magnitude of the reductions is accurately stated, even when due consideration is given to the major changes that were made in the course of the 1990s to the methodology applied in these countries to collect and properly attribute statistical data on science and R&D. R&D statistics for the CEECs, which used to be not very uniform (cf. Meske, 1990), are now increasingly conforming to the measurement of scientific and technological activities as laid down in the 1993 Frascati Manual (OECD, 1994), which is standard in the OECD countries and generally more strictly defined. However, because the curves of the general trends in R&D personnel in the CEECs (cf. Figure 19.1 and the methodological note) do not all show the absolute change, but rather extrapolate the percentage changes where radical organizational changes have taken place, the changes in methodology have affected the shape of the curve only slightly. This view is also confirmed by the changes in East German R&D personnel, which in this respect can be confirmed and itemized separately.

At the beginning of 1990, acting on a suggestion made by the Stifterverband, an umbrella organization for a number of foundations engaged in the promotion of science, the GDR statistics on R&D expenditure and personnel were recalculated in accordance with OECD methodology, with use being made of detailed primary documents (cf. Stifterverband, 1990). As a result, the changes that have taken place in East Germany since 1989 can generally be broken down accurately, with a distinction being made between changes attributable to methodology and the real changes that took place from 1990

Table 19.1 CEECs: trends in R&D intensity (GERD/GDP; %)

	1989	1990	1991	1992	1993	1994	1995	1996	1997	1998	1999	2000	Relations of min. and max. values, %	GERD in 2000 to 1989/90, %
Russia	.	**2.03**	1.43	*0.74*	0.77	0.84	0.79	0.90	0.99	0.93	1.06	1.09	36	33
Ukraine	**3.10**	2.50	2.50	1.60	0.70	0.60	0.60	0.50	0.50	0.40	0.40	*0.30*	10	4
Belarus	.	**2.27**	1.43	0.82	*0.78*	0.80	0.89	0.88	0.85	0.82	1.09	0.82	34	30
Estonia	**1.30**	1.10	0.60	*0.40*	.	.	0.63	0.60	0.60	0.61	0.76	0.70	31	43
Latvia	.	**1.60**	.	.	0.48	*0.42*	0.52	0.46	0.43	0.45	0.42	0.48	26	19
Lithuania	(0.5–0.6)*	.	.	.	*0.35*	0.52	0.48	0.52	0.57	0.57	0.52	0.60*	58	76
Poland	0.90	**1.10**	0.60	0.84	0.87	0.82	0.74	0.76	0.76	0.72	0.75	*0.58*	53	82
Czech Republic	**4.08**	2.14	2.02	1.71	1.22	1.13	*1.04*	1.08	1.18	1.28	1.25	1.37	25	33
Slovakia	**3.88**	1.75	2.25	1.88	1.53	0.96	0.98	0.97	1.13	0.82	*0.68*	0.69	18	18
Hungary	**1.96**	1.61	1.09	1.08	1.00	0.93	0.75	*0.67*	0.74	0.70	0.70	0.82	34	44
Romania	**2.60**	.	.	.	0.82	0.68	0.80	0.71	0.58	0.49	0.41	*0.37**	14	11
Bulgaria	**2.63**	2.38	1.53	1.64	1.18	0.88	0.62	*0.52*	0.52	0.59	0.59	0.55	20	14
FR Yugoslavia	*0.84*	1.11	1.15	1.20	0.86	1.17	1.11	1.26	1.29	1.24	**1.40**	.	[167]	81
Slovenia	.	1.80	.	**1.90**	1.60	1.77	1.71	1.44	*1.42*	1.48	1.50	.	75	96

*provisional/estimated data; **max.**, *min.*
Source: Own compilation based on national statistics and country chapters.

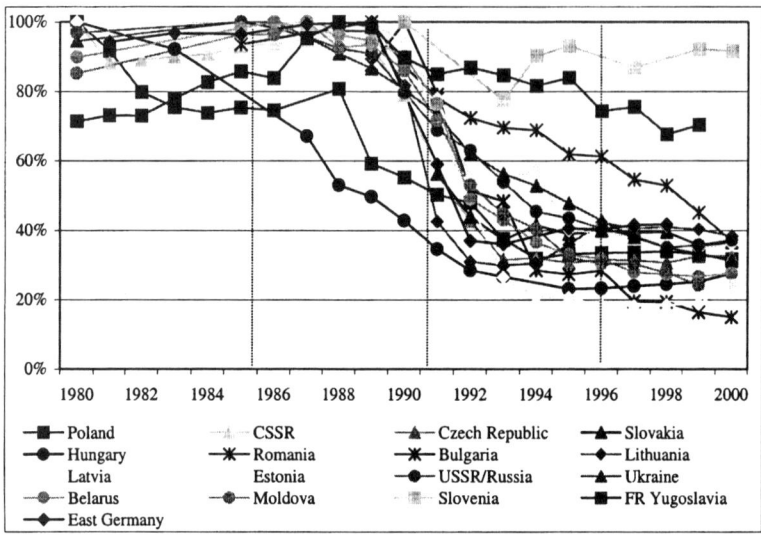

Fig. 19.1: CEECs: general trends in R&D personnel
Source: Own compilation based on national statistics and country reports.
Methodological note: The highest value in the time series for each country mentioned in this Figure and in Figures 19.3 – 19.6 has been set at 100 per cent. These time series merely indicate trends and are not based in all countries on a consistently uniform time series. In the case of breaks in the statistics as a result of changes in the data collection methodology, the changes have been extrapolated using the percentage changes rather than the absolute data.

onwards (cf. Figure 19.2). When calculated according to OECD methodology, there were approximately 30 per cent fewer people working in R&D in 1989 than GDR statistics indicated. In 1989, even according to calculations that allowed for comparability, the overall share of R&D personnel in the total employed population was equivalent to the level in the FRG; the share of the individual sectors, however, was different, because the GDR had a higher proportion of non-university research (especially in the academies) and a lower proportion of R&D in the higher-education sector and in the industrial sector than in the FRG (cf. Meske, 1993, Table 1). Nevertheless, between 1989 and 1993 the number of R&D personnel in East Germany fell by approximately two thirds, i.e. to 36 per cent of the former level; by 1995, personnel levels had returned to about 40 per cent of the 1989 level and have remained there ever since. The curve for East Germany shown in Figure 19.1 is based on these data; if the GDR statistics for 1989 were used to calculate the reduction, the cutback would have been as high as 75 per cent. Nevertheless, the East German

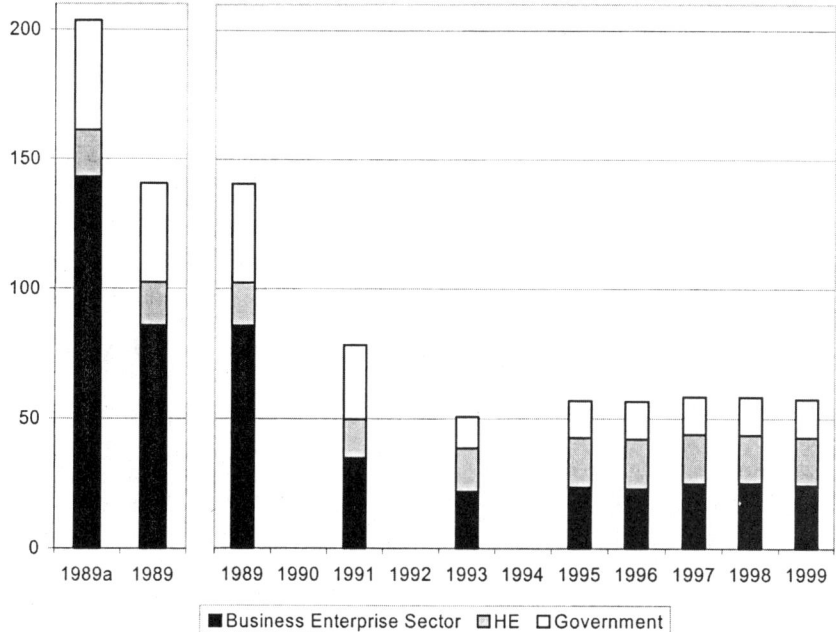

Fig. 19.2 East Germany: methodological and real changes in R&D personnel (1 000 FTE) 1989a: Statistics of the GDR, all other data according to OECD methodology. Source: Own compilation, based on Stifterverband, 1990; BMBF, 2002, p. 427.

curve would still lie within the CEEC interval of between 20 and 40 per cent (25 per cent of the 1989 maximum in 1993 and around 28 per cent from 1995 onwards).

Analysis of the changes in East Germany focused attention at an early date on the fact that the changes in science and research were taking place not uniformly but in ways that differed considerably both between and within the three major sectors of the S&T system. This was then confirmed in the course of a comparative analysis of the transformations within the S&T systems of other former socialist countries. One of the striking findings of this analysis was that substantially concurrent patterns often emerged despite differences of approach to transformation and the course of events associated with it in the individual countries. In particular, there were conspicuous differences between the three major *sectors* of the science system, i.e. higher education (HE), the governmental sector (which in most CEECs was and is centered around the Academy of Sciences/AoS) and industrial R&D (now the business enterprise sector/BES).

As with the evolution of the number of R&D personnel, however, and over and above any dominant trend towards reduction, there were also considerable differences between the countries in respect of the extent and timing of the changes; consequently, we have analyzed the changes in R&D personnel over the whole of the 1990s based on information drawn from the chapters in Part II and from other sources. This analysis reveals differences not only between the countries but also between the three major S&T sectors (public R&D/AoS; universities/HE; industrial R&D) and in the evolution of the structure of R&D personnel, in particular regarding the share of scientists in total R&D personnel and its age structure.

2. COUNTRY-RELATED ANALYSIS OF R&D PERSONNEL

2.1. THE EUROPEAN CIS COUNTRIES

In the USSR, a gradual reduction in R&D personnel had already begun at the end of the 1980s; it accelerated markedly following the dissolution of the USSR into its various successor states between 1991 and 1992.

In the four European CIS countries, this process of reduction was relatively slow from the mid-1990s onwards. In the Ukraine and Moldova it has not yet leveled out, whereas in Russia and Belarus an upturn began in 1998/99 (cf. Figure 19.3).

Russia: Against the background of a general reduction in R&D personnel between 1990 and 1993 to 67.7 per cent of the starting level (and to about 44 per cent by 1998), there were relatively slight structural changes between the three major sectors. Between 1994 and 1999, the share of the governmental sector (AoS and some former branch institutes) rose from 25 per cent to 28 per cent (cf. Gokhberg and Mindeli, 2001, p. 28, 33), while the share of R&D personnel in the HE sector (11 per cent in 1994) remained relatively constant at between 10 and 11 per cent. From 1994 onwards, the share of the business enterprise sector (now recorded in official statistics, but also including independent (former branch) R&D institutes and a wide range of very diverse forms of organization) fell slowly from 64.4 to 61.3 per cent in 1999, but the number of enterprises with their own R&D departments fell from 449 in 1990 to 240 in 1998 (cf. Gokhberg et al., 2000, p. 81). Although the number of R&D personnel employed in the private non-profit sector rose from about 300 to 1 600 (FTE) by 1999, its share, at between 0.1 and 0.2 per cent, remains vanishingly small. These data confirm the impression gained from the analysis of the organizational change that the reduction of R&D personnel in Russia was much less a reflection of fundamental structural reforms than a consequence of reduced state funding and the 'survival strategies' that all organizations had to adopt as a result. This included, in particular, retaining scientists, the R&D es-

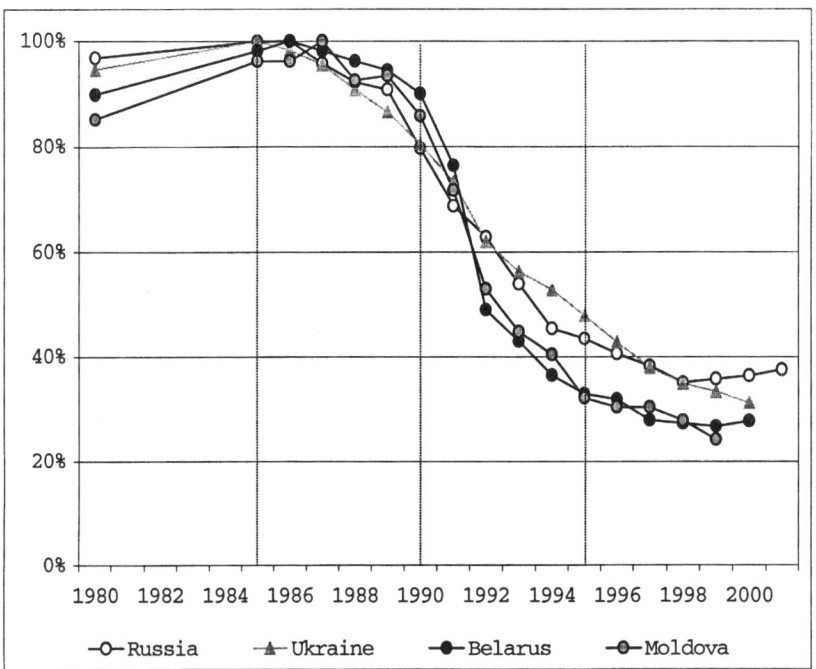

Fig. 19.3 European CIS countries – R&D personnel
Source: Own compilation, based on data from Nadiraschwili, 1994, p. 76; Gokhberg and Mindeli, 2001, p. 139; Fig. 4.2 and Tables 5.11 and 6.2 in this volume.

tablishments' most important resources. As a result of the emigration of many internationally known experts, the share of researchers in R&D personnel fell slightly up to 1994, from 52 to 47.5 per cent, but has been relatively stable since then at 48 per cent. However, within the researcher category, the share of individuals with the degree of Candidate of Science (CoS) rose from 13.4 per cent in 1991 to 20.1 per cent in 1999, while the share of those with the degree of Doctor of Science (DoS) rose from 1.8 per cent in 1991 to 5.0 per cent in 1999. As a result, the share of these most highly qualified groups within the researcher population rose overall quite considerably, from 15 per cent in 1991 to 25 per cent in 1999 (or to 12 per cent of the total R&D personnel) (Gokhberg and Mindeli, 2001, p. 28, 36). In Russia, consequently, the main adjustment in the R&D system was to the reduced state funding, which also explains the slight recovery with the renewed increase in R&D personnel from 1998 onwards. This is in line with the 'patterns of preservation, restructuring and survival' which have characterized the S&T policy in Russia in the post-Soviet era (cf. Radosevic, 2003). However, functional adjustment to the new demands made of R&D establishments by the economy and society at large in a context of increased international openness has, at best, probably only just begun.

Ukraine: The trends here are similar to those in Russia. The number of R&D employees fell between 1990 and 1993 to 70 per cent of its initial level and to only 39 per cent of that level by 2000. Unlike in Russia, however, it has not yet been possible to put an end to the fall in R&D personnel levels, since the Ukraine's economic decline had not halted until the year 2000. The share of scientists and engineers in R&D personnel remained practically unchanged at between 62 and 63 per cent, although here too, as in Russia, the share of the CoS category rose from 11 to 15 per cent and that of the DoS category from 1 to 3.5 per cent. The two categories combined increased their share from 11.7 per cent in 1990 to 13.7 per cent in 1994 and to 18.6 per cent of all scientists and engineers (or to nearly 12 per cent of the R&D personnel) in the year 2000. Here too, the sectoral structure remained relatively unchanged. The share of the AoS sector in all scientists and engineers engaged in R&D rose significantly from 18 to 27 per cent by the year 2000, not least because some parts of the former all-union (branch) R&D institutes located in the Ukraine had been incorporated into the AoS; that of the HE sector remained relatively constant at 9 or 10 per cent. The share of the branch institutes fell from 74 to 52 per cent; in-house R&D increased its share of total R&D personnel from 8.6 to 9.7 per cent. These last changes suggest that functional adjustment has also begun in industrial R&D.

Belarus: In Belarus, the number of R&D personnel fell between 1990 and 1994 to 40 per cent of its starting level and by 1999 had fallen to 30 per cent of that level; it was not until the year 2000 that the first increase, of 3.6 per cent, was recorded (cf. Table 6.2). As in Russia, this increase is obviously attributable to the recovery of GDP. Here too, the share of scientists and engineers rose from 55 per cent in 1990 to about 60 per cent (from 1994 onwards). Within this group, the share of the CoS category rose from 10 per cent in 1990 to 20 per cent in the year 2000 and that of the DoS rose from 0.9 to 4.2 per cent. Thus the two categories combined increased their share in the total number of scientists and engineers engaged in R&D from 11 per cent in 1990 to 20 per cent in 1994 and 24 per cent in the year 2000 (or more than 14 per cent of total R&D personnel).

Thus three common trends can be identified in the CIS countries. First, a drastic reduction in R&D personnel in the first half of the 1990s was followed by a slower rate of reduction and a slight recovery in line with the evolution of GDP. Second, the structure of the S&T system has changed only slightly, with industrial R&D having seen the main reductions and the governmental sector expanding in relative terms. Consequently, the S&T system is still oriented more towards science rather than to innovation in the enterprise sector. The role of the HE sector in R&D remains weak. Third, the increase in the share of the most highly qualified scientists shows that the core of the science personnel was largely maintained in the 1990s. With the necessary resources, however, actual research capability was probably severely restricted, which is reflected, for example, in the renewed decline in publications since the mid-1990s (cf. Chapter 20).

2.2. THE BALTIC STATES

In the Baltic states, the number of R&D employees fell to between 20 and 40 per cent of the former level, as was the case in the European CIS countries; here, however, the reduction began earlier and took place more quickly, for the most part between 1990 and 1992 (cf. Figure 19.4). Moreover, it went hand

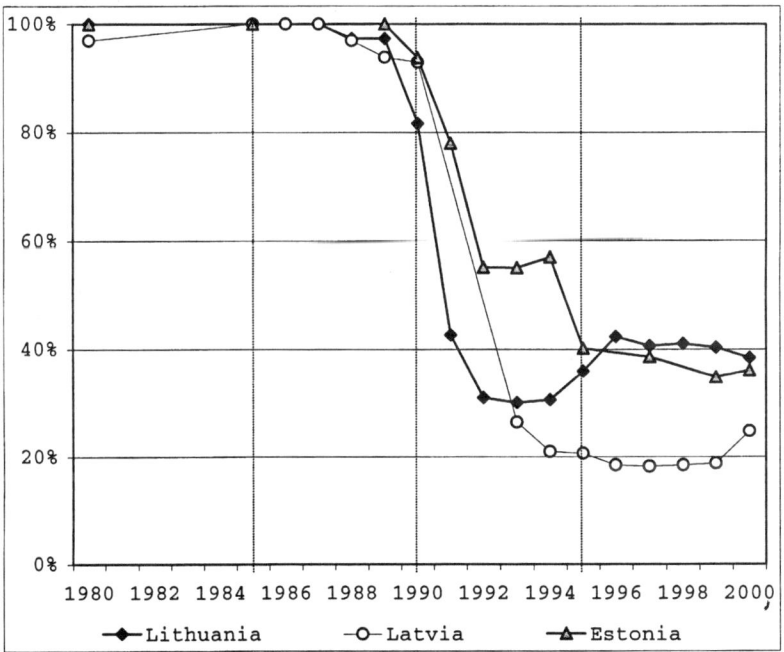

Fig. 19.4 Baltic states – R&D personnel
Source: Own compilation based on data from Nadiraschwili, 1994, p. 76; Martinson, 1995, p. 71; Tables 7.2, 8.1 and 9.3 in this volume.

in hand with extensive organizational changes and changes in the system of gathering statistics, for which reason comparable sets of figures are not always available for the 1990–1994 period.

Estonia: The number of R&D personnel fell between 1990 and 1993 to 59 per cent of its previous level, although the number of researchers fell to 'only'

63 per cent of that level. The number of researchers in the HE sector remained virtually unchanged, with the result that its share in the total researcher population rose from 40 per cent in 1990 to about 60 per cent in 1994. At the same time, the share of the AoS rose from 17 to almost 22 per cent (despite a fall in the absolute number of researchers), while the share of the remaining (mostly branch) institutes fell from 43 to 22 per cent as a result of the rapid reduction in personnel. After 1994, following further restructuring and reduction of scientific institutions and the incorporation of the AoS institutes into the university system, the share of the HE sector in the total researcher population rose to 80 per cent by 1999 (14 per cent of whom were in incorporated research institutes). The governmental sector had a share of only 19 per cent, while the private non-profit sector had a less than 1 per cent share. It should be noted that data on the business sector are not separately identified in Estonian statistics; however, the country's remaining industrial R&D capacities are likely to be relatively limited (see Table 7.2; cf. Martinson, 1995, p. 76–79; and Paasi, 1998). In 2000 only 507 (out of a total 4 570) researchers were employed by firms and only 9 per cent of R&D was financed by the business enterprise sector (Martinson, 2002, p. 6).

Lithuania: The number of R&D personnel fell between 1989 and 1993 to 31 per cent of its previous level. Since then, the HE sector's share in the total researcher population has been about 70 per cent. The public sector (state research institutes) has a share of about 25 per cent, other institutes a share of about 5 per cent (cf. Tables 9.3 and 9.4; Martinson, 1995, p. 78).

Latvia: Against the background of a reduction in R&D personnel between 1990 and 1994 to nearly 20 per cent of its former level (the largest fall in all the countries!), the share of the HE sector rose to 32 per cent while that of other non-profit research institutes (public sector) was 40 per cent. Here, however, the share of the business sector was still 28 per cent (cf. Table 8.1 and Martinson, 1995, p. 78). After a drop in R&D personnel that continued until 1997, moderate growth was achieved from 1998 onwards, and in the year 2000 there was even a 32 per cent increase.

In all the Baltic states the sharp decline in R&D personnel was associated with considerable organizational changes in favor of the HE sector. This created the conditions for improving the integration of teaching and research in the public sector. It remains to be seen whether public-sector R&D will be able to compensate for the severe decline in industrial R&D in Estonia and Lithuania.

2.3. THE CENTRAL EAST EUROPEAN COUNTRIES

Despite a general trend towards a reduction in R&D personnel in the five, now six independent CEECs, there were differences from country to country in the 1980s as well as in the 1990s, which reinforces our view that even under socialism there was some diversity (cf. Figure 19.5). Hungary and Poland were

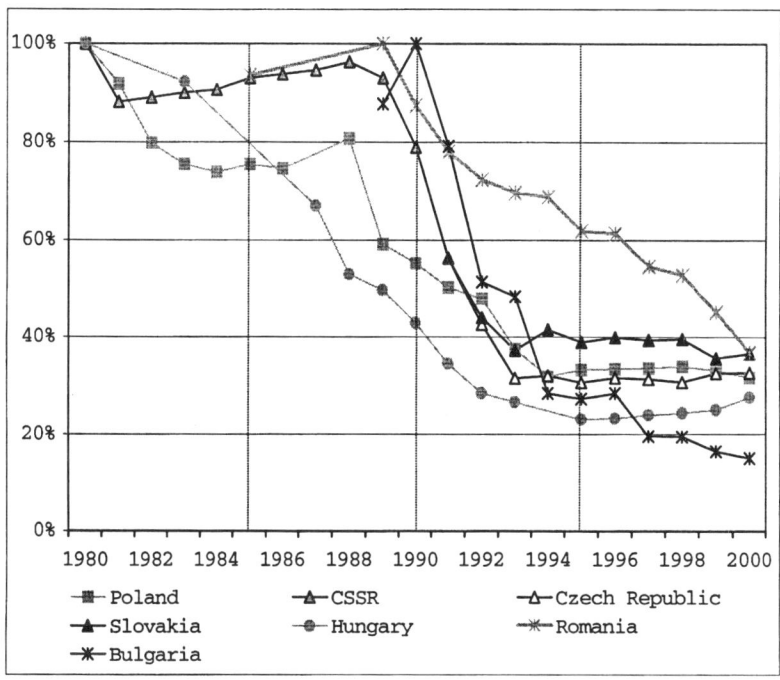

Fig. 19.5 Six CEECs – R&D personnel
Source: Own compilation based on national statistics, country reports, and Tables 10.2, 11.1, 11.2, 12.2, 14.1, 15.8 and Figure 13.1 in this volume.

exceptions in that their peak staff levels were reached as early as 1980, whereas in all the other countries, the increase in personnel levels continued until the end of the 1980s. In Hungary and Poland, personnel levels were more or less halved during the 1980s. Thus the further cutbacks from 1990 onwards resulted in a drop to 'only' about 50 per cent of 1990 levels rather than to 20–40 per cent as in the other countries. Nevertheless, levels in all these countries are now very similar, standing at between 20 and 40 per cent of the all-time high (with Romania, too, which has been slow to restructure, reaching that level in 2000, whereas Bulgaria had fallen below the 20 per cent level since 1997).

As already shown in Chapter 18 and as is evident from the path of the curves, this similarity in the level of reduction hitherto attained conceals very diverse changes in the individual countries.

Poland: R&D personnel levels fell between 1988 and 1994 to less than 40 per cent or to about 30 per cent of their former peak level und have remained at around this level ever since.

However, the impact of the cutbacks on the various sectors differed considerably, with the share of the HE sector, for example, rising from 24 per cent in 1988 to 41 per cent in 1994 and that of the AoS rising from 6 to 10 per cent. On the other hand, the branch institutes' share fell from 45 to 38 per cent, while that of in-house R&D fell by more than half, from 24 to 11 per cent; the practically oriented S&T service units virtually disappeared, with their share falling from 1.5 to 0.1 per cent (cf. Table 10.2). This means that industrial R&D bore the brunt of the reductions; as early as 1994, the public sectors (HE plus AoS) had increased their share in total R&D personnel from 30 per cent in 1988 to 50 per cent. The HE sector even began to grow again in absolute terms subsequently, while the industrial sector suffered further reductions. As a result, the relative positions of public and industrial R&D were virtually reversed between 1988 (30 and 70 per cent respectively) and 2000 (62 and 38 per cent respectively). After a minor increase starting in 1995 (only in the HE sector), in 1999 and 2000 the number of R&D employees fell again, especially in industrial R&D (branch institutes and in-house R&D). In Poland, the transformation was associated with a strengthening of the (traditional) HE sector and the maintenance of AoS capacities as part of the public research infrastructure. As the ability of successive governments to take action has improved as GDP has increased, these sectors have been further strengthened. Developments in industrial R&D have gone in exactly the opposite direction. After the deep cuts in the 1980s, efforts were made initially to retain the branch institutes' core capacities as far as possible; the objective now is to ensure that they are adapted much more closely to actual needs and hence to financial demand. However, since the privatization of the economy has been a slow process in Poland and is not yet complete (on this point see Mense-Petermann, 2002, p. 236; and Szomburg, 2001, p. 87), its ability to absorb R&D results is still very limited. The upshot is that, in contrast to public R&D, there has been a further decline in industrial R&D.

In conjunction with these structural shifts and assisted by funding criteria that, among other things, stipulate a minimum number of highly qualified scientists among the employees of scientific institutes, the number of scientists and researchers declined by much less than the overall personnel level, with their share in the total R&D workforce increasing in consequence. Within the total employment of S&T institutions, the share of PhDs (CoS) increased from 23.0 per cent in 1997 to 26.1 per cent in 2000 and the share of PhDs with a further research degree (DoS) rose from 7.0 per cent to 7.8 per cent – taken together an increase from 30 to 33.9 per cent within three years (data by Kozlowski from the Statistical Yearbook of Poland). This is attributable to the structural changes that have taken place at the expense of industrial R&D, while at the same time reflecting a functional restructuring of the S&T system, similar to those that have taken place in other countries in transition, towards more basic science and less experimental development and service.

Czech Republic: Between 1989 and 1994, the number of R&D personnel declined to about 30 per cent of its peak level, with the sharpest reduction taking place between 1990 and 1992. The absolute number of university scientists actually increased during this period (although the number of researchers working in this sector declined to one third of its peak level) and the number of AoS personnel declined by 'only' 50 per cent. The reduction, driven by neoliberal thinking, took place largely at the expense of industrial R&D personnel. Between 1990 and 1995, the number of industrial R&D personnel declined to about 30 per cent of its peak level and in manufacturing to as little as 12 per cent (cf. Chapter 11, in particular Tables 11.1 and 11.3). As a result, the share of the business enterprise sector in total R&D personnel had declined to 50 per cent by 1995; the governmental sector (including AoS) had a share of about 34 per cent and the HE sector about 16 per cent. In 1999/2000 the number of R&D personnel increased once again, especially in the HE sector, reaching 107 per cent of its 1995 level, and a 22 per cent share of the total in 2000. The share of the BES, on the other hand, had declined to 48 per cent and that of the governmental sector to less than 30 per cent (cf. Table 11.2). These changes were accompanied by a slight increase in the relative number of researchers from 53 to 56 per cent of total R&D personnel and in their qualification levels (cf. Schneider, 1998, p. 41).

The organizational changes in the Czech Republic have primarily involved functional and structural adaptations to the new (market) conditions. The HE sector is being strengthened by state funding, while industrial R&D, which has been severely slimmed down, is now largely concentrated in (private) manufacturing companies or in independent firms providing scientific and technical services. Although here too the number of R&D personnel was severely reduced in the early 1990s, the Czech Republic still has (together with Slovenia) the highest R&D density of all the CEECs because of its high starting level (cf. Table 19.1).

Slovakia: Although the initial conditions here were similar to those in the Czech Republic, the personnel reduction took place somewhat more slowly because of policy differences (no liberal, laissez-faire policy); to date, however, it has not proved possible to halt the cutbacks in the BES. The number of R&D personnel fell to 40 per cent of its peak value between 1989 and 1993, with a reduction in the HE sector to 78 per cent, in the AoS to 54 per cent and in industrial R&D to 28 per cent. This led to a reduction in the industry's share from 54 to 37 per cent, while the HE sector's share rose from 6 to 13 per cent and that of the AoS from 14 to 19 per cent (cf. tables in Chapter 12). Between 1994 and 2000, the number of R&D personnel fell again to less than 90 per cent of its 1994 level. It remained more or less unchanged in the AoS at 97 per cent of its 1994 level, increased further in the HE sector to 118 per cent and fell even further in industry. In the year 2000, R&D personnel increased by 2.5 percent – but only in HE and the AoS. It decreased further in manufacturing and in the whole business enterprise sector (BES) by about 10 per cent, with the result that by 2000 BES had only a 34 per cent share in total R&D personnel.

At the same time, PhDs in all three sectors were far less affected by the reductions than the R&D workforce as a whole. Consequently, between 1989 and 1993, scientists with a PhD increased their share in the R&D workforce from 11 to 20 per cent and to 29 per cent by 2000 (calculated by OECD methods).

Hungary: The reduction in R&D personnel that had begun in the 1980s continued until the mid-1990s. Just in the seven years between 1988 and 1995, the number declined further to 45 per cent of its starting level. The greatest reduction was in the business sector (and particularly in-house R&D), whose share in 1995 was still only 38 per cent, while the HE sector had seen its share rise to 33 per cent. The governmental sector (including AoS) maintained its share of just under 30 per cent (cf. Imre, 1998, p. 72; Mosoni-Fried, 1997). Between 1995 and 2000, the number of R&D personnel rose again to 120 per cent of its 1995 level (= 28 per cent of the 1980 level), with the business enterprise sector in particular increasing its share at the expense of the governmental sector (cf. Figure 13.1 and Table 13.2). This increase is attributable largely to the activities of foreign subsidiaries that not only produce in Hungary but have also in some cases taken over or built up and extended domestic R&D capacities. Thus in contrast to other CEECs, this increase is due less to the stabilization of state R&D funding than to the increasing integration of R&D into the activities of firms operating internationally.

Over the period as a whole, the share of scientists and engineers in total R&D personnel rose from 47.5 per cent in 1988 to 52 per cent in 1993 and 59 per cent in 1999. Between 1991 and 1997, the number of scientists with PhDs working in R&D units increased to 124 per cent and those with the DoS to 117 per cent of the starting level, reaching together a share of 26 per cent of total R&D personnel by 1997 (cf. IRO-AoS, 1999). This increase is here not due solely to the structural shift in favor of the HE and Academy sectors (as is the case in the other countries), since the industrial R&D facilities in the foreign investment enterprises operate to international standards with good financial resources and a high share of scientists.

Romania: From 1989 onwards, there was a constant but relatively slow decline in the number of R&D personnel to 62 per cent in 1995 and 37 per cent in 2000. Between 1993 and 2000, numbers fell to 53 per cent (of the 1993 level), with the sharpest decline in the enterprise sector (to 43 per cent); the governmental sector saw a decline to 73 per cent, while the HE sector actually recorded an increase to 184 per cent. However, since the latter sector had only a 2.7 per cent share in total R&D personnel in 1993, compared with the enterprise sector's 77 per cent and the governmental sector's 20 per cent share, no fundamental change in the overall structure has taken place to date, despite the changed dynamic. In the year 2000, the enterprise sector's share was still almost 62 per cent (primarily in the form of state-owned facilities in the former branch institutes), while the governmental sector had increased its share to 28 per cent and the HE sector to nearly 10 per cent.

A comparison of Romania with Hungary reveals common features as well as fundamental differences in the evolution of R&D personnel. As Figure 19.5 shows, both countries' personnel curves had come very close to each other in the year 2000 at about 30 per cent of the peak level, despite the path of those curves being very different. In Hungary, after a relatively severe reduction, the number of R&D personnel has now begun to rise in internationally competitive companies (which have a share of more than 40 per cent of the total potential) and in the HE sector. Consequently, R&D can be regarded as a healthy basis for the further strengthening of Hungary's national innovation system. In Romania, on the other hand, the reduction in R&D personnel has been extremely slow and has not yet come to an end. Since this reduction took place without any fundamental organizational or functional changes to the R&D system and industrial R&D still has little connection with firms, the adjustment of R&D to market economic conditions will continue and probably lead to a further reduction in R&D personnel.

Bulgaria: The number of R&D employees rose until 1990, after which there was a very sharp reduction to 28 per cent in 1994. This reduction was concentrated almost wholly in the branch sector, while the HE sector actually grew in absolute terms. Accordingly, between 1989 and 1993, the branch sector's share of scientists fell from 42 to 24 per cent, while the AoS increased its share from 15 to 17 per cent and that of HE went up from 43 to no less than 59 per cent. After a short 'respite' in 1994–1996, by the year 2000 R&D personnel figures had dropped once again by almost 50 per cent. This represented a fall to a level of just 15 per cent of the 1990 all-time high. After 1993, cutbacks were concentrated, as they had been previously, in the enterprise sector, which by the year 2000 had a share of only 15 per cent of total R&D personnel (cf. Table 15.8). Since 1997, therefore, Bulgaria has become the country with the largest proportionate losses of R&D personnel. This is attributable primarily to the fact that, after a relatively severe reduction in R&D as a predominantly 'communist inheritance' at the beginning of the 1990s, the country's own political instability and the effects of the decline in neighboring Serbia meant that economic recovery started later and more slowly than in most of the other CEECs (cf. Figure 18.2). Consequently, state funding for R&D is still lacking in Bulgaria and demand in the economy is also low. This situation has to date prevented any fundamental functional changes in the S&T system. This is also reflected in the structure of qualifications. The number of researchers rose between 1994 and 1996, but then fell again, reaching 75 per cent of the 1994 level by the year 2000 (cf. Table 15.6). Since 1994 the relative shares of researchers have shifted in favor of the state sector, which had a 68 per cent share in 2000, compared to 57 per cent in 1994; the figures for the HE and enterprise sectors were 20 and 12 per cent respectively, compared to 27 and 15 per cent in 1994. In contrast to these trends, the number of scientists in the HE sector remained virtually unchanged, something that can be attributed to an increase in teaching activity accompanied by a reduction in research. Furthermore, the absolute and relative decline in all three sectors in the share of researchers with scientific degrees in the total body of researchers (cf. Table 15.7) can be viewed as an indicator of the continuing insecurity of research activity and of its continuing loss of attractiveness in Bulgaria, which contrasts with developments in most other countries.

2.4. THE SUCCESSOR STATES TO THE SFR YUGOSLAVIA

The successor states to the SFR Yugoslavia deviate somewhat from the basic CEEC pattern. Thus the FR Yugoslavia and Slovenia recorded only relatively slight reductions of about 20 per cent in R&D personnel between 1990 and 1994, while Croatia and Macedonia actually increased their R&D personnel (cf. Figure 19.6). A fundamental reason for this particularity seems to lie in Yugoslavia's looser coupling to the USSR and its avoidance of full integration into the 'socialist world system', with its overestimation of S&T. As a consequence, in the former SFR Yugoslavia there were some fundamental quantitative and institutional deviations from the 'Soviet model' of STS. In particular, the universities rather than industry were mainly responsible for R&D. The 'branch R&D institutes' did not constitute a major sector and in-house R&D was less developed. There were only a few independent R&D institutes whose main source of income was public contracts. It was possible for them to survive transition with reduced government funding.

FR Yugoslavia: The number of R&D personnel fell between 1989 and 1999 to 70 per cent of its starting level; the first reduction took place in 1990/1991, but it was not until 1996 that the second round of cuts began. The main factor here was the economic decline caused by internal and external conflicts, up to

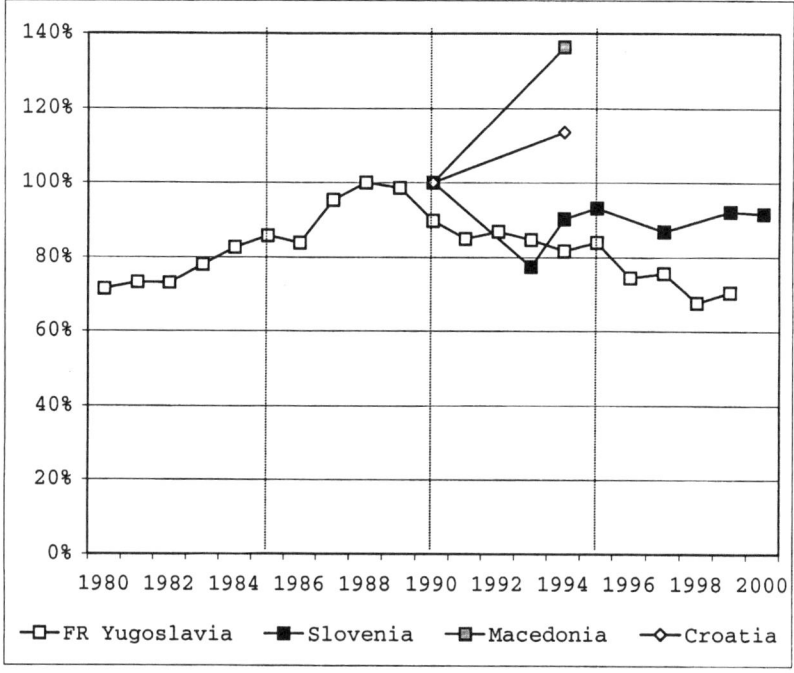

Fig. 19.6 Successor states of the SFR Yugoslavia – R&D personnel
Source: *Own compilation based on data from Kutlača, 1996; Stanovnik, 1998; SURS, 2001, p. 28; and Table 16.6 in this volume.*

and including the NATO bombing, which caused a further fall in GDP after 1998 (cf. Figure 18.2).

The reductions were greatest in state independent R&D institutes, where personnel levels fell to 70 per cent of the starting level between 1990 and 1999; in the HE sector and in R&D units in industry, the reductions were to 84 and 83 per cent of the previous level respectively. As a result, the HE sector further increased its relatively high share in total R&D personnel from 52 per cent in 1990 to 56 per cent in 1999, to the detriment of the independent institutes, whose share declined from 40 to 36 per cent. R&D institutes in industry retained their low share of about 8 per cent. In the HE sector, the number of researchers actually increased in absolute terms to 114 per cent of the starting level, while it remained virtually the same in industry. In the course of the 1990s, therefore, the share of researchers in total R&D personnel rose from 40 to 50 per cent (cf. Figure 16.4), which meant that skilled personnel were for the most part retained in R&D.

Slovenia: Between 1990 and 1993, the number of R&D employees fell to

77 per cent of its initial level, but then rose again by 20 per cent until 1995, and has stagnated since then at 92 per cent of the 1990 level. The number of researchers rose in absolute terms from 1990 onwards and reached 130 per cent of its starting level by the year 2000, bringing this group's share within total R&D personnel from 38 per cent in 1990 to 54 per cent in 2000. Between 1993 and 2000, there was a particularly marked increase in the number of R&D employees in the business enterprise sector, which went up to 143 per cent of the 1993 level, while in the HE sector it rose to 119 per cent and in government institutes to 102 per cent of its former level (cf. Table 17.1). The basis for this was the relatively stable growth of GDP from 1993 onwards and a strengthening, albeit very gradual, of innovation activity in the business enterprise sector. Consequently, the business sector's share in total R&D personnel rose from 33 to 40 per cent, while the shares of the other two sectors fell, from 30 to 26 per cent in the case of the governmental sector and from 37 to 34 per cent in higher education. The share of researchers rose in all sectors. In 1999, the proportion of PhDs within total R&D personnel was 21 per cent and among researchers 37 per cent (cf. Table 17.2).

3. COMPARATIVE ANALYSIS OF R&D PERSONNEL CHANGES

A comparative analysis of the changes in R&D personnel in all the CEECs under investigation here reveals the following trends.

3.1. BASIC TRENDS

In virtually all the countries, the first half of the 1990s saw a considerable reduction in the number of R&D personnel, in most cases to between 20 and 50 per cent of the peak level in the 1980s (cf. Figure 19.1). The successor states to the SRFY are an exception in this respect, since they recorded only slight declines, of between 10 and 20 per cent, or even increases.

In the second half of the 1990s, the reductions slowed down considerably in all the countries. In some countries, the reductions then continued, albeit at a slower pace (Ukraine, Lithuania, Romania, Bulgaria, FR Yugoslavia); in most countries, however, numbers stabilized, and they began to rise again slightly towards the end of the decade in a few countries (Russia, Belarus, Latvia, and Hungary). Clearly, these changes in R&D personnel in the 1990s were influenced mainly by economic circumstances, that is the changes in GDP (cf. Figure 18.2). After the sharp decline in GDP and R&D personnel in the first half of the 1990s in all the countries, those countries whose GDP stagnated or continued to decline in the second half of the 1990s were unable to halt the reduction in R&D personnel. On the other hand, those countries whose economies began to recover were able to build up their R&D personnel again, particularly if organizational and functional changes to the R&D system had taken place at the same time.

3.2. DEVELOPMENTS IN THE THREE MAJOR S&T SECTORS

The changes in the three S&T sectors were even more differentiated.

a) In one group of countries, the reductions were (relatively evenly) spread across all three sectors; in other words, there were no essential structural shifts despite the fact that the industrial sector had the largest absolute losses. This was the basic trend in the European CIS countries in particular, as well as in Romania (cf. Figure 19.7). In these countries, the transforma-

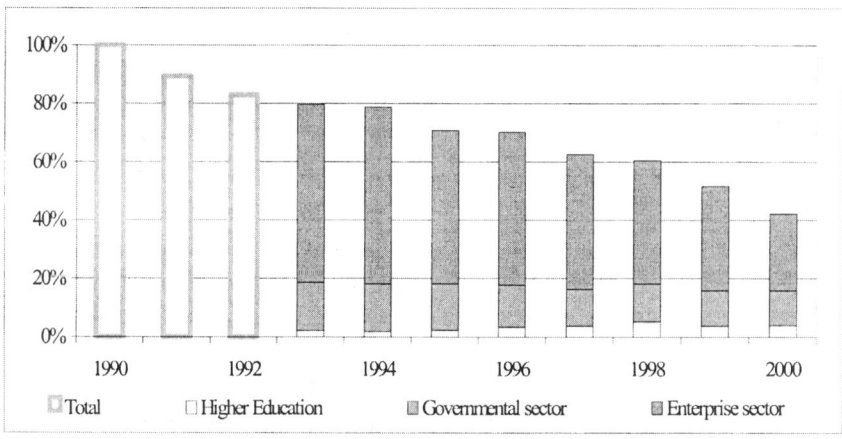

Fig. 19.7 Romania: R&D personnel by sector (%)
Source: Own compilation based on Table 14.1.

tion proceeded falteringly. Economic decline meant there was no money available to fund R&D and because of the particularly pronounced 'Soviet nature' of the S&T system even functional and structural changes were ineffectual. Thus in these countries – and only in these countries – industrial R&D maintained its dominant share of between 50 and 70 per cent of total R&D personnel, who are still largely concentrated in the former branch institutes. Thus the high share of industrial R&D in these countries is essentially a problematic inheritance from the socialist period rather than an effective precondition for new innovations in the years to come. This view is reinforced by the fact that the share of in-house R&D capacities in these countries is below 10 per cent. At the same time, the AoS have been able to maintain and even increase their significant 20 per cent share of total R&D personnel. Although the HE sector has increased its (previously very low) share of total R&D personnel in all these countries, it was still only 10 per cent at most in the year 2000.

b) In another, relatively large group of countries, public-sector R&D (HE and government) increased its share of total R&D personnel, in some cases considerably, despite a reduction in absolute numbers in most countries.

This group includes Poland (cf. Fig. 19.8), the Baltic states, the Czech Republic, Slovakia, Bulgaria and the FR Yugoslavia. Here, the largest (absolute and relative) losses of R&D personnel were in industry.

These losses in industrial R&D particularly affected the former branch institutes and, contrary to all expectations, also had a very serious impact on the relatively highly developed in-house R&D capacities in Poland and in the Czech Republic. The above-average reduction in industrial R&D should be regarded primarily as an adjustment to market economic conditions and to a demand that has largely disappeared.

In some countries (Bulgaria and the Czech Republic, for example), R&D in the HE sector remained relatively weak, while in Estonia, Lithuania and Poland it was considerably strengthened. In Estonia, Lithuania and the FR Yugoslavia the governmental sector was considerably reduced in favor of the HE sector, while in Bulgaria and Slovakia it maintained or even increased its share. All the countries in this group have maintained their public R&D sector as far as possible, giving preference to either the HE or the governmental sector (organizational centers of gravity that also exist in OECD countries) depending on their various traditions and political orientations.

c) A special group consists of Slovenia and Hungary. In Slovenia, there were, at the beginning of the 1990s, only relatively small changes in total R&D personnel and its sectoral structure. This is attributable primarily to the specific starting situation of this country, which did not have a Soviet-influenced R&D system. But, in the second half of the 1990s in Slovenia, the number of industrial R&D personnel increased as a consequence of and precondition for the favorable evolution of domestic enterprises. Its share in the R&D personnel reached 40 per cent. A similar renewed growth of industrial R&D can be observed in Hungary, too, to a high degree based on FDI (cf. Fig. 19.9). These two countries are examples that a strengthening of this sector cannot be expected until there is an upturn in innovative activity among competitive firms.

These sectoral analyses of R&D personnel changes show that it is usually the *HE sector* that has experienced the lowest cutbacks; in some countries it has even grown in absolute terms. In all countries this expansion of the HE sector is attributable especially to an increase in teaching activity, while research

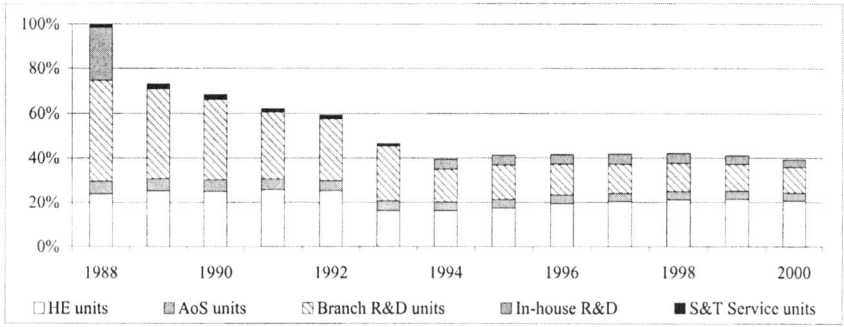

Fig. 19.8: Poland: R&D personnel by sector (%)
Source: Own compilation based on Chapter 10.
Methodological note: No data available for 'In-house R&D' between 1989–1993; therefore the total height of the columns for these years is not comparable with those for the years before and after; data since 1995 according to OECD methodology, therefore not comparable to previous years.

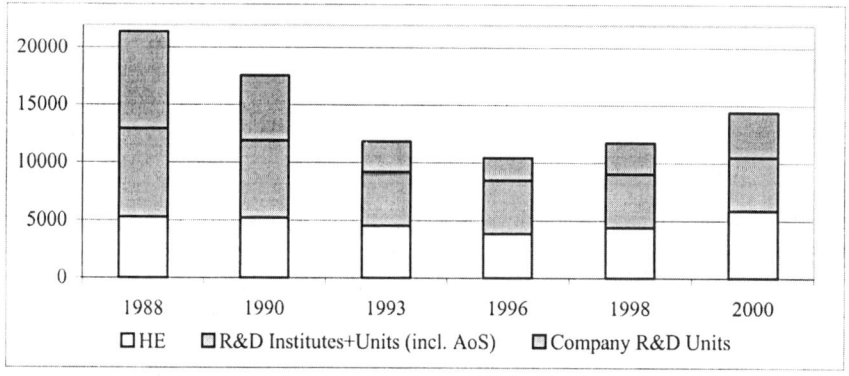

Fig. 19.9 Hungary: scientists and engineers by R&D sector (calculated FTE)
Source: Own compilation based on Chapter 13.

has tended either to be reduced or to be strengthened at the expense of the former Academy sector, which has seen institutes transferred to the universities. In practically all of these countries HE institutions are becoming increasingly diverse, with private and regional establishments emerging; this increasing diversity has been accompanied by an expansion of student numbers and often of faculty as well. In the *governmental sector* there are widely varying tendencies. In some countries the former Academy institutes or the public research institutes, as the case may be, have been maintained and strengthened, while in other countries they have faced sizeable reductions or in some cases been made subordinate to the universities, in particular in Estonia.

On the whole, it is *industrial R&D* that has consistently suffered the greatest absolute and – in most countries – also relative losses. That is the case in countries in which such capacities were concentrated primarily in the independent branch R&D institutes. However, it is surprising that in-house capacities, which some countries did indeed have, have also undergone considerable reduction, because most of the major enterprises in the modern branches of the economy, in which such capacities were concentrated, have by now been privatized, eliminated or heavily downsized. Clearly, we are dealing here with several concurrent trends that were, it is true, more or less pronounced in the individual countries but which ultimately all tended in the same direction, that is towards the dismantling of industrial R&D capacities. The economic decline after the collapse of the socialist bloc led to a decline in output and to the closure of whole enterprises and their R&D facilities, particularly in the armaments industry. Moreover, inadequate funding meant that rationalization measures had also to be taken in virtually all R&D establishments; one of the main objectives here was to retain the core scientific personnel. The former branch institutes' efforts to adjust to the demand of the market economy by increasing the share of contract research remained futile, since firms were hardly in a financial position to innovate or preferred to do so through technology transfer from the West. True, the migration of industrial R&D personnel to the business enterprise sector has strengthened innovative SMEs, but it has reduced R&D capacities. Despite the stability and, in some cases, the economic recovery that has now been achieved, the best that has proved possible in most countries is for the dismantling of R&D to be held up rather than halted; only in Hungary and Slovenia has industrial R&D begun to expand again in a way that is functionally linked to firms' innovation activities. The comparison of Figures 19.7–19.9 shows these divergent trends in different CEECs.

3.3. CHANGES BY SCIENTIFIC FIELD

These sectoral and functional shifts toward teaching and away from (industrial) research activity were accompanied by changes that varied from one scientific field to another. In this area there is a clearly evident trend towards a strengthening of the humanities and social sciences, in particular such disciplines as economics, management science and jurisprudence, which become more important in democracies subject to the rules of the market economy, as opposed to the natural sciences and, in particular, engineering, which once predominated in the socialist countries and are now in the doldrums or in some countries in decline in absolute terms. By virtue of their former pre-eminence, however, their position among R&D activities often remains fairly strong relative to Western countries. One example is the evolution of post-graduate student numbers in Belarus (cf. Table 6.4).

These macro- and meso-level changes in the organization and activities of scientific facilities have had a great influence on the remaining staff and especially on the qualification, function and age structure.

3.4. CHANGES IN THE QUALIFICATION STRUCTURE

In virtually all countries, the reduction of the number of researchers and/or scientists and engineers was smaller than that of total R&D personnel. This was the case in East Germany, where R&D personnel declined from 1989–93 to 36 per cent of the peak level, while the number of scientists 'only' went down to 43 per cent. As a result of the transfer of West German scientists into top positions (professors, directors of institutes) the number of scientists began to rise again from 1995 onwards, reaching almost 60 per cent of the 1989 level, while total R&D personnel rose only to approximately 40 per cent (cf. Meske, 2001). An increase in the proportion of researchers and/or scientists and engineers as well as of scientists with PhDs in the total R&D workforce is a phenomenon that is typical of the CEECs (cf. section 2 in this chapter). In particular, the number of the most highly qualified personnel (with PhD; the degrees of Candidate of Science/CoS and DoS) fell only slightly and in some cases not at all. As a result, the share of these most highly qualified categories within total R&D personnel increased, and in some countries their number even increased in absolute terms. The main reason for this was the structural changes in favor of public R&D, where the share of scientists is generally higher than in industrial R&D. This trend was reinforced by internal rationalization measures and the exclusion of non-research activities from R&D establishments. This is also confirmed by an analysis that found that the overcapacities within the GDR's research personnel in comparison to the FRG "do not relate so much to the professional category of scientists as to auxiliary research personnel. In the FRG there are 1.3 subordinate staff members per researcher, in the GDR 2.44 such staff members" (Meyer, 1990, p. 39; own translation). Since these trends were more or less pronounced in the individual countries, it is hardly surprising that, by the end of the 1990s, numbers in the most highly qualified personnel categories varied from country to country. Thus the share of researchers with PhD/CoS/DoS in total R&D personnel climbed as high as 45–50 per cent in countries such as Estonia and Slovakia, where industrial R&D now has a less than 10 per cent share. Even in countries such as Hungary and Poland, where industrial R&D had a share of up to 40 per cent, the most highly qualified categories had a share in excess of 20 per cent. In the CIS countries, on the other hand, where industrial R&D still had a 50–60 per cent share and the processes of organizational restructuring had proceeded slowly, these categories had a share of only 12–14 per cent, despite the rise that took place here too.

3.5. CHANGES IN THE AGE STRUCTURE

However, the retention of above-average numbers of the most highly qualified scientists and engineers in R&D was associated with major changes in the age structure. Against the background of considerable reductions in the overall number of R&D personnel, it was primarily young and middle-aged scientists who left the scientific establishments, whether voluntarily or because they were forced to do so. At the same time, very few young people were prepared to embark upon a relatively poorly paid and above all extremely uncertain scientific career. As a result, by the end of the 1990s, there was just a tiny fraction of scientists under the age of 30, and even the share of those under 40 had declined dramatically (in Lithuania from almost 24 to under 13 per cent from 1993 to 1999; it only began to increase again in 2000; cf. Figure 19.10).

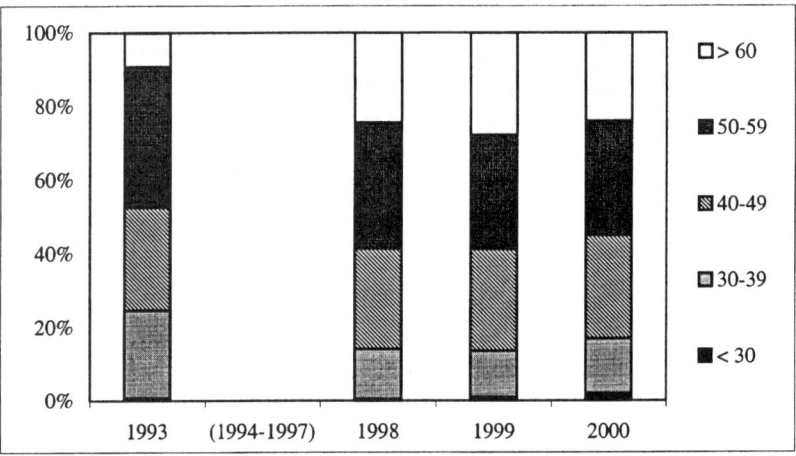

Fig. 19.10 Lithuania: age structure of scientists (%)
Source: Own compilation based on Table 9.6.

On the other hand, the share of the over-50s and over-60s rose significantly (in Estonia, the share of researchers aged over 50 rose from 34 to 44 per cent between 1991 and 1999; in Romania, the over-40s had a 62 per cent share by 1999; cf. Chapters 7 and 14). Only in the successor states to the SFR Yugoslavia were effective measures taken to counter this trend, a particularly problematic one for the future development of science and technology in the CEECs (cf. Nesvetailov, 1998), by instituting a number of special support programs for young people. As a result, the average age of researchers in Slovenia fell from 44–46 years in the 1980s to 38–40 years at the end of the 1990s (cf. Chapter 17). In the other countries, it is only in recent years that there have been signs that interest in PhD courses and careers in science is beginning to grow once again (in Belarus, for example; cf. Table 6.4).

REFERENCES

BMBF (2002) *Faktenbericht Forschung*. Berlin: Bundesministerium für Bildung und Forschung (BMBF).
EBRD (2000) *Transition Report 2000*. London: European Bank for Reconstruction and Development (EBRD).
Gokhberg, L., and L. Mindeli (Eds.) (2001) *Russian Science and Technology at a Glance: 2000. Data Book*. Moscow: Centr issledovanii i statistiki nauki (Center for Science Research and Statistics – CSRS).
Gokhberg, L., A. Gudkova, L. Mindeli, L. Pipiia, and A. Sokolov (2000) *Organizacionnaia Struktura Rossiiskoj Nauki* (Organizational structure of Russian science). Moscow: CSRS.
Imre, J. (1998) "S&T in Hungary: Past, Present and Future." *Transforming Science and Technology Systems – The Endless Transition? NATO Science Series 4: Science and Technology Policy – -Vol. 23*. Ed. W. Meske, et al. Amsterdam: IOS Press, 69–82.
IRO-AoS (Institute for Research Organization of the Hungarian Academy of Sciences) (Ed.) (1999) *Science in Hungary*. CD-ROM. Budapest: Akadémiai Kiadó Budapest.
Kutlaca, D. (1996) "The Transformation of the S&T System in Yugoslavia directed to a New Innovation System (Country Report)." Unpublished paper. Berlin: Wissenschaftszentrum Berlin für Sozialforschung (WZB).
Martinson, H. (1995) *The Reform of R&D System in Estonia*. Tallinn: Estonian Science Foundation (ESF).
Martinson, H. (2002) "Impact of internationalisation on building S&T capacity in an small country." Paper presented at the 4th Triple Helix Conference, November 6–9, 2002, in Copenhagen/Denmark and Lund/Sweden.
Mense-Petermann, U. (2002) "Kontinuität und Wandel – Zum Erklärungspotenzial institutionalistischer Ansätze in der Transformationsforschung." *Berliner Journal für Soziologie* no. 2: 227–242.
Meske, W. (1990) "Industrie-FE in der DDR – Umfang, Strukturen, Tendenzen." *IGW-Report 2 über Wissenschaft und Technologie in der DDR und anderen RGW-Ländern*. Erlangen: Institut für Gesellschaft und Wissenschaft (IGW), 19–33.
Meske, W. (1993) "The restructuring of the East German research system – a provisional appraisal." *Science and Public Policy* 20, no. 5: 298–312.
Meske, W. (2001) "Wissenschaft in Ostdeutschland – Eine ambivalente Zwischenbilanz des deutsch-deutschen Einigungsprozesses nach 10 Jahren. "*Autonomie oder Anpassung? Die Vernetzung von Wissenschaft, Staat und Gesellschaft gestalten. Dokumentation der 20. GEW-Sommerschule*. Ed. F. Bretschneider and G. Köhler. Frankfurt a.M.: Gewerkschaft Erziehung und Wissenschaft (GEW), 249–274.
Meyer, H. (1990) "Wissenschaftspolitik, Intelligenzpolitik – Das Personal für Wis-

senschaft, Forschung und Technik in der DDR." *Intelligenz, Wissenschaft und Forschung in der DDR*. Ed. H. Meyer. Berlin: de Gruyter, 1–51.

Mosoni-Fried, J. (1997) "Transformation of the R&D System in the Transition Economies: The Changing R&D System in Hungary (Country Report)." Unpublished paper. Berlin: WZB.

Nadiraschwili, A. (1994) *Die Transformation der Wissenschaft in den Ländern der ehemaligen UdSSR*. WZB-Paper P 94–401. Berlin: WZB.

Nesvetailov, G.A. (1998) "Compromised Futures: The Consequences of an Aging Research Staff." *East European Academies in Transition*. Ed. R. Mayntz, et al. Dordrecht: Kluwer, 93–106.

OECD (1994) *Proposed standard practice for surveys of research and experimental development – Frascati Manual 1993*. Paris: OECD.

Paasi, M. (1998) "The Emergence of Innovative Firms in Estonia." *Transforming Science and Technology Systems – The Endless Transition? NATO Science Series 4: Science and Technology Policy – Vol. 23*. Ed. W. Meske, et al. Amsterdam: IOS Press, 211–221.

Radosevic, S. (2003) "Patterns of Preservation, Restructuring and Survival: Science and Technology Policy in Russia in the Post-Soviet era." *Research Policy*, forthcoming.

Schneider, C. (1998) "Institutional Transformation in the Industrial R&D Sector – Changes in Organizational Structures, Functions, and Interrelations. Analysis by Country: Poland (Country Report)." Unpublished paper. Berlin: WZB.

Stanovnik, P. (1998) "The Slovenian S&T Transition." *Transforming Science and Technology Systems – The Endless Transition? NATO Science Series 4: Science and Technology Policy – Vol. 23*. Ed. W. Meske, et al. Amsterdam: IOS Press, 98–107.

Stifterverband (1990) *Forschung und Entwicklung in der DDR – -Daten aus der Wissenschaftsstatistik 1971 bis 1989*. Essen: SV-Gemeinnützige Gesellschaft für Wissenschaftsstatistik m.b.H. im Stifterverband für die Deutsche Wissenschaft.

SURS (2001) *Slovenija v številkah 2001* (Slovenia in figures 2001). Ljubljana: Statistièni Urad Republike Slovenije (SURS – Statistical Office of the Republic of Slovenia).

Szomburg, J. (2001) "Ostdeutschland – Polen. Zwei Transformationsansätze – zwei Entwicklungswege. " *Zehn Jahre Deutsche Einheit – Bilanz und Perspektiven. Tagungsband*. Halle: Institut für Wirtschaftsforschung Halle (IWH).

20. PUBLICATION ACTIVITY IN CEECS DURING THE 1990S

Werner MESKE [1]

The organizational restructuring and the changes in the resources made available for R&D, and particularly in the number and structure of R&D personnel, are supplemented and, to a certain extent, confirmed by output data. The severe decline in the industrial R&D sector is also reflected in the small number of patent applications from the CEECs in the early 1990s, a trend that only began to rise gradually in some countries towards the end of the 1990s (cf. Havas, 2002, Table 10). On the other hand, the number of publications by CEEC scientists appearing in internationally distributed journals has risen in most countries during the 1990s (cf. Table 20.1), despite the reductions in personnel.

There are differences between the individual countries, in both the increase in publications and in its timing (cf. Figure 20.1), that tally fairly well with the dynamic of scientist numbers and other resources devoted to R&D already outlined.

a) The pattern of increase in the six Central East European countries was relatively continuous, with an increase in their publications included in the Science Citation Index (SCI) from 12 622 in 1989 to 14 889 (118 per cent) in 1995 and 17 742 (140 per cent) in 1999 (cf. Table 20.1). Only in Bulgaria, where there was an initial increase until 1993, has the number of publications fallen again below the 1989/90 level, while Slovakia has experienced fluctuations but no sustained increase. In all these countries, one of the major factors influencing the growth in publications in the 1990s was the rapid rise in international cooperation: SCI publications written in collaboration with scientists from EU member states have increased considerably

---

[1] All tables and figures in this section are the author's own compilations, based on data supplied by H.-J. Czerwon, using the SCI Data Base (cf. Czerwon, 2000). Data source: Science Citation Index 1985–1999 (CD-ROM versions, 1985–1999). The following publication types ('citable items') were taken into consideration: (1985–1996) articles, letters, notes, reviews, (1997–1999) articles, letters, reviews.
Methodological hint regarding the assignment of papers to countries: A paper was assigned to a CEEC in those cases where at least one author (co-author) came from this country. International cooperation: A paper is deemed internationally co-authored if at least two corporate addresses are from different countries; in many cases more than two countries contributed to internationally co-authored publications.

Table 20.1 CEECs: publication activity (1989–1999; number of SCI publications)

	Year										
	1989	1990	1991	1992	1993	1994	1995	1996	1997	1998	1999
The six Central East European countries											
Poland	5 128	4 854	4 876	5 247	5 052	5 361	6 160	6 216	6 108	6 705	7 176
Hungary	2 318	2 169	2 371	2 477	2 378	2 432	2 582	2 631	2 848	3 134	3 367
CSFR	3 217	3 332	3 241	3 702	3 849						
C. R.						2 710	2 612	2 981	2 964	3 222	3 222
Slovakia						1 352	1 378	1 449	1 385	1 519	1 330
Romania	581	434	460	673	671	889	899	1 170	1 183	1 267	1 335
Bulgaria	1 378	1 407	1 398	1 462	1 645	1 323	1 258	1 245	1 327	1 297	1 312
total	**12 622**	**12 196**	**12 346**	**13 561**	**13 595**	**14 067**	**14 889**	**15 692**	**15 815**	**17 144**	**17 742**
The Baltic states:											
Estonia				250	233	286	334	371	377	437	454
Latvia				286	230	236	225	231	235	275	271
Lithuania				273	211	261	261	275	319	355	430
total				**809**	**674**	**783**	**820**	**877**	**931**	**1 067**	**1 155**
The four European CIS countries											
Russia				26 556	21 857	24 138	22 988	21 679	21 914	21 699	21 020
Ukraine				4 695	3 331	3 427	3 469	3 223	2 956	3 102	3 165
Belarus				1 122	821	1 133	887	766	753	791	776
Moldova				267	193	193	183	183	169	155	145
USSR/ total	34 328	33 615	32 356	32 640	26 202	28 891	27 527	25 851	25 792	25 747	25 106
Follower to the former SFR Yugoslavia:											
FR Yugoslavia				834	618	637	633	693	677	747	759
Croatia				662	732	652	726	675	693	696	739
Slovenia				500	526	588	622	649	791	825	938
SFRY/ total	1 587	1 939	1 789	1 996	1 876	1 877	1 981	2 017	2 161	2 268	2 436
All CEECs	**48 537**	**47 750**	**46 491**	**49 006**	**42 347**	**45 618**	**45 217**	**44 437**	**44 699**	**46 226**	**46 439**

Source: Czerwon, 2000.

quicker than the overall number of publications. The share of publications co-authored with scientists from EU member states has risen as a result from about 10 per cent (only Poland and Hungary already had a share of 16 per cent in 1989) to 30 per cent and more. Germany has the highest share of such co-authorships, accounting for more than 10 per cent by itself. If publications jointly co-authored with scientists from the USA (the share is usually 10 per cent and more) and other countries (e.g. Japan) are included,

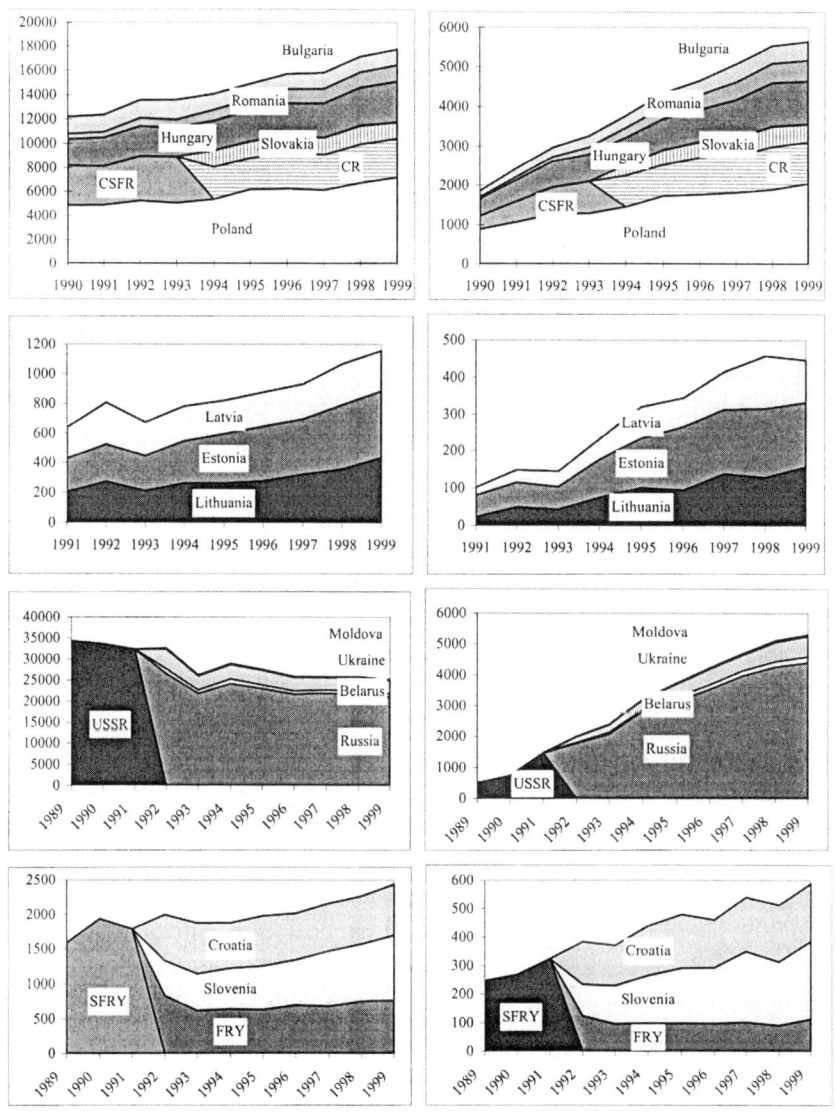

a) Publications (SCI)   *b) Co-authored papers with EU 15 countries*
Fig. 20.1 *CEECs: publication activity (SCI) and co-authored papers with EU15 countries, 1989–1999*
Source: Own compilation based on Czerwon, 2000.

virtually 50 per cent of all SCI publications by the scientists of these six countries are the result of international collaboration.

b) The trends in publication activity in the three Baltic states are very similar. Here however, there was a sharp increase in 1992 and a decline in 1993.

Both these developments can be explained by the specific situation of these countries, where there had been strong bottom-up initiatives taken by scientists even before the collapse of the USSR, while the ensuing turbulence and decline had a delayed effect on the publications in 1993. The number and share of publications co-authored with scientists from abroad have increased even more rapidly in these countries than in the six Central East European countries, particularly as a result of the close cooperation with Finland and Sweden. In Estonia, for example, both these countries had a 15 per cent share in international co-authorships in 1999, putting them ahead of Germany at the top of the table (Czerwon, 2000, Table 27).

c) The evolution of publication activity in the four European CIS countries was considerably more differentiated. Here too, there was an increase in the number of publications in 1993 after the opening-up to the outside world. After a short-lived rise in 1994/95, presumably attributable largely to the publication of earlier research findings, the number of publications stagnated in Ukraine and Belarus and even fell slightly in Russia and Moldova. The reasons for this decline probably lie in the emigration of scientists, unfavorable working conditions and, despite a rapid overall increase, considerably lower shares of joint publications with scientists from EU member states (approximately 20 per cent, the exception being Moldova with over 30 per cent) than in the groups analyzed before.

d) The evolution of publication activity in the successor states to the SFR Yugoslavia was even more diverse. In Slovenia, there was a constant increase from 1992 onwards, underpinned by a growing share of international co-authorships. There was an increase in the share of publications jointly authored with scientists from EU member states from 20 to 30 per cent of all SCI publications, while a further 11 per cent were jointly authored with scientists from the USA. In Croatia, publication activity stagnated, with a slight increase in the share of co-authorships with scientists from EU member states to almost 30 per cent. In the FR Yugoslavia, the number of SCI publications fell until 1995 before rising slightly again, despite the share of co-authorships with scientists from EU member states remaining more or less unchanged at around 15 per cent.

The following conclusions can be drawn from these data.

1. The fact that the number of SCI publications mostly rose or at least remained approximately the same supports the argument that a relatively large core of productive scientists has been maintained in the CEECs, despite the considerable reduction in R&D personnel.
2. The rise in the absolute number of publications despite the reduction in the number of scientists in most countries shows that scientists have been pub-

lishing more than hitherto, which suggests that they are now giving higher priority to internationally regarded publications as an outlet for their research findings. This may be attributable to an adjustment to international modes of behavior (acquisition of individual scientific reputation) but also to the disappearance of prohibitions or other obstacles in the CEECs and to the sharp reduction in applied R&D. The contrary trends that emerged again in the CIS countries after 1995 (and, to a certain extent, in Bulgaria, too) suggest, however, that the reductions in personnel and the deterioration in research conditions have had a much more serious effect than in the other CEECs. A large proportion of scientists have been unable, for linguistic, financial and other reasons, to offset the disadvantages they suffer in their own country by engaging in international cooperation, and in any case in Russia "neither active involvement in international actions nor the converse is in itself a conclusive indicator of a researcher's professional effectiveness" (Mirskaja, 1998, p. 43).

3. The increased number and high share, mostly in excess of 30 per cent and in some cases more than 50 per cent, of work co-authored with foreign scientists in total publications in the EU applicant countries (cf. Figure 20.2) suggest that 'Western' colleagues have made a significant contribution as 'boundary spanners' to this increase, as was also the case with East German scientists and research institutes (cf. Meske et al., 1997). "The willingness of advanced countries to support research in the countries under reform and to help their scientists integrate into world science, coincided with the liberation of the initiative of individual scientists and research groups not only of political but also bureaucratic restrictions." (Mirskaya, 1997, p. 303). In this way, (English) language difficulties were overcome, presentation adapted to international standards and access to reputable journals facilitated.

4. These developments also helped to break down Western scientists' prejudices concerning the efficiency of their East European colleagues. Such prejudices had a significant role to play, especially in 1990 prior to the incorporation of the GDR into the FRG (cf. Meske, 2002), but also from that time on in proposals relating to the transformation and renovation of science in other former socialist countries (cf. OECD, 1992). Prof. Zacher, president of the renowned Max-Planck-Gesellschaft, for example, spoke to this effect when he referred to a "research desert in the GDR", notably in the humanities and social sciences (Frankfurter Allgemeine Zeitung no. 141 of June 21, 1990, p. 31; own translation). Later, more thorough investigations of selected areas of research in the former GDR with respect to the cognitive and methodological capacities they reflect show that scientific standards were adhered to, not only in natural sciences and technology but even in the social sciences, despite their orientation to the prevailing system

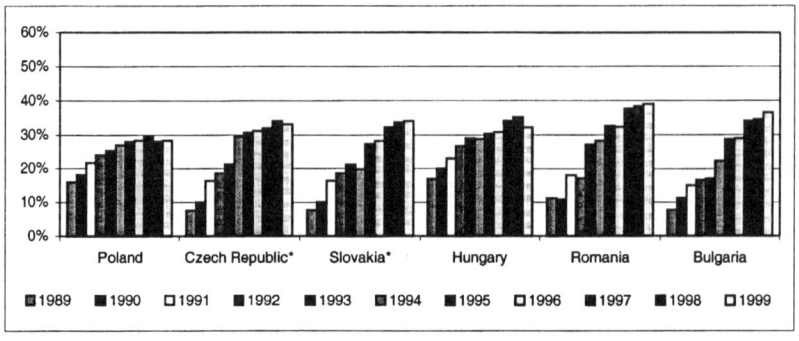
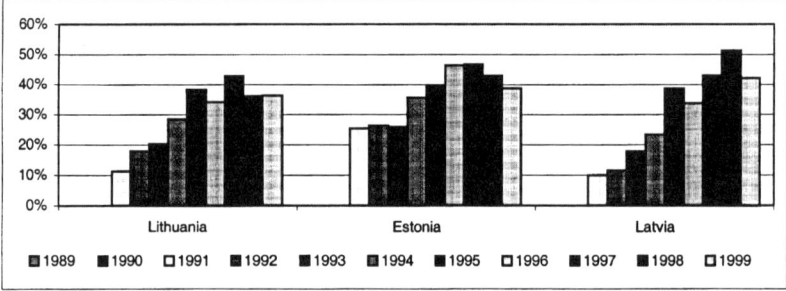
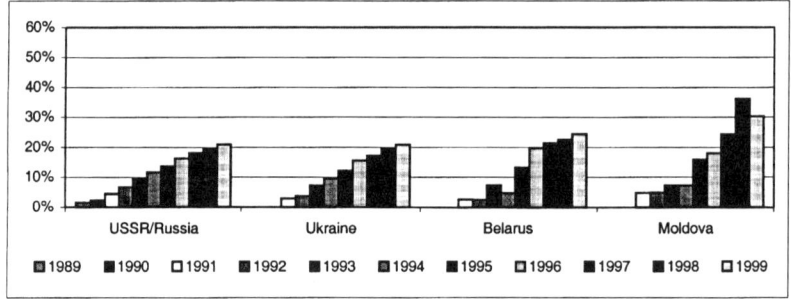
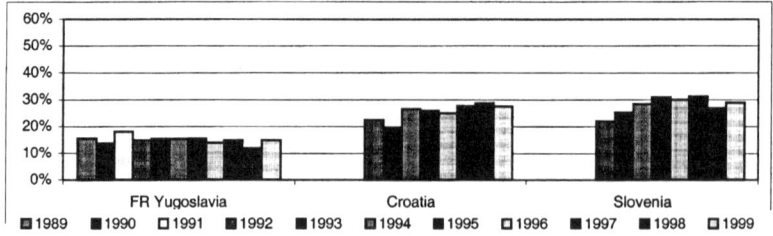

Fig. 20.2 CEECs: share of papers co-authored with scientists from EU15 countries
*;data for the years 1989-1993 refers to Czechoslovakia

(cf. Kocka and Mayntz, 1998; Mayntz, 1998). Evaluations of science in the Baltic states that were carried out by Danish and Swedish experts in 1992 reached similar conclusions.

However, the positive changes that took place during the 1990s and are reflected in the number and structure of publications by CEEC scientists should not blind us to the fact that these developments were influenced by a number of specific factors and should not be regarded as a trend that will necessarily be sustained in the future. Thus the sharp increase in the number of SCI publications in the first half of the 1990s is attributable to a large extent to the publication of an enormous number of results produced in the 1980s. The subsequent slumps in research work caused by the emigration of internationally renowned scientists and the deterioration in the conditions for research work led to a stagnation and decline in the number of SCI publications in the 4 CIS countries, Bulgaria and the FR Yugoslavia in the second half of the 1990s. In the other countries, these difficulties, which also existed there, were concealed by the development during that period of international cooperation and the resultant increase in co-authored publications. Thus the 6 Central East European countries did indeed increase the volume of SCI publications between 1992 and 1999 by a total of 4 181 (from 13 561 to 17 742); however, this increase was achieved solely as a result of publications jointly authored with collaborators from abroad, since the results published jointly with EU15 scientists rose by 2 671 and those published with scientists from the USA, Japan, Canada and Switzerland by 1 770. In other words, the results published jointly with scientists from other countries, which increased by a total of 4 441 publications, exceeded the total increase in SCI publications. The same applies to the Baltic countries. Among the EU applicant countries, Slovenia is the only one in which the increase in publications is not attributable solely to the increase in publications jointly authored with foreign scientists. In the other countries, the decline in 1999 relative to 1992 would have been even greater if it had not been offset by the rise in international co-authorships.

The trends highlighted here can be said to reflect the fact that the lack of resources for R&D in most CEECs has already begun to have an impact not only on their research activities but also on their results and published output. Consequently, improvement of science funding and infrastructure in the CEECs is imperative for the maintenance and, in the long term, improvement of the performance of their S&T systems. It is not only the production of internationally respected research results that depends on this, but also, among other things, international collaborations on an equal footing, with mutual exchanges of scientists, the retention of leading scientists in the country, the recruitment of young scientists and the rapid implementation of results. "Publications in international journals [...] play a particular role in the integration of national scientists and institutes into international science. In some cases they are just a prelude – a first step to an international reputation, which can encourage the author's further advancement in international collaboration. In other cases, they point to the author's involvement in the process of international interaction between researchers and, chiefly, present the results of joint work to the international scientific community." (Mirskaya, 1997, p. 306). According to Mirskaya's investigations, the evolution of publication activity in the 1990s is also differentiated, inconsistent and often ambivalent. Thus cooperative links with EU15 and other OECD countries have hitherto usually taken the relatively one-sided form of individually organized visits by East European scientists to these countries. "In the initial phase of transformation, with the old background of pre-planned research on particular topics continuing over decades, this new kind of collaboration may have a positive impact and induce productive changes. However, if not supplemented with institutional structures, it will support only the most advanced scientists and research, with a strong similarity to the West. Since it is not orientated towards the preservation of specific national problem areas and the bringing-up of a new scientific generation, such collaboration in fact is in opposition to these objectives" (Mirskaya, 1997, p. 307).

Similar conclusions were reached in a study carried out in the early 1990s by Rudolph of visiting scientists from the CEECs at German research institutes: "At that time the mobility was virtually all in one direction (that is from East to West); this was due not only to the difference in living and working conditions but also to a number of funding and promotional programs set up expressly for this purpose". The main beneficiaries of these programs were the West German institutes, which were able to use the visiting scientists as "a skilled and relatively cheap source of flexibility [...] While the short-term value of visiting scientists from CEE/CIS countries in German non-university research establishments seems to be evident to all concerned, medium-term risks are not to be excluded". For the country of origin, there is "the risk that research programs will not be geared primarily to national priorities but that research areas and topics will be selected on the basis of whether or not they can be linked to foreign programs, thereby increasing the chances of scientists being invited to visit research institutes abroad [...] For individual researchers, their professional future at the end of their stay abroad is often unclear. According to their own statements, many of them are somewhat uncertain whether their institutions and jobs will even exist when they go back and even less certain about their chances of being promoted on the basis of their experiences abroad. [...] For these researchers, the good working conditions in the West (remain) the main motivation for their move and, for some, the possibility of finding further employment once they are there." (Rudolph, 1994, p. 46–47; own translation).

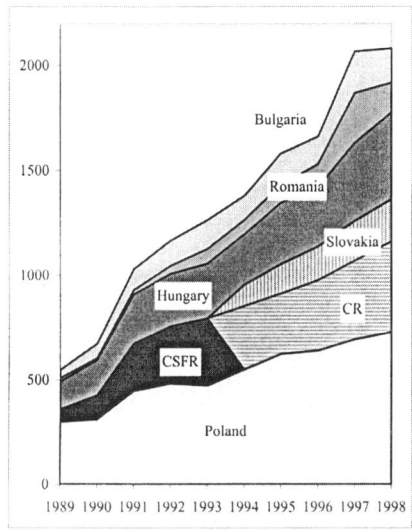 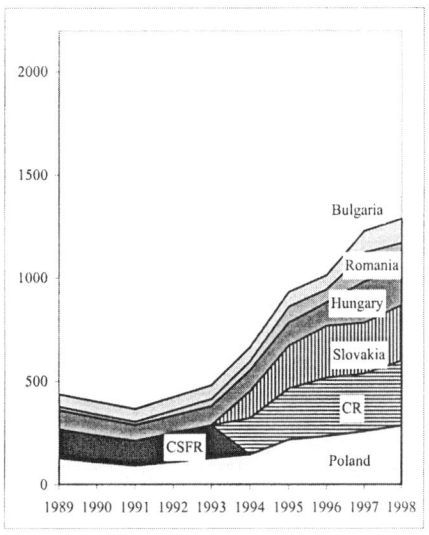

a) with German scientists *b) with the 6 CEECs scientists*
Fig. 20.3 Six Central East European countries: papers co-authored with German scientists and scientists from these six CEECs
Source: Czerwon, 2000.

The sharp increase in CEEC scientists' international links during the 1990s undoubtedly played a significant part in the 'survival' of scientific activities and in the transformation processes in these countries. However, the nature of the interactions to date raises doubts as to whether a durable basis for cooperation on an equal footing has in fact been created. Moreover, the stagnation in the volume and share of publications in the CEECs jointly authored with scientists from OECD countries observed towards the end of the 1990s (cf. Fig. 20.2) leads one to suppose that the phase of integration into international communities has now been successfully completed. Future developments will certainly depend on whether and how science in the CEECs succeeds in building up and extending its own basis for performance-enhancing research. "The preservation and productive transformation of Russian science requires primarily purposeful state support and an efficient national science policy aimed at rational reformation of the organizational system of science. Only under these conditions can contemporary international interactions turn into genuine cooperation that would proceed on equal terms and would be interesting and mutually beneficial to all the participants" (Mirskaya, 1998, p. 44).

The fact that jointly authored publications involving scientists from the six

Central East European countries themselves also rose constantly during the 1990s must be seen as a step in this direction (cf. Figure 20.3). However, with a share of 5 per cent in total publications, the extent of these joint authorships, and hence the cooperation between these countries, is still very modest. It should be borne in mind that Germany and the USA alone each have a share of approximately 10 per cent or more of the co-authored publications of the individual CEECs.

REFERENCES

Czerwon, H.-J. (2000) "International scientific cooperation of EIT countries: a bibliometric study." Unpublished paper. Berlin: Wissenschaftszentrum Berlin für Sozialforschung (WZB).

Havas, A. (2002) "Does Innovation Policy Matter in a Transition Country? The case of Hungary." *Journal of International Relations and Development* 5, no. 4: 380–402.

Kocka, J., and R. Mayntz (ed.) (1998) *Wissenschaft und Wiedervereinigung – Disziplinen im Umbruch*. Berlin: Akademie-Verlag.

Mayntz, R. (1998) "Die Folgen der Politik für die Wissenschaft in der DDR." *Wissenschaft und Wiedervereinigung – Disziplinen im Umbruch*. Ed. J. Kocka and R. Mayntz. Berlin: Akademie-Verlag, 461–483.

Meske, W. (2002) *Science in Formerly Socialist Countries – Asset or Liability within New Societal Conditions?* WZB-Paper P 02–401. Berlin: WZB.

Meske, W., J. Gläser, G. Groß, M. Höppner, and C. Melis (1997) "Die Integration von ostdeutschen Blaue-Liste-Instituten in die deutsche Wissenschaftslandschaft. DFG-Forschungsbericht." Unpublished paper. Berlin: WZB.

Mirskaya, E. (1997) "International scientific collaboration in the post-communist countries: modern trends and priorities." *Science and Public Policy* 24, no. 5: 301–308.

Mirskaya, E. (1998) "The role of international interactions in contemporary science in Russia." *Science and Public Policy* 25, no. 1: 37–45.

OECD (Ed.) (1992) *Reviews of National Science and Technology Policy: Czech and Slovak Federal Republic*. Paris: OECD, Centre for Cooperation with the European Economies in Transition.

Rudolph, H. (1994) *Ex Oriente Lux? Gastwissenschaftlerinnen und Gastwissenschaftler aus Mittelosteuropa und der ehemaligen UdSSR an deutschen Forschungsinstituten*. WZB-Paper P 94–105. Berlin: WZB.

Part IV

Science and technology in CEECs at the beginning of the 21st century

21. A PROVISIONAL APPRAISAL: THE TRANSFORMATION OF S&T DURING THE 1990S AND THE CHALLENGES OF THE 21st CENTURY

Werner MESKE

Comparative analysis of the changes in S&T in the individual CEECs reveals that approaches to the transformation of the 1990s varied greatly, as did the course it took and the (interim) results achieved. At the same time, there are a number of common features and similarities, particularly in the sequence, nature and 'direction' of institutional, structural and quantitative changes. Taking into account experiences in East Germany as well, these findings were used to construct a 'three-phase' model of S&T transformation in CEECs (cf. Meske, 1998 and 2000) that generalizes significant characteristics of the transformation process and can be used as a yardstick in characterizing and assessing the course it has taken to date in individual countries and the results it has produced. Thus this model is not designed to be used for 'ranking' the countries in transition, but rather, taking a characterization of the current situation as a starting point, for identifying the stage the transformation of the S&T system has reached in each country and the fundamental problems and tasks that remain as those systems develop further in future.

1. THE 'THREE-PHASE MODEL'

The entire system of S&T in the individual CEECs underwent significant changes under the influence of the endogenous preconditions and internal activities necessitated by the countries' initial situations and the effects of the national and international environment. These changes were both top-down, that is driven by political and economic decisions, and bottom-up, that is the result of the activities of S&T actors, which were a product of the democratization process and the greater autonomy enjoyed by organizations and individual scientists.

Dealing with these individual processes took different lengths of time; however, they were also interrelated, so certain changes could not be successfully concluded until several processes had been successfully completed.

Thus the process of transforming the S&T system (STS) can be divided into three main phases (cf. Figure 21.1), each one dominated by particular kinds of changes.

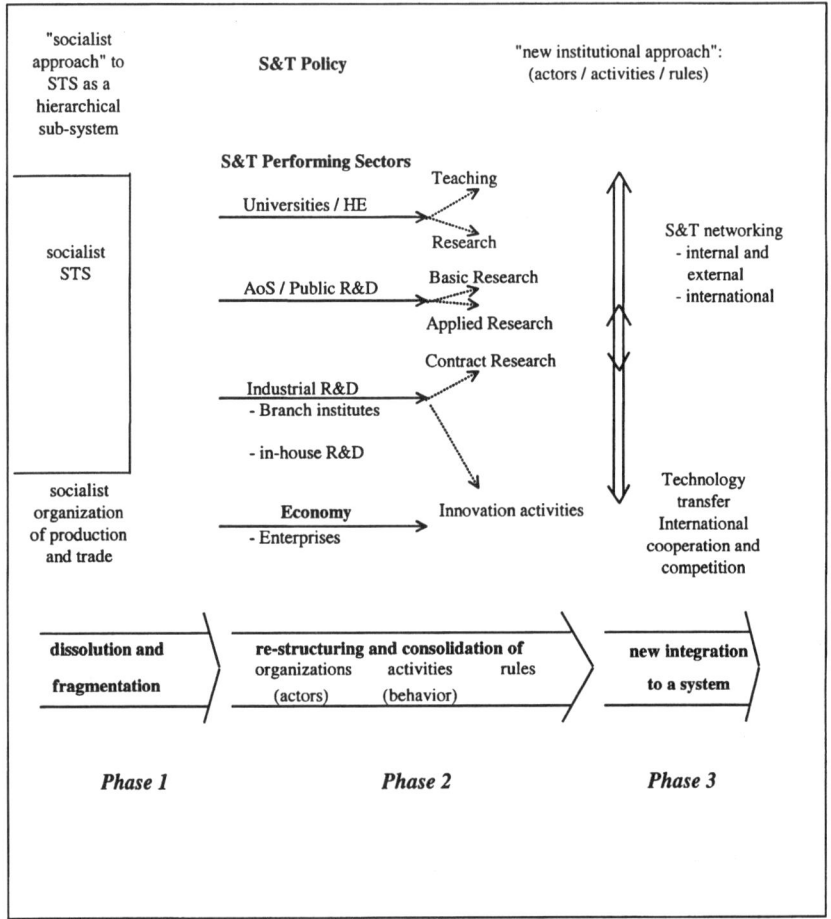

Fig. 21.1 Phases in the transformation of the S&T system

Phase 1 included the dissolution and fragmentation of the old (socialist) STS, both internationally and nationally.

Phase 2 included the reorganization of S&T, during which two fundamental tasks had to be accomplished if the transformation of the S&T system was to proceed successfully.

a) S&T institutions had to be reorganized by creating new actors (in particular by fundamentally restructuring old or creating new bodies and organizations in politics, industry and science) and introducing and enforcing new rules governing their activities and behavior.
b) These newly created individual actors in politics, industry and science had subsequently to be consolidated, that is they had to be able to find their specific place in the new societal system and to maintain it over the long term.

In *Phase 3* the key feature is the emergence/building of a new S&T system as a balanced configuration of the different actors, activities and rules. The aim of this phase is the development of an effective and internationally competitive national innovation system through relatively stable relationships and a sustainable 'dynamic balance' within the STS as well as between it and other areas of society. This phase also constitutes the completion and end of the transformation process.

We start from the assumption that all CEECS have in principle to go through these three phases in transforming their S&T systems. However, the form and duration of each phase may vary from country to country, and overlaps between the phases are probably the rule rather than the exception.

2. THE DISSOLUTION OF THE SOCIALIST S&T SYSTEM IN ALL CEECs IN THE EARLY 1990s (PHASE 1)

In the early 1990s, all the CEECs were having to deal with Phase 1, that is the dissolution and fragmentation of the former socialist STS. This involved, firstly, the break-up of the former socialist 'world system' with its relatively one-sided international relations. The dissolution of this international system began in 1989 with changes of government in several countries and the temporary opening of borders in Hungary. It gathered pace considerably with the economic and political incorporation of the GDR into the FRG in 1990 and was completed with the break-up of the Warsaw Pact and the CMEA/COMECON in the middle of 1991 (cf. Lavigne, 1995, p. 91–112).

Even at this stage, however, there were differences between the countries. While these international processes were in some cases the consequence of changes in the political systems of nation states (e.g. in Poland, the CSSR, Hungary and the GDR), in other countries they were the prelude to subsequent changes of system (e.g. in Romania and Bulgaria) or the break-up of confederations of states and the emergence or re-emergence of independent states (as a result of the break-up of the USSR and the SFRY in 1991 and of the CSFR in 1993). In the first group of states, the dissolution of the old international and national systems took place simultaneously, while in the second group they were successive processes. In the countries that had regained their independence, the difficulties caused by the delays in the break-up of the old system were further compounded by the problems associated with the emergence of political actors and bodies keen to make their mark for the first time, particularly in the Baltic States and other former Soviet republics. In the Czech Republic and, above all, in Slovakia, the dissolution processes lasted until the two republics parted company in 1993; in the successor states to the SFRY, they took very different courses, ranging from the rapid breakaway and re-orientation of Slovenia as early as 1991 to a transitional phase marked by political instability that even in the year 2000 was not complete in the FR Yugoslavia, Macedonia und Bosnia-Herzegovina.

The processes leading to the dissolution of the political system impacted directly on S&T policy. Ministries and other bodies responsible for S&T were broken up; legislation and contractual and other arrangements were annulled. However, while existing bodies and arrangements were abolished or suspended, new ones were not always created to take their place; as a result, there were often lengthy periods of uncertainty and unclearness between the break-up of the old system and the constitution of new actors.

For these reasons, the first phase of the transformation was characterized primarily by the dissolution and fragmentation of the old system. In HE and the AoS in particular, organizations were not completely dissolved; rather, they remained intact to a large extent, although individual establishments or parts of them were closed. Generally, however, existing organizations were scrutinized with regard to their rights and obligations and in many cases transformed in stages and 'reengineered' to give them a new profile.

In most CEECs, industrial R&D went through a similar transitional phase. Thus branch institutes maintained a de facto existence as quasi-state establishments long after the dissolution of the ministries to which they answered; similarly, R&D departments in not yet privatized or restructured state enterprises remained in existence. Precisely because of these frequently uncertain periods of transition between the break-up of the old system and the restructuring and reengineering of its fragments in the various sectors, which often lasted for variable periods of time even in the same country, it is virtually impossible to say exactly when this first phase ended in the individual countries (cf. the detailed surveys in the country chapters in Part II of the present volume). In East Germany, for example, GDR science policy was completely eliminated in October 1990, when the GDR became part of the FRG and its Ministry of Science and Technology was dissolved and not replaced. Most state, non-university R&D establishments, particularly the research institutes run by the GDR research academies, continued to exist, however, until they were dissolved at the end of 1991. In the HE and industrial R&D sectors, the dissolution or replacement of the old structures lasted until at least 1993 because the reengineering and privatization processes lasted longer in these areas. One factor that played a role here is that in Germany the individual *Länder* are responsible for education, including higher education. After unification, these *Länder* had first to be established in East Germany and then made operational (cf. Meske, 1991). In this respect, therefore, the situation was similar to that in other CEECs. These differences in the dissolution of the various sectors of the GDR's S&T system meant that the process of reducing personnel levels also took a different course in the various sectors. Consequently, what was in overall terms a rapid and severe reduction in R&D personnel in East Germany took place in stages and lasted until the end of 1993 (cf. Figure 19.2).

The differences in the dissolution of 'old' institutions in East Germany are largely confirmed by comparative analysis of the changes in institutional structures and personnel in the CEECs. This also applies to the finding that the extent of the reduction in R&D personnel is a relatively good indicator of the dissolution and fragmentation of the old STS, both over time and in terms of the depth of the associated cuts in individual sectors and countries.

The path of the personnel reduction curve in Figure 19.1 shows, for example, that the transformation started early in Poland and Hungary and that in most countries the dissolution processes were concentrated in the period between 1990 and 1993/94. Figure 19.3 shows that, in the European CIS countries, this first phase started as early as the late 1980s under Gorbachev but did not gather pace until after the disintegration of the USSR in 1991 and gradually ran out of steam after 1995. In the Baltic States, the dissolution got under way in 1989/90, but was completed shortly after the demise of the USSR, in 1992 in Estonia and Lithuania and 1993/94 in Latvia (cf. Figure 19.4). The curves in Figure 19.5 show that five Central East European countries had completed the decisive changes by 1993/94; Romania, on the other hand, had indeed made some fundamental changes by 1994/95 but had not followed them through consistently to a successful conclusion. Figure 19.6 also shows that, of the successor states to the SRFY, only Slovenia had completely implemented the first phase of S&T transformation by 1993.

Consequently, it can be estimated with a relatively high degree of certainty that the first phase of the transformation, that is the dissolution and fragmentation of the old socialist STS, took place in most of the countries investigated here largely between 1990 and 1993/94. However, it had not been completed in all CEECs by that time; in the CIS countries and in Slovakia, Romania and the FR Yugoslavia, it continued in some cases at least into the second half of the 1990s. This statement is also supported by findings on the institutional changes in the S&T sectors and policies in the individual CEECs during the first half of the 1990s (cf. Tables 18.1 and 18.2).

3. THE REORGANIZATION OF S&T IN INDIVIDUAL SECTORS AND COUNTRIES UP TO THE END OF THE 1990S (PHASE 2)

The second phase began with the emergence of new and powerful actors in S&T. The differences that already existed in the chronology of the dissolution of the old system, combined with the influence of the more or less pronounced fragmentation that ensued, resulted in a further reinforcement of the differences between the individual countries in the course of the reorganization processes that followed. In the course of the GDR's accession to the FRG, the creation of new actors in East Germany was thoroughgoing and rapid due to the 'transfer of institutions' from West to East Germany (cf. Lehmbruch, 1992; on the approaches adopted in different S&T sectors cf. Hilbert, 1994; Mayntz, 1994 and 1994a; Labrousse, 1999). Other countries undergoing transformation had to seek out and implement their own solutions with regard to actors and rules, a process that was necessarily lengthier.

Despite the resultant differences in the nature and pace of the reorganization, very similar trends in the changes in the 3 main sectors of the S&T system can be discerned. In most cases, phase 2 lasted considerably longer than phase 1, because the establishment of new science organizations and then making them actually able to function are processes rather than single formal acts. Moreover, it took a long time, often several years, to consolidate them, particularly under new political and (market) economic conditions.

3.1. HIGHER EDUCATION

In all countries and virtually all segments of the 'new elites', there was a widespread consensus that higher education plays an important role in every modern state as part of the education and science system. For this reason, HE institutions were to be retained or even strengthened. In order to ensure continuity of educational provision, only marginal organizational and quantitative changes in personnel had taken place in the first phase. This trend continued in the second phase: from the beginning, HE establishments had been not only relatively strong and stable actors in the science system but also frequently in a position to extend their influence to the political level as well. Consequently, both state and private resources (particularly tuition fees) have been used to maintain and strengthen the sector, which has also been supported by foreign aid programs. There has been a relatively high degree of continuity in both establishments and personnel, and new establishments – some of them private –

have been set up at regional level. In most countries, consequently, the second phase of the transformation began relatively early in this sector with the granting of considerable autonomy for individual institutions and continued with reforms of academic staff, curricula, administration etc. Because of limited financial resources, prolonged structural changes and a lack of international experience among academic staff, not all institutions always meet high international quality and performance standards. Nevertheless, in all CEECs HE institutions emerged as independent and relatively strong actors in the formation of a new S&T system, in some cases as early as the first half of the 1990s and at the latest from the middle of that decade onwards. Since most institutions continued to make a mark for themselves in the second half of the 1990s, with a few being closed down, phase 2 was largely completed in this sector by the end of the decade with the consolidation of the surviving institutions.

In most of the countries, however, the universities continue to concentrate primarily on teaching, and the EU candidate countries in particular "perform favorably compared to the EU for the share of the working-age population with tertiary education (with Bulgaria, Cyprus, Estonia and Lithuania equal or above the EU mean)" (EC, 2002, p. 3). On the other hand, research in the HE sector has not been strengthened to any significant extent except in those cases in which institutions from the former AoS sector have been assigned to the HE sector, as in Estonia and Latvia, for example. The persistent weakness of research in HE is in part a legacy of the past, therefore, but is due primarily to the inadequate financial resources available for R&D.

3.2. PUBLIC RESEARCH ORGANIZATIONS

The evolution of non-university academic science in the individual CEECs is considerably more varied than in HE. One reason for this lies in the very diverse and often widely diverging views of the new political actors responsible for the sector, but is also attributable to the influence of higher education, which in this phase in particular acted as a competitor. Another factor was the general lack of financial resource in all the countries, which led to a sharp reduction in the public (and private) resources available for R&D. As a result, the processes of establishing new actors in the public research organization sector, and in particular the reengineering and consolidation processes, lasted a very long time in most CEECs. However, by the second half of the 1990s at the latest, clear decisions had been taken in all CEECs in this sector as well, leading to public non-university research organizations remaining intact or being recreated following varying degrees of restructuring.

The actual shape of the sector is very different in the individual CEECs. This applies above all to the reorganization of the AoS sector. Whereas AoS establishments in the European CIS countries remained in place and were even strengthened as the leading research institutions, in Estonia and Latvia they were completely dissolved as research organizations in favor of the universities and retained only as learned societies. In the former SFRY, there were no Soviet-type research academies. However, in its successor states there were and still are a number of 'independent' public research institutes. In the other Central East European countries and in Lithuania, the AoS has been preserved, although the distinction between the Academy as a traditional learned society and its research establishments, most of which have been accorded greater autonomy within associations of institutes, has been further reinforced. In some cases, the public research sector has also been strengthened through the incorporation of former branch R&D institutes from the industrial sector, particularly in the form of departmental research organizations in individual ministries.

In all CEECs the public non-university research institutes have been preserved as a strong pillar of the public research system, alongside the university sector, often as part of the AoS. In a number of countries, however, they have lost their previous dominant position in the S&T system to the universities. The amount of core institutional funding these research institutes receive has been very considerably reduced in all countries. A competitive system of public funding for individual projects has been introduced in most cases, but this has not fully compensated for the reduction in core funding. Attempts to obtain increased funding from business and other sources have met with little success to date.

At the beginning of the 1990s, funding for most of the institutes was wholly inadequate. In most countries, there was an improvement in funding in the second half of the decade, which led to some degree of consolidation among the institutes, in accordance particularly with the (very variable) rates of growth in GDP in the individual countries. Thus in most CEECs, phase 2 in this sector as well was virtually concluded at the end of the 1990s with the consolidation of the reorganized public research institutes. In the light of the generally positive economic trends, there are good prospects of this happening in the other countries, such as Ukraine, Romania and Bulgaria, in the early years of the new century.

3.3. INDUSTRIAL R&D

Compared with the institutional changes that have already taken place and the progress that has been made with consolidating institutions in the HE and public research sectors (which are reflected in, among other things, publication activity, cf. Table 20.1), the situation in industrial R&D in nearly all the CEECs must be regarded as extremely problematic. In most countries, virtually no progress has been made in building up the in-house R&D potential that is the essential precursor for 'independent' innovation activity in firms. Even in those countries in which some enterprises had their own R&D departments during the socialist period, the first phase of transformation at the beginning of the 1990s saw a sharp reduction in numbers and personnel levels and little attempt has been made to build them up again since. Thus most of the range of S&T institutions crucial to innovation that exist in the leading OECD countries still exist or have been put in place in the CEECs, but they are still far too underdeveloped. This situation had scarcely changed in most countries by the end of the 1990s, as our analyses and the results of the European Innovation Scoreboard of November 26, 2002 show (EC, 2002). Thus business expenditure on R&D (BERD) relative to GDP in the candidate countries (CCs) lies between only 6 and 65 per cent of the EU average and is hence far below the level that has been achieved in public expenditure on R&D (calculated as GERD minus BERD) relative to GDP (16–102 per cent of the EU average) (cf. Table 21.1).

Table 21.1 Candidate countries: level and trends in R&D expenditure relative to EU average (= 100%)

	Level relative to EU average (%)		Trends relative to EU average (%)	
	Public R&D/GDP	Business R&D/GDP	Public R&D/GDP	Business R&D/GDP
Bulgaria	61	9	13	-43
Czech Republic	81	64	28	7
Estonia	79	12	-1	21
Hungary	67	28	12	21
Lithuania	79	6	20	-36
Latvia	43	15	-13	78
Poland	67	20	8	-19
Romania	16	24	-32	-49
Slovenia	102	65	-9	4
Slovakia	36	35	-25	-36

Source: Own compilation based on EC, 2002, p. 16–27.

Comparison of the trends (using data for the last 3 or 4 available years in each case) also shows that most CEECs are actually falling further behind, particularly in business R&D. The only countries in which the evolution of R&D expenditure is unambiguously positive are Hungary and the Czech Republic and, in the case of business enterprise R&D expenditure, Estonia, Latvia and Slovenia, which overall has the highest level of R&D expenditure among the CCs. Even in countries that have largely completed the process of enterprise restructuring, R&D activities have not moved ahead very much because technology (both hardware and know-how) has been mainly imported and domestic R&D has either not been required or could not be utilized for financial reasons.

Despite this generally unfavorable situation compared with the HE and public research sectors, there are still considerable differences in industrial R&D between the individual CEECs (cf. Figures 19.7–19.9). Whereas in some countries industrial R&D is continuing to decline and in others stabilization at a low level has been achieved and consolidation begun, enterprises in Slovenia and Hungary have recently begun to expand their internal R&D capacities again.

Besides R&D departments within enterprises, a large number of new, innovative SMEs have emerged during the transformation process as a result of spin-offs or start-ups. These new scientific and technological service providers or R&D-oriented companies are filling a gap that used to exist in the CEECs' innovation systems. However, since the market for R&D remains undeveloped, these firms are increasingly turning to production and non-scientific services or are simply struggling for survival. Some of these new firms developed out of the branch R&D institutes. In all CEECs, this sector, which used to dominate industrial R&D, has been reduced considerably in quantitative terms and has also been reorganized. However, the planned reorientation of these institutions towards applied R&D and predominantly contract-based financing has not advanced beyond its early stages because of inadequate demand for their services and results. Thus the consolidation of most of these now independent industrial R&D firms is not yet complete. Their future is even to a large extent undecided, since they have played little part to date in firms' innovation activities. In contrast to the HE and public research sectors, therefore, the industrial R&D sector in most countries did not see the completion of phase 2 in the course of the 1990s. The main exceptions are Slovenia, Hungary and the Czech Republic, where there are clear signs of stabilization and renewed growth in industrial R&D as well, albeit at a relatively low level compared with the previous situation. Thus it must be assumed that only these three countries, the ones most advanced in the transformation of S&T, had completed phase 2 in all S&T sectors by the end of the 1990s. The other countries are still in this phase, since in some cases, for example the CIS countries and Romania, the restructuring of the former branch institutes and, in particular, the consolidation of the remaining industrial R&D capacity is not yet complete.

4. S&T AT THE TURN OF THE CENTURY

In all the CEECs, considerable changes in the scope, structure, organization and funding of S&T have been made within a decade. Great progress has been made in converting the public sector, particularly HE institutions and public R&D institutes, to modern forms of management, organization and funding through the abolition of hierarchical management systems and the granting of greater autonomy to institutions and scientists. The opening-up of S&T systems to international science has led in particular to a strengthening of cooperation with scientists from the EU, the USA and other OECD countries. On the other hand, the reduction and ageing of the scientific workforce, combined with cuts and/or tight constraints on funding, equipment and other preconditions for exacting research work, has often led to the abandonment and/or interruption of proven and successful lines of research and called into question the international competitiveness of research findings. What is more, the old management hierarchies have been abolished but not replaced by other effective management instruments. In most CEECs, this has led to the fragmentation of research topics, difficulties in adapting R&D to each country's needs and aims and the disruption of important cooperative links both within and outside the science system.

There are particularly serious problems with industrial R&D in all CEECs, since the reorganization is still not concluded in this sector or has been achieved only by greater than average reductions in the number of establishments and in personnel levels. Although the scientific and technological infrastructure for innovations has at the same time been developed and strengthened in a great diversity of ways in all countries, little progress has been made in exploiting R&D results or incorporating them into companies' innovations and in many cases the scope of in-house R&D capacities has fallen far below earlier levels.

As a result of these very diverse changes, a number of existing imbalances in S&T have remained and been reinforced, while new ones have emerged. This applies both to individual scientific institutes and their activity profiles and personnel and resource structures and, particularly, to relations between the individual sectors. Since applied research and industrial R&D have become particular weaknesses, the relationship between science and the economy has also been damaged. The problems created by the largely uncoordinated reorganization of the individual elements of the S&T system have been further compounded by the fact that, for a long time and for diverse reasons, science policymakers were unable or unwilling to take on a coordinating role. It is only since the end of the 1990s or 2000/2001 that fresh attempts have been made in the CEECs to exert greater political influence on the science and research system. If national innovation systems are to be developed, priorities should be determined and strategies developed. Most importantly, innovation activities in enterprises should be actively promoted, with particular efforts being made to make use of the results produced by national science systems.

"Most of the PACs [pre-accession countries] have to realize that, without sound priorities and concentration of means, future prosperity will be uncertain," declared the vice-president of the Slovenian Academy of Science and Arts at the 'Bled Forum' in December 2001 (Kralj, 2002, p. 72). However, there is "a tendency in the candidate countries to remain close to the stated policy objectives of the EU and its present member states, without taking into account the restrictions of the government budget and the need to be innovative" (Brandsma et al., 2002, p. 3).

Since most of the candidate countries are relatively small and less powerful economically than some of the regions of the larger EU member states, such an approach is likely to be problematical. However, the development of the European Research Area means they will have to address the task of developing their own, restricted research profiles as a matter of urgency (EC, 2001). If science in the CEECs is not to be reduced to a mere appendage of the leading international research centers, it must consider its own strengths and, taking full account of national needs and priorities, make its own specific contributions to world science. This applies as of now to the candidate countries, which are now fully integrated, for example, into the EU research and technical development framework programs. However, the other CEECs also need to develop their own identities in this area, since they will soon be neighbors and partners of an ambitious EU that is gaining strength in the scientific and technological sphere (cf. EC, 2002a).

It is in the public sector in particular that there have been some early attempts in the CEECs to put in place a new system by establishing functionally and regionally differentiated HE organizations and linking them to the public, non-university R&D institutes. These links have been established primarily for teaching purposes but are increasingly being extended to cooperation in research. However, cooperation between these public-sector organizations and the industrial sector, and within industry itself between R&D organizations and firms, is less well developed.

Scientific and technical cooperation between the various types of firms is also still very underdeveloped.

Without wishing to underestimate or ignore the progress that has been made or the considerable differences in the course and results of the transformation of S&T in the individual countries, dispassionate analysis of the situation at the end of the 1990s shows that the development of new, efficient and internationally competitive national innovation systems (NIS) in all CEECs constitutes an as yet unresolved problem. This applies even to the countries that have made most progress, which indeed have reorganized and consolidated all three S&T sectors but without having overcome the weaknesses that have emerged in the business enterprise sector in particular. Thus the development of new national science and innovation systems, that is phase 3, is at best under way in the countries in transition but is as yet nowhere near completion. Although there are very different problems and tasks to be addressed in the short term in the individual CEECs, in the medium and long term they will all have to undertake the strategic task of developing new national S&T and innovation systems against the background of the increasing international openness of science and greater economic globalization. Dealing successfully with this task is the most important challenge facing CEECs as they seek, at the beginning of the 21st century, to develop their S&T systems.

All the data suggest that the greatest problems currently facing CEECs as they seek to reshape their national S&T systems concern the economy, and are reflected in particular in the weakness of industrial R&D, the most important mediator between industry and academic science. The various indicators show that all the countries in transition, irrespective of the stage they have reached in their institutional changes, are experiencing inadequate demand from business enterprises for both internal and external R&D activities. Even in countries such as Hungary, Slovenia and the Czech Republic, in which in-house R&D activities are beginning to expand again, the share of R&D personnel in industry is only 40 per cent of the total, which is far below the 60 per cent share in the leading OECD countries (cf. Eurostat, 2001, p. 22 and 27). To judge from the East German experience, this problem is likely to persist in the transitional economies for a long time to come. In East Germany, as is well known, the *institutional* adjustment to West German conditions started a long time ago and has been largely successfully concluded. Thus "in the past ten years the Fraunhofer Institutes in the new *Länder* have developed very rapidly indeed. From the fragments of a disintegrating research landscape there have arisen ultra-modern, vital competence centers with an efficient infrastructure. The East and West German institutes are now virtually indistinguishable from each other, even in their financial structures" (Erfolg im Osten, 2002, p. 8; own translation). The situation in other public research institutes and in the universities is judged to be similar (Stifterverband, 2002). Despite this, East Germany has not seen the emergence of a new (regional) innovation system that in any way matches that of West Germany. The higher than average decline in industrial R&D reduced its share in total R&D personnel from more than 60 per cent to slightly more than 40 per cent between 1989 and 1993 (cf. Figure 19.2). At the outset, there were some very optimistic statements, with former federal research minister Krüger, for example, declaring in August 1993 that the number of industrial R&D personnel in the new *Länder* would rise again in the medium term and level off near the old figure (quoted in the *Deutsche Universitätszeitung* no. 17 in 1993, p. 6). Since then, much effort and money has been expended on promoting R&D at federal and *Land* level (cf. Hornschild, 1998;

Ruprecht and Becher, 1998; Tamasy, 1998). Despite all this, the share of industrial R&D in total R&D personnel in East Germany has been somewhat over 40 per cent since the mid-1990s, compared with 65 per cent in West Germany (cf. BMBF, 2000, p. 75). Despite some positive developments in R&D personnel in East German industry (cf. Hermann-Koitz et al., 2002), this gap between the two Germanys, which did not arise until after 1990, has not narrowed again in recent years, since "industry in the new *Länder* has not taken part in the enormous drive towards R&D intensification that seized West Germany in the second half of the 1990s" (Legler et al., 2002, p. 51; own translation). The fundamental causes of this, which cannot be eliminated in the short term, lie in the fact that research-intensive industries play a significantly smaller role in East Germany than in West Germany. Moreover, small firms are predominant in the East, which means that the large firms that account for the overwhelming share of R&D personnel in the leading countries are almost completely absent or the fairly recently established subsidiaries of large West German or foreign groups with only limited R&D capacities of their own (ibid., p. 57; cf. also Spielkamp et al., 1998). These differences are demonstrated by a comparison of R&D structures in East and West Germany since the mid-1990s (cf. Figure 21.2).

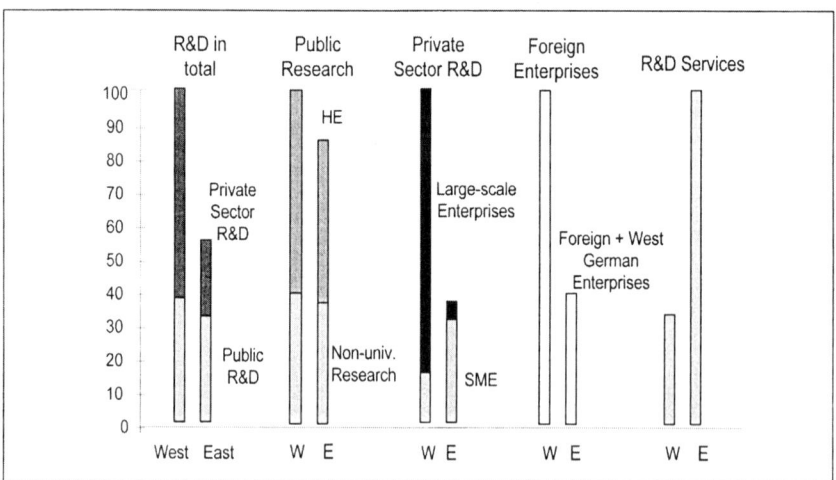

Fig. 21.2 Comparison of R&D structures in West and East Germany (comparable relations in staff levels)
Source: Meske, 1998, p. 302. Note: R&D services are enterprises that fulfill R&D tasks for private and public contractors. In East Germany they were principally created from the former GDR branch R&D institutes. The data for East Germany were multiplied by 4.6 for this comparison, in accordance with the employee ratio of 1:4.6 between East and West Germany. All data for the year 1996.

Since 1996 there has been a proliferation of views that hold the original concept of rapid economic catch-up to be unsuccessful in East Germany, meaning that for this part of Germany new and independent solutions and development opportunities must be sought. Under-secretary Ludewig (the former Chancellor Kohl's spokesman on East German economic affairs) warned as early as 1996 against reducing the economic development of East Germany to a "mere race to catch up with the West": "We need an independent, self-sufficient economic structure in East Germany" (quoted in the *Berliner Zeitung*, Nov. 25, 1996, p. 16; own translation). In 2000 President of the Bundestag Thierse of the SPD advanced his thesis that East Germany is 'tottering on the brink', provoking a new debate, which still continues (cf. AG Perspektiven für Deutschland, 2001). As an expert opinion provided by the leading economic research institutes states, the prospects for the East German economy can at best be "assessed with guarded optimism. In the medium term the recovery process will get back into gear [...] One must nevertheless avoid succumbing to illusions: the road that the East German economy has yet to travel is still long" (Kurzexpertise, 2001, p. 34; own translation).

This conclusion must necessarily be considered just as valid for industrial R&D and for the total science system, because the deficiencies in this respect can no longer be explained solely in terms of old-fashioned patterns of behavior. Thus it is true that there are few innovative large companies in East Germany, but R&D intensity (GERD/turnover) in SMEs is significantly higher than in West Germany (Stifterverband, 2001, p. 23). An "empirical study shows that there are no deficiencies in cooperation frequency or cooperation continuity in East Germany compared to West Germany, and differences in cooperation partner priorities only reflect the given structural differences between the two regions. With respect to the outcomes of innovation cooperation it could be shown that among the cooperating firms there are clearly more firms that carried out innovations and market novelties than among non-cooperating firms. [...] Nevertheless, these positive circumstances do not translate into a better productivity of cooperating enterprises in East Germany while cooperating enterprises in West Germany show the expected effect of higher productivity. [...] Thus, the productivity gap between East and West German enterprises engaged in innovation cooperation rather supports the assumption that co-operations in East Germany are not (yet) fully functioning, especially with respect to the commercialization of new products. [...] It is indeed the case that R&D intensive branches are less productive than non R&D intensive branches within East Germany. In 2000, East Germany's non R&D intensive branches reached nearly 70 per cent of the West German productivity level while R&D intensive branches reached only 60 per cent" (Günther, 2003, p. 17–18).

5. THE END OF 'REMEDIAL MODERNIZATION' – THE RE-ENGINEERING OF SCIENCE AND INDUSTRY AS A NEW CHALLENGE IN THE 21st CENTURY

It is clear from the experiences in East Germany and the findings from the CEECs that the challenge facing all CEECs as they enter the third phase of the transformation is no longer to overcome or eliminate the outdated and unsuitable institutions, structures and behaviors they inherited and to replace them with ready-made 'models' through processes such as institutional transfer or 'remedial' modernization. The latter process undoubtedly played an important and mostly positive role in the reorganization of individual institutions and the establishment of new institutions in all CEECs during the 1990s. In future, however, it will become increasingly necessary actively to develop new and in some cases even internationally novel profiles in industry and science. This is a crucial qualitative difference in the transition from the second to the third phase of the transformation. The need now is to identify the prospects for S&T in the individual CEECs against the background of globalization, and in particular of EU expansion and integration, and to develop the structures and profiles in science and industry to which existing strengths in S&T are best suited.

This re-engineering of S&T in order to make new demands match the available (quantitative and qualitative) resources undoubtedly requires considerable feats of creativity. And it will be the success or failure of CEECs' efforts in this direction that will reveal whether and to what extent the now-reorganized S&T potential inherited from the socialist era continues to be primarily a (financial) burden or can be used as a valuable asset for social development in an environment characterized by international competition and globalization. The relative success of CEECs' attempts to meet changing demands in HE, the reorganization of R&D facilities and the emergence of the modern infrastructures required to support innovation, together with the increase in publications of international standard, should all be seen as evidence in favor of the capacities now available in CEECs. However, it has also been demonstrated that the inherited institutional preconditions in science and research, along with individual knowledge and levels of educational attainment, are not assets and advantages in and of themselves. They must constantly be seen in relation to the overall societal framework. To that extent they are potential preconditions that only take effect and turn into real assets when they are able to adapt to new social circumstances, or when they help to shape these circumstances, or when they contribute to the fulfillment of the ever-unfolding new needs, possibilities and demands of society. Accordingly, the fundamental question does not seem to be whether the capacities inherited from the socialist era are assets or liabilities. Instead, the problem seems to be one of how to employ, change and utilize the personnel and other capacities inherited from the socialist era and to combine them with the additional abilities, skills, experiences etc. that now exist in such a way that they can be used as assets in meeting the new challenges (cf. Meske, 2002).

However, the third phase of S&T transformation, which is now under way, will also take the CEECs out of their special 'system transformation' role and increasingly into the ranks of the OECD countries, which have long recognized the need to modernize their R&D and innovation systems and are introducing fundamental reforms to that end. This was very clearly expressed at international level as early as 1989 at the conference on "The research system in transition" (cf. Cozzens et al., 1990). Mayntz even takes the view that the two societal models that used to prevail in the East and the West respectively, both of which were based on a linear innovation model and "relied either on central control or spontaneous self-management, [...] have been replaced, both in theory and in practice, by an approach that depends on a combination of political control and societal self-regulation" (Mayntz, 1997, p. 142; own translation). In Mayntz's view, this observation applies in particular to innovation systems. Since then, various theoretical concepts (cf. in particular the debate on the shift to a new mode of knowledge production, Gibbons et al., 1994; or the triple helix concept, Etzkowitz and Leydesdorff, 1997) and practical measures have been deployed in order to implement such reforms. Thus in Germany, the Science Council submitted "Proposals for the future development of the science system", which were "intended to play a decisive role in guiding the debate on reform over the next 10 to 15 years. These proposals are not intended primarily for implementation in the short term but rather to stimulate a thoroughgoing process of debate and reform in the science system" (Wissenschaftsrat, 2000, p. 4; own translation). At the same time, a start was made on evaluating the most important research organizations (cf. BLK, 1999; Wissenschaftsrat, 2001; 2001a) and implementing a number of changes to the HE system (cf. Wissenschaftsrat, 2001b).

Since it often went unrecognized that the Western societal system also faced new challenges, and also because of the pressure of time, the process of system transformation in the CEECs was strongly influenced by the notion of 'remedial modernization' and proceeded on the basis of more or less uncritical acceptance of various Western models, many of them already out of date. Thus Etzkowitz noted that "having abandoned central planning, some Eastern European countries took a laissez-faire stance and severely limited their role in S&T policy. On the one hand, Western countries in which government traditionally played a quite limited role in innovation are moving closer to the former socialist system, adopting a variant of the branch science model but with the important difference of a lack of centralized controls. The U.S. government, for example, enlarged its role in S&T policy, enacting laws during the 1980s to encourage universities and national laboratories to engage in technology transfer. In addition to this indirect industrial policy, government programs have been established during the past decade to directly support civilian technological innovation, especially in niche priority areas. [...] This is a transition going in the opposite direction from Eastern Europe, where the state is reducing its role in industry. The U.S. transition suggests that there is an important role for the state in S&T policy, a principle that has come under severe questioning in the East." (Etzkowitz, 1998, p. 57 and 65). The transfer of institutions from West to East Germany not only brought about the renewal of science in terms of structure, personnel and expertise, but also introduced some of the serious shortcomings of the Western system into the East and eliminated some of the positive aspects of the old East German system. These included good individual supervision and advice for students, completion of studies within the prescribed time, a very high student success rate and well-established forms of distance learning suitable for working people (cf. Ehrhardt, 2002). In comparison, the student drop-out rate in Germany is now 27 per cent, the main cause being the poor conditions students have to endure (BMBF, 2003). Thus the transfer of institutions and personnel has not necessarily led to the mobilization of East Germany's capabilities. Szomburg (2001) makes a similar point in his comparison of the transformation process in East Germany and Poland. He suggests that, besides the structural differences referred to above, one of the reasons for the unintended effects of the transformation process to date is that traditions and patterns of socialization from the socialist era continue to make themselves felt. The behavior of the actors in the newly created institutions is influenced by these hangovers from the socialist past, but not in the same ways as in similar institutions in the 'Western' system, which has evolved historically. Taking experiences in East Germany (compared with West Germany) as a starting point, Mayntz argues "that only when both levels, the institutional macro level and the micro level of individual behavior, are considered simultaneously and in terms of their interdependence does a central problem in the actual transformation process become discernible, namely the tendency for the processes to fall apart at these two levels. [...] If the tension between institutions (in the sense of formal structures) and the micro world of individual actions and behavior is one of the fundamental problems in the transformation process, then some fascinating theoretical questions are raised as a result. If the objective is to initiate societal change and to channel it in a particular direction, does it make any sense at all to start with the institutions and then expect everything else to fall into place by itself? To what extent do institutions shape individuals' actions and behavior, to what extent, conversely, do they depend on predefined patterns of behavior acquired in other ways?" (Mayntz, 1994b, p. 23; own translation). The pertinence of these questions is particularly confirmed by the preceding analyses of the transformation of S&T and the very different (interim) results that have been achieved in the individual CEECs to date.

It should not be assumed that different patterns of behavior are always simply a disadvantage, a 'legacy' to be overcome. Why should they not also have specific strengths and offer new opportunities for creative thinking that can be used positively to cope with new challenges? Some examples of the positive aspects of the legacy for East German science are listed above; to the list might be added such qualities of GDR scientists as a talent for improvisation and a tremendous capacity for acting on their own responsibility (cf. Syrbe, 2002, p. 7). It should be easier for the CEECs than for East Germany to capitalize on their own good experiences. Such an approach is also urgently required because the structural shift towards modern systems of innovation and a knowledge-based society is of major significance for the further growth of the economy and employment. The future of the CEECs as members or immediate neighbors of the EU can only be safeguarded by developing a knowledge-based economy, the development of which is a declared aim of the EU (cf. Cadiou et al., 2000). The current situation in S&T in all CEECs, which is at least problematic and often downright unsatisfactory, raises questions above all about the opportunities and means of overcoming this situation in order to build up effective innovation systems. In this connection, the interaction between science and industry is of crucial importance.

After the collapse of the Soviet bloc, it became necessary for the CEECs to re-engineer themselves, a process that at national, regional and local level is in many cases only in its initial stages. Developing national capacities further, to the point where R&D-intensive products and services that can compete in European and international markets are being developed, is both a condition and a challenge for the development of S&T in all CEECs, and there will be no generally applicable or uniform solutions (cf. Chapter 22).

It is probable that such development will lead to even greater regional diversity and differentiation than in the EU15 countries. This is a trend that is already becoming discernible within the EU candidate countries as well as between these countries and the other CEECs. It should not be forgotten that the reengineering required by the science systems in all these countries has two aspects. Firstly, the systems must ensure that advances in international science are fully incorporated into national education systems. The new knowledge has first to be 'translated' into each national language and then imparted and diffused by the education system as well as by new forms of 'lifelong learning'. At the same time, however, the science systems in CEECs, as part of the international scientific community and, in particular, of the emerging European Research Area (EC, 2000), must develop their own specific research profiles and make their own contributions to the international stock of knowledge. This will not be achieved without tensions between national and international interests and new problems will emerge with regard to the individual mobility of leading scientists and of the up-and-coming generation of scientists, which will take the new form of 'brain circulation' (cf. A.S. Saxenian, quoted by Seeling, 2002).

In East Germany, the debate on the future of science and industry has also reinvigorated the search for real medium-term development goals and the strategies best suited to achieving them. Alongside a shift in government funding, which is now being increasingly targeted towards innovative activities within *enterprises*, there is also an observable tendency for funding in East Germany to be concentrated more and more within high-priority regions. Although in the beginning hopes for a rapid economic upturn in East Germany and of its catching up with West Germany in the short term had led to economic assistance being dispersed fairly evenly over the whole region (referred to as the 'watering can principle'), actual developments in the 1990s did away with illusions of evening out differences quickly. Since overall development is only making slow progress and the catching-up process has actually got into a rut, the new strategy is based upon concentrating most government funding on selected priority regions with favorable conditions for accelerated growth. The regions selected, because of the high concentration of enterprises, but also of R&D institutes, higher education, skilled labor and transport within them, are expected to develop and utilize the synergy effects that these factors make possible. Some examples are the historically well-established industrial and technologically advanced urban concentrations of Dresden (microelectronics), Jena (optics, devices technology) and Berlin (biotechnology, transport). These centers are meant to become pre-eminent 'islands' of growth with high innovative propensities, and they are also supposed to co-operate with one another and ultimately to fuse economically to the point that new regional clusters are formed, which will diffuse benefits to the less developed areas in their surroundings.

This strategy squares with practical experience gained in the EU in the course of its development. Recent data on the economic and social situation in the EU15 regions shows that, while the gap of per capita GDP between the member states has indeed declined, the gaps between regions fail to exhibit such a trend and have remained more or less constant. As far as unemployment is concerned, the gaps at the regional (as well as the national) level have even widened (see EC, 1999, Graph 1, Table 3 and Graph 5, Table 3). This practical experience has also been corroborated and understanding deepened through scientific studies. These show that regions constitute 'local production systems' (Crouch et al., 2001). They make out some important characteristics of regions undergoing especially dynamic development, such as clusters of SMEs in the region and a regional environment that is 'rich' in 'common commodities' such as well-developed business services, skilled labor, joint R&D projects, superior higher education and R&D institutions. Of interest is also the finding that there are different types of such clusters and networks of SMEs based either on the horizontal co-operation of very similar SMEs (the many small high fashion firms around Treviso/Italy, for example) or on hierarchically structured co-operation of SMEs around big enterprises (as in the region around Stuttgart in Baden-Württemberg where the multinational Daimler-Chrysler Corporation is at the core).

In all of these cases, the state and the local authorities have played a great role in organizing the networks and developing the necessary (scientific, educational, etc.) infrastructure. The German Federal Ministry of Education and Research (BMBF) has developed in pursuit of this aim a support program for East Germany called 'Inno-Regio'. Its purpose is "to create new jobs by aggregating the existing potentials of educational and research institutions, as well as businesses and civil administration. For this purpose, marketable products and services are to be developed through the increased cooperation of research institutes with enterprises" (BMBF, 2001; own translation).

In view of the fact that most of the Central and Eastern European countries themselves and virtually all their firms are small and that even the large countries (particularly CIS members) have to deal with the problem of regional differentiation, the experience East Germany has acquired in supporting regions and SMEs could provide some important lessons for the development of new S&T systems in all these countries. Joint initiatives by academia, research institutions, enterprises and regional authorities should be supported, in order to develop such networks of formal and informal relationships as have helped to make many regions in the EU15 competitive. Obviously, this should involve the FDI-based firms in particular, because of their huge economic and technological importance in the CEECs.

REFERENCES

AG Perspektiven für Ostdeutschland (Ed.) (2001) *Ostdeutschland – eine abgehängte Region? Perspektiven und Alternativen.* Dresden: Junius-Verlag.

BLK (1999) *Forschungsförderung in Deutschland. Bericht der internationalen Kommission zur Systemevaluation der Deutschen Forschungsgemeinschaft und der Max-Planck-Gesellschaft.* Bonn: Bund-Länder-Kommission für Bildungsplanung und Forschungsförderung (BLK).

BMBF (2000) *Bundesbericht Forschung 2000*. Bonn: Bundesministerium für Bildung und Forschung (BMBF).
BMBF (2001) Pressemitteilung No. 155/2001 of October 10, 2001. Bonn: BMBF.
BMBF (2003) Pressemitteilung No. 028/2003 of March 6, 2003. Bonn: BMBF.
Brandsma, A., S. Ertel, and P.-A. Farrer (Eds.) (2002) *Proceedings – Forum Bled 2001. Enlargement Futures Project*. Enlargement Futures Report Series 07. Seville: Institute for Prospective Technological Studies (IPTS).
Cadiou, J.-M., et al. (2000) *Synthesis Report – The IPTS Futures Project*. Seville: IPTS.
Cozzens, S.E., P. Healey, A. Rip, and J. Ziman (Eds.) (1990) *The Research System in Transition*. Dordrecht/Boston/London: Kluwer Academic Publishers.
Crouch, C., P. le Galès, C. Trigilia, and H. Voelzkow (2001) *Local Production Systems in Europe: Rise or demise?* New York: Oxford University Press.
EC (1999) *Sixth Periodic Report on the Social and Economic Situation in the Regions*. Brussels: European Commission (EC).
EC (2000) *Towards a European Research Area*. Communication from the Commission no. 6/2000. Brussels: EC.
EC (2001) *Towards a European Research Area – Key Figures 2001. Special edition: Indicators for benchmarking of national research policies*. Internet: europa.eu.int/comm/research/area/benchmarking2001.pdf.
EC (2002) *European Trendchart on Innovation – 2002 European Innovation Scoreboard: Technical Paper No 2 – Candidate Countries*. Internet: trendchart.cordis.lu/Reports/Documents/report2.pdf.
EC (2002a) *More Research for Europe – Towards 3% of GDP*. Communication from the Commission no. 499/2002. Brussels: EC.
Erfolg im Osten (2002) *10 Jahre Fraunhofer-Gesellschaft in den neuen Bundesländern*. O. O.: Fraunhofer-Gesellschaft.
Ehrhardt, M. (2002) "Der Erneuerungsprozess – Stärken und Schwächen." *10 Jahre danach – Zur Entwicklung der Hochschulen und Forschungseinrichtungen in den neuen Ländern und Berlin*. Essen: Stifterverband für die Deutsche Wissenschaft, p. 6–9.
Etzkowitz, H. (1998) "The Endless Transition: Designing a Triple Helix of Academic-Industry-Government Relations in the United States." *Transforming Science and Technology Systems – The Endless Transition? NATO Science Series 4: Science and Technology Policy – Vol. 23*. Ed. W. Meske, et al. Amsterdam: IOS Press, p. 57–66.
Etzkowitz, H., and L. Leydesdorff (1997) *Universities in the Global Economy: A Triple Helix of University-Industry-Government Relations*. London: Cassell Academic.
Eurostat (2001) *Forschung und Entwicklung: jährliche Statistiken – Ausgabe 2001*. Luxembourg: Europäische Gemeinschaft.
Gibbons, M., C. Limoges, and H. Nowotny (1994) *The new production of knowledge. The dynamics of science and research in contemporary societies*. London: Sage Publishers.
Günther, J. (2003) *Innovation cooperation in East Germany – only a half-way suc-*

cess? Discussion paper no. 170. Halle: Institut für Wirtschaftsforschung Halle (IWH).

Hermann-Koitz, C., W. Horlamus, and T. Konzack (2002) *Strukturelle Analyse der Entwicklung von FuE-Potenzialen im Dienstleistungssektor und verarbeitenden Gewerbe in den neuen Bundesländern.* Berlin: EuroNorm Gesellschaft für Qualitätssicherung und Innovationsmanagement mbH.

Hilbert, A. (1994) *Industrieforschung in den neuen Bundesländern – Ausgangsbedingungen und Reorganisation.* Wiesbaden: Deutscher Universitäts-Verlag.

Hornschild, K. (1998) "FuE-Förderung in Ostdeutschland durch das Bundesministerium für Wirtschaft – Ergebnisse aus einer Wirkungsanalyse." *Innovationen in Ostdeutschland: Potentiale und Probleme.* Ed. M. Fritsch, et al. Heidelberg: Physica-Verlag, 327–356.

Kralj, A. (2002) "Comment." *Proceedings – Forum Bled 2001. Enlargement Futures Project.* Enlargement Futures Report Series 07. Ed. A. Brandsma, et al. Seville: IPTS, 71–73.

Kurzexpertise des DIW, IWW, and IWH (2001) "Gesamtwirtschaftliche und unternehmerische Anpassungsfortschritte in Ostdeutschland: Kurzexpertise." *Zehn Jahre Deutsche Einheit – Bilanz und Perspektiven.* Tagungsband, Sonderheft 2/2001. Ed. Institut für Wirtschaftsforschung Halle (IWH). Halle: IWH, 7–34.

Labrousse, A. (1999) "Der komplexe Wandel von Institutionen und Organisationen in der ostdeutschen Transformation." *BISS public* Vol. 27 (Halbjahresband I/1999): 105–131.

Lavigne, M. (1995) *The Economics of Transition – From Socialist Economy to Market Economy.* New York: St. Martin's Press.

Legler, H., H. Belitz, B. Gerke, C. Grenzmann, and R. Marquardt (2002) *Industrieforschung in Deutschland – Positionen im internationalen Vergleich.* Materialien zur Wissenschaftsstatistik, Heft 12. Essen: SV-Gemeinnützige Gesellschaft für Wissenschaftsstatistik m.b.H. im Stifterverband für die Deutsche Wissenschaft.

Lehmbruch, G. (1992) "Institutionentransfer im Prozeß der Vereinigung: Zur politischen Logik der Verwaltungsintegration in Deutschland." *Verwaltungsintegration und Verwaltungspolitik im Prozeß der deutschen Einigung.* Ed. W. Seibel, et al. Baden-Baden: Nomos Verlagsgesellschaft, 41–46.

Mayntz, R. (1994) *Deutsche Forschung im Einigungsprozeß. Die Transformation der Akademie der Wissenschaften der DDR 1989 bis 1992.* Frankfurt a.M.: Campus Verlag.

Mayntz, R. (Ed.) (1994a) *Aufbruch und Reform von oben. Ostdeutsche Universitäten im Transformationsprozeß.* Frankfurt a.M.: Campus Verlag.

Mayntz, R. (1994b) "Die deutsche Vereinigung als Prüfstein für die Leistungsfähigkeit der Sozialwissenschaften." *BISS public* no. 13: 21–24.

Mayntz, R. (1997) "Forschung als Dienstleistung? Zur gesellschaftlichen Einbettung der Wissenschaft." *Berichte und Abhandlungen Bd. 3.* Ed. Berlin-Brandenburgische Akademie der Wissenschaften. Berlin: Akademie-Verlag, 135–154.

Meske, W. (1991) "Zum Wissenschaftssystem in den neuen Bundesländern und dessen Veränderungen im Zeitraum 1990/91." *Soziologen-Tag Leipzig 1991 – Soziologie in Deutschland und die Transformation großer gesellschaftlicher Systeme.* Ed. H. Meyer. Berlin: Akademie Verlag, 679–688.

Meske, W. (1998) "Öffentliche Forschung als notwendige Infrastruktur für Innovationen in Ostdeutschland." *Innovationen in Ostdeutschland: Potentiale und Probleme.* Ed. M. Fritsch, et al. Heidelberg: Physica-Verlag, 293–312.

Meske, W. (2000) "Changes in the innovation system in economies in transition: basic patterns, sectoral and national particularities." *Science and Public Policy* 27, no. 4: 253–264.

Meske, W. (2002) *Science in Formerly Socialist Countries – Asset or Liability within New Societal Conditions?* WZB-Paper P 02–401. Berlin: Wissenschaftszentrum Berlin für Sozialforschung (WZB).

Ruprecht, W., and G. Becher (1998) "Der Netzwerk-Ansatz der FuE-Förderung für die neuen Bundesländer – Das Beispiel des Programms Auftragsforschung West – Ost." *Innovationen in Ostdeutschland: Potentiale und Probleme.* Ed. M. Fritsch, et al. Heidelberg: Physica-Verlag, 357–378.

Seeling, S. (2002) "Vereint auf Jahre." *Deutsche Universitätszeitschrift* no. 13: 18.

Spielkamp, A., G. Becher, W. Meske et. al. (1998) *Industrielle Forschung und Entwicklung in Ostdeutschland. Untersuchung im Rahmen des Forschungsvorhabens 'Analyse der Situation, der Probleme und Perspektiven der FuE in der Ostdeutschen Wirtschaft'.* Baden-Baden: Nomos Verlagsgesellschaft.

Stifterverband (2001) *FuE-Datenreport 2001 – Forschung und Entwicklung in der Wirtschaft 1999–2000.* Essen: SV-Gemeinnützige Gesellschaft für Wissenschaftsstatistik m.b.H. im Stifterverband für die Deutsche Wissenschaft.

Stifterverband (2002) *10 Jahre danach – Zur Entwicklung der Hochschulen und Forschungseinrichtungen in den neuen Ländern und Berlin.* Essen: Stifterverband für die Deutsche Wissenschaft.

Syrbe, M. (2002) "Da wurde nicht lange geklagt." *Erfolg im Osten – 10 Jahre Fraunhofer-Gesellschaft in den neuen Bundesländern.* O. O., 6–7.

Szomburg, J. (2001) "Ostdeutschland – Polen. Zwei Transformationsansätze – zwei Entwicklungswege." *Zehn Jahre Deutsche Einheit – Bilanz und Perspektiven.* Tagungsband, Sonderheft 2/2001. Ed. IWH. Halle: IWH, 74–94.

Tamasy, C. (1998) "Technologie- und Gründerzentren als Instrument der Technologiepolitik in Ostdeutschland." *Innovationen in Ostdeutschland: Potentiale und Probleme.* Ed. M. Fritsch, et al. Heidelberg: Physica-Verlag, 313–326.

Wissenschaftsrat (2000) *Thesen zur künftigen Entwicklung des Wissenschaftssystems in Deutschland.* Köln: Wissenschaftsrat.

Wissenschaftsrat (2001) *Systemevaluation der Blauen Liste – Stellungnahme des Wissenschaftsrates zum Abschluß der Bewertung der Einrichtungen der Blauen Liste.* Köln: Wissenschaftsrat.

Wissenschaftsrat (2001a) *Systemevaluation der HGF – Stellungnahme des Wissenschaftsrates zur Hermann von Helmholtz-Gemeinschaft Deutscher Forschungszentren.* Köln: Wissenschaftsrat.

Wissenschaftsrat (2001b) *Personalstruktur und Qualifizierung: Empfehlungen zur Förderung des wissenschaftlichen Nachwuchses.* Köln: Wissenschaftsrat.

22. WHAT FUTURE FOR S&T IN THE CEECs IN THE 21st CENTURY?

Slavo RADOSEVIC

The economic and institutional transformation of the former socialist Central and Eastern European countries (CEECs) started in the 1990s, at a time of profound technological and institutional transformation in the world economy driven in particular by the further development and diffusion of information and communication technologies (ICT). Internet technology became the most visible sign of the underlying process of ICT development that had been going on since the mid-1970s.

The end of the Cold War has been followed by an historically unprecedented period of US dominance of the world economic and S&T system. This has been accompanied by radical changes in the world economy, characterized by the liberalization of trade and, in particular, of capital flows. China's entry into the global trading system and the emergence of global production networks have radically changed the terms of competition in the global economy. The increasing multiplicity of markets and production locations has further intensified cost-based competition, which in many sectors has become global. The fragmentation of trade and production systems across large number of countries has pushed the terms of competition for developed countries further towards non-price factors, in particular knowledge, design and innovation. The modes of entry into world markets and the catching-up processes based on the combination of protected domestic markets and aggressive export expansion that were possible for South-East Asian economies have become impossible in the new conditions of reciprocal trade and capital openness. This new global political economy has made it more difficult for developing countries and CEECs to catch up (Rodrik, 2000).

1. DUAL TRANSFORMATION OF CEECs

Under these conditions, CEECs face a formidable dual task. First, they need to transform themselves institutionally into market economies. Second, they need to grow and integrate themselves into global production and technology networks. This dual transformation, both institutional and technological, is essential if we are to understand the context in which the transformation of S&T

systems in CEECs is taking place. At the beginning of the 21st century, the countries of the former socialist bloc are now 'tiered' in several regards, in particular in terms of growth, trade and foreign direct investment (FDI) integration (Chavance and Magnin, 1998). A few central European economies (Poland, Czech Republic, Slovenia, Slovakia and the Baltic states) have integrated relatively quickly with the EU economies, both as markets and, increasingly, through production networks. Bulgaria and Romania, although laggards in this process, have become part of the accession group with good prospects for a further deepening of integration with the EU. The south-east European economies of Albania and the former Yugoslav republics of Croatia, Bosnia and Herzegovina, Serbia, Montenegro and FYR Macedonia have fallen behind in this process and their prospects of joining the EU in the near future are poor. Russia, Belarus, Ukraine and Moldova, which were declining for most of the 1990s, have started to recover. However, in institutional terms, they seem to be becoming different types of market economies with pronounced post-Soviet features, particularly in relation to FDI and state-business relationships (Dyker, 2000; Hanson, 1997).

While some central European economies have embarked on a path of sustained recovery, others are still struggling with the 'transformational recession'. Economic divergence and increasing disparities in per capita incomes among CEECs persist, whether measured over the whole period 1989–2000 or just for the period after 1993 (see UN ECE, 2000, chart 5.4.4, p. 186) [1].

This 'tiering' of the CEECs corresponds to some extent to national differences in the institutional transformation of S&T systems (Meske et al., 1998; Chapter 18 in this volume). However, there is no simple relationship between the transformation of these economies and the transformation of their S&T systems. Progress in institutional transformation as defined by transition economics is not linearly related to growth, recovery and an increased role for S&T. S&T systems are governed by public and private actors and are not mere reflections of levels of development and progress towards the market economy. A variety of supply and demand factors, coupled with historical characteristics and political decisions, makes understanding of the future role of S&T in CEECs a complicated issue in which extrapolations are of limited use.

The various contributions to this volume reveal the patterns and determinants of the transformation of S&T systems in CEECs since the end of the socialist period. S&T as economic and military activity was essential to the rise and subsequent demise of the centrally planned systems. Future prospects for the development of S&T are equally essential to the economic growth and

---

[1] However, within individual CEE sub-regions, convergence rather than divergence has been the prevailing trend, especially in the period 1993–2000 (ibid.).

social welfare of the CEE economies in the 21st century. From that perspective, the evidence presented in this book is very pertinent to an understanding of the former socialist bloc's potential for catching-up and not just of the future patterns of S&T transformation.

This chapter builds on the evidence from country studies and inter-country comparisons presented in the previous chapters in order to analyze what the future holds for S&T in the CEECs in the new century. In particular, we address the following questions. What are the technological and institutional factors that are shaping the transformation of S&T in CEECs at the beginning of the 21st century? How will differences in economic and institutional transformation across the different tiers of CEECs affect the future of S&T in those countries? How will EU integration affect the role of S&T in CEECs economies? Unlike the country chapters in this book, which are focused on 'narrow' national systems of innovation (NSI)[2], our analysis is framed within the 'broad NSI' and growth perspective.

Our main argument is that the future role of S&T will differ significantly in the various CEECs. In this respect, we may expect to see increasing differentiation across the region or cases of simultaneous catching up and falling behind, which can be partly attributed to the role that S&T systems will play in these economies. Our argument is that national differences in this respect will depend on the interaction between 'narrow NSI' (S&T system) and 'broad NSI' or the overall institutional setup that influences innovation. The current exclusive focus on the various aspect of transition (privatization, liberalization, corporate governance, banking and finance system reform, etc.) will help us little in understanding which economies are likely to grow in the long term (Frydman et al, 2000). In a knowledge-based economy, growth will depend essentially on a strong S&T system or the 'narrow NSI' and how that system is embedded within the wider economy. This alignment between S&T and the economy depends not only on market reforms, which dominated the policy agenda during the 1990s, but much more on the development of public – private interfaces within and between the 'narrow' and 'broad' systems of innovation.

We begin in the second section by presenting a few stylized facts about the transformation of S&T and growth during the 1990s in the CEECs. In the third section, we outline the main external challenges that stem from the mode of growth based on the knowledge economy (3.2.) and the increasing integration of the world economy (3.1.). How CEECs will cope with these two challenges is the key issue if they are to grow in the next 15 to 30 years. The fourth section

---

[2] The NSI in a narrow sense embraces those institutions which are directly involved in R & D and the dissemination of the results of R & D.

analyzes two key emerging features of the transformation of S&T in the region and the trade-offs that they imply. These are the post-Soviet R&D model and the Europeanization of research, technology and development (RTD). In conclusion, we summarize the main arguments and discuss the policy implications of our findings.

2. GROWTH AND S&T IN THE POST-SOCIALIST TRANSFORMATION: REALLOCATION VS. ACCUMULATION

Evidence on growth shows that by 2001 only four CEECs had managed to surpass their 1989 GDP levels (Poland, Slovakia, Slovenia and Hungary) (EBRD, 2001). Instead of a short journey through a 'vale of tears', as the transition has been popularly characterized, most of the CEECs still find themselves at levels of development below those they had reached in 1989. In most of the CEECs, growth, the curve of which was expected to be J-shaped, came much later and at lower rates than expected. Nevertheless, all the CEEC economies are now growing and most of them have undergone tremendous institutional changes and increased both productivity and exports.

However, the CEECs' sources of growth during the 1990s seemed to be at odds with the increasing emphasis elsewhere on knowledge as the main factor driving growth (Dosi et al, 1994). Technology was not the force driving recovery and growth in the CEECs during the 1990s.

Differences in factor endowments in labor, capital and knowledge, or other variables, which are usually used in growth accounting exercises, cannot explain the differences in economic growth between CEECs during the 1990s (Havrylyshyn et al., 1998). Factor expansion is not significantly linked to growth in the transition period. It seems that reallocation rather than accumulation was the driving force behind recovery and growth in CEECs during the 1990s. Intra-firm, intra-sector and inter-sectoral reallocations were pervasive features of the CEECs, and those economies *other things being equal* that performed better in this respect managed to recover earlier.

The socialist legacy left huge scope for efficiency improvements and growth based on reallocations with minimal investment. Among the most important are reallocations of labor from industry to services, from loss-making, state-owned enterprises to new, private-sector companies, from domestic to foreign-owned firms and from unprofitable to profitable lines of businesses. The perception of post-socialist transformation as entirely a problem of reallocations and static allocative efficiency is based on the implicit assumption that achieving static allocative efficiency is sufficient for long-term growth. A 'progress in transition' framework based on a few simple criteria (price and trade liberalization, privatization, banking reform, etc.) derived from this perspective is then used to measure how far any particular country has advanced in its transition (EBRD, 2000).

The notion that growth is a function of 'progress in transition' comes up against the problem that there is no consensus on the *specific* factors that might explain recovery and growth. Economists consider initial conditions, macroeconomic policies and structural reforms in the transition period as the major determinants of recovery during that period. True, each of these factors is in itself positively related to growth but the major problem is how they are related to each other. For example, policy choices are influenced by different initial conditions and, hence, cannot be considered as independent factors. Many institutional factors are omitted due to data problems.

An exclusive focus on allocative efficiency issues during the transition period stems from a critique of socialism from the same perspective. From a different point of view, the main deficiency of socialism was not allocative inefficiency but the inability to generate and diffuse technical change and to adapt institutionally (adaptive efficiency) (Grabher, 1997; Grabher and Stark, 1997). From this perspective, the criteria of adaptive efficiency and the capacity of institutions to promote innovation and technical change are used to assess post-socialist transformation. As a result, many but not all the institutional changes will be assessed differently. For example, institutions such as large state-owned enterprises and business groups are not seen as a priori 'inefficient' (Amsden and Hikino, 1994; Granovetter, 1995; Khanna and Palepu, 1997; 1999). From a mainstream perspective, they are regarded as problems because of improper corporate governance and the possible distortion of allocative efficiency. Their role as generators of demand for innovation and R&D and as agents that can compensate for deficient capital and skill markets is not taken into account. From this alternative perspective, the achievement of static allocative efficiency does not guarantee growth. Rather, growth is driven by an accumulation of technology and an institutional framework conducive to technical change (Dosi et al, 1994).

From this perspective, the relationship between S&T and growth has certain features that are unique to the post-socialist transition. Firstly, *recovery and growth have been unrelated to domestic technology and R&D*. The sources of growth in CEECs have not so far been directly linked to R&D but to the acquisition of knowledge in the production process and through different forms of firm-based learning (Dyker and Radosevic, 1999). This suggests that the main capability acquired is production capability or the capability to produce in accordance with the standards of efficiency and quality required in export markets. The technology capability or the ability to generate change seems to be much less significant. Technical change embodied in imported equipment has had a greater impact, but the mere import of such equipment does not of itself guarantee that it will be used to generate innovations in the absence of indigenous technological capability. There have been visible improvements in export competitiveness in several central European economies, driven mainly by foreign direct investments (Hunya, 2000). However, these improvements were not produced by the formal S&T system.

In particular, technical change involving domestic R&D is absent. Growth has not been accompanied by increased investment in R&D (Radosevic and Auriol, 1998). In fact, in six out of ten CEECs, growth has been accompanied by a decreasing GERD/GDP ratio. Moreover, this ratio is negatively correlated to growth in three out of the four fastest growing CEECs (Poland, Hungary and Slovenia). In all CEECs, public R&D systems were marginalized during the 1990s by budget cuts and a lack of demand for R&D.

Secondly, there has been *significant productivity growth but little technological development*, except in sectors with high levels of FDI. Given the abundance of idle capacity and the considerable potential for efficiency gains, the expansion of output during the transformation has been based mainly on non-investment sources of growth. As the Polish case shows, early expansion coupled with structural shifts and a decline in employment is likely to have been caused by unprecedented efficiency gains (Zukowski, 1998).

Once growth in CEECs resumed, it was accompanied by rising labor productivity in industry. In the early stages of transition, developments in productivity throughout the region were dominated by the decline in output, as many firms initially avoided large-scale layoffs even though demand for their products had collapsed. Firms often hoarded labor and financed current production through inter-industry credit. This decline in measured productivity began to be reversed in most countries after 2–3 years, primarily as a result of labor shedding. More recently, some of the central European economies, primarily Hungary, have entered a phase of rapid productivity growth, driven by foreign direct investments. However, strong fluctuations in rates of labor productivity growth in other CEECs suggest that improvements are being driven more by layoffs, the closure of unproductive lines of businesses and reactive restructuring than by continuous technological improvements (EBRD, 1999; Stephen, 2000).

A slow decline in productivity growth in all CEECs may indicate that this 'mode of growth' is coming to an end and that any further productivity gains will require new investment in tangible and intangible assets. The ability to generate investments for a prolonged period of time is essential for the CEECs to catch up and grow (Brzeski and Colombatto, 1999). Now that the growth potential based on reallocations and idle capacities is being exhausted, the accumulation of sector and technology-specific knowledge will become distinctive features of growing CEE economies. However, recovery and growth in post-socialist CEECs will involve more than a shift from growth based on unused capacities and reallocations within firms and within and between sectors to growth based on capital and technology accumulation.

The short history of post-socialist transformation has shown that growth is a complex, non-linear phenomenon in which nationally specific alignments between factor endowments and external and domestic political factors explain growth much better than a few seemingly unrelated variables in growth regressions. Growth is not a simple function of 'progress in transition', as transition economics tends to suggest, but a much more idiosyncratic phenomenon. A narrow understanding of what constitutes growth has led to some misconceptions. The most influential of them has already been alluded to, namely the idea that growth follows from 'progress in transition' defined in terms of a few simple criteria that constitute a properly functioning market economy (EBRD, 1999). This has resulted in a series of misjudgments regarding growth prospects. For example, Czech macro-economic stability and rapid privatization in the early 1990s subsequently turned out to be illusory rather than sustainable. On the other hand, the Polish economy, which before 1990 was considered a 'basket case', turned out to be the fastest growing economy in the region. Its growth has been attributed to initial 'shock therapy', although Polish progress in transition does not seem distinctively different from several other central European economies. The Polish economy, which for ten years grew on very limited technology accumulation, has now reached its threshold level in terms of industrial upgrading with slim prospects for further robust growth (Kubielas, 2003). Russia, which immediately after the financial crisis of August 1998 was also considered a hopeless case has suddenly come to be regarded as an economy with respectable growth potential. Estonia which, together with the other two Baltic economies, was considered to be among the leading economies in terms of progress in transition has managed just to keep pace in terms of recovery with Belarus, which can be considered as still very much the post-Soviet economy. Slovenia, an economy that was slow to privatize and with limited FDI, has managed to grow faster than Hungary, the CEE economy with the highest level of FDI. All this suggests that the narrow conceptual framework of 'progress in transition', the key elements of which are criteria related to static allocative efficiency, is too weak to explain the long-term growth prospects of CEECs.

Growth is not simply a continuous and incremental process of accumulating capital, technology and knowledge but rather a systemic process with complementarities and synergies operating between various growth factors once they reach specific threshold levels. This explains why industrial upgrading is a non-linear process and why it is difficult to understand the relationship between S&T and growth (Radosevic, 1999; Ernst, 1998). Industrial upgrading is a function of other, more general factors, such as education, science, general purpose tangible investments (e.g. telecommunications), a diversified knowledge base, as well as of specific factors like firm-specific skills, know-how, training and sector-specific equipment (Porter, 1990). Growth in the CEECs during the 1990s shows that general factors of production, like an educated labor force, are not sufficient for growth. Firm and sector-specific knowledge requires a variety of institutional preconditions if it is to develop. For example, the development of local subcontractors in the automotive industry in CEECs requires a variety of activities to match each other. Efforts by assemblers and first-tier suppliers have to dovetail with the activities of the still weak local authorities, the higher education system and state support for vocational training.

In summary, ten years of post-socialist transformation have shown that growth has been unrelated to domestic technology accumulation. Equally, growth cannot be equated with progress in transition. While it is true that CEE economies have adopted more of the institutional features of the market economy as they have grown, these factors do not have the power of independent factors but are combined with a variety of historical and developmental features. There are signs that the sources of productivity growth, which have been mainly in the realm of 'reallocations', are now coming to an end and that the CEECs will have to grow on the basis of technology accumulation. This brings back the issues of S&T systems, their role in CEE economies and in particular the links between 'narrow' and 'broad' national systems of innovation.

Countries may grow for some time based on an historically specific constellation of economic and socio-political factors which may then exhaust their potential once the underlying mode of growth in the world economy changes. In that respect, our knowledge of growth in post-socialist CEECs is still quite limited. This chapter continues with an analysis of two of the main features of the mode of growth to which CEECs will have to adjust if they are to grow. These are, firstly, the increasing knowledge intensity of production systems or the increasing knowledge base of growing economies and, secondly, the increasing financial, production and technological integration of the global economic system within which CEECs operate. What demands do these two key features of the mode of growth place on CEECs?

3. TECHNOLOGICAL INTEGRATION AND THE KNOWLEDGE-BASED ECONOMY: TWO KEY EXTERNAL CHALLENGES TO S&T AND GROWTH IN CEECs

Economic growth in the 21st century takes place within an increasingly interdependent world economy in which the generation, utilization and diffusion of knowledge are playing an ever more important part. In this section we analyze what effects these two factors will have on S&T systems in the CEECs.

3.1. THE CEECs: FROM PRODUCTION TO TECHNOLOGICAL INTEGRATION?

In an interdependent global economy, industrial upgrading occurs through technology accumulation mediated by various forms of international inter-firm linkages (alliances, subcontracting etc.). Autonomous technology accumulation or technological development within national boundaries is untenable due to the increasing complexity and pace of technical change. An increasing variety of modes of integration into regional and global production networks is a key precondition for technology accumulation in transitional economies. Protectionism and learning behind national borders are becoming increasingly difficult and unviable options. This experience is even more relevant in the case of the previously closed socialist economies, which were often forced to 'reinvent the wheel' in sectors closed to foreign trade in technology.

Economic liberalization in the CEECs has given enterprises in these countries an opportunity to integrate into the global economy, while their S&T systems are now able to integrate fully into international science. In terms of financial and trade integration, all CEECs are now full members of the global economy as markets. In terms of production integration, specifically in terms of FDI and subcontracting linkages, there are great differences between individual CEECs (Lankes and Venables, 1996). As far as globalization is concerned, the CEECs can already be divided into several tiers. Broadly speaking, there are significant differences in the scale and features of international integration between the central European, Balkan and European CIS economies. The trade patterns of CEECs, which reveal a multiplicity of different adjustment patterns, suggest that the roles of specific sectors or countries in Eastern Europe in production and technology will increasingly diversify (Guerrieri, 1999; Hotopp et al., 2002).

The increased financial, trade and production integration of the CEECs has highlighted the emerging problem of technology catch-up and weak or non-existent technological integration. While consumer catch-up seems quite advanced, the CEECs are lagging far behind in technological terms. Financial and trade integration does not automatically lead to production integration, while production integration does not lead automatically to technological integration. Case study evidence shows the difficulties involved in deepening production integration and points to the discontinuous nature of technological integration and the emerging structural barriers Eastern European firms face after initial productivity improvements [3] (see Radosevic, 1999; 1999a).

The CEECs are reintegrating into the international economy at a time when trade patterns are being strongly shaped by the complex integration strategies of multinational companies (MNCs) that are constructing international production networks. Their penetration into CEECs, especially in central European, has exposed how organizationally and technologically weak domestically controlled firms in CEECs are and how dependent they are on global production networks. On the other hand, in countries that have received large volumes of FDI and investors are not only market seeking but also efficiency (export) and knowledge seeking, the effects on productivity, employment and technology transfer are strongly positive (Hunya, 1997; 2000; Djankov and Hoekman, 1999). For example, Eichengreen and Kohl (1998, p. 39) conclude that "the correlation between FDI flows and increase in unit values, shifts in factor intensity and the other measures of trade performance [...] strongly suggest that FDI is integrating the more advanced economies into multinational production networks, shifting these economies towards R&D-intensive and human capital intensive products". FDI brings capital but also transfers assets to more efficient owners. This latter aspect is very important in CEECs, where foreign owners have advantages in terms of corporate governance as well as easier access to capital markets and technology. The result is a big difference in terms of productivity between domestic and foreign owned firms.

Our research on East – West industrial networks shows that the weakest actors in this process are national networks (domestic firms and their linkages to S&T organizations) (Radosevic, 2003a). The integration of the CEECs S&T systems into international science is quite advanced in some countries, such as Russia and Hungary. However, international linkages in S&T are in the CEECs mainly a substitute for inadequate domestic funding rather than a sign of a well integrated and developed national S&T system. This crisis-driven internationalization of S&T systems has its limits in the sense that it

---

[3] From a sample of 90 Hungarian subcontractors, Szalavetz (1996) showed that close cooperation with foreign partners brings considerable productivity improvements. In her sample, all processing firms received a transfer of technology or equipment and half the firms benefited from investment or working capital finance provided by the foreign partner. However, after the initial push the learning process gradually slowed and finally stopped completely. Szalavetz (1996, p. 5) points out that "Once the Hungarian company had undergone sufficient restructuring to ensure that cooperation can go smoothly, foreign partners abandon any further developmental effort". This occurred even in those cases where foreign partners had decided to increase their equity in the Hungarian company (ibid., p. 53).

leads to the integration of R&D groups into international networks of scientific excellence but further widens gaps with domestic industry. Particularly in those CEECs in which foreign controlled MNC networks have become quite well established, the weakness of domestic 'narrow' NSI has become an obstacle to further production and technology integration. Elsewhere, we have pointed to emerging structural barriers to growth in several industrial sectors in the more advanced CEECs in which the constraints are eminently domestic and should be addressed by policymakers (Radosevic, 1999a). As the cases of some sectors suggest, industrial upgrading is a continuous process and today's specializations may not be sustainable or economically profitable in the medium or long term. The case of the car industry, considered the most advanced in terms of restructuring in CEE because it integrated early and successfully, shows the type of problems that are emerging. These are no longer company-internal problems to which FDI was a sufficient and necessary answer. The emerging problems are systemic in nature and FDI alone may not provide the solution. The successes of the last 10 years in trade and production integration have brought with them new structural barriers, one of which is the weakening of S&T systems. The continued marginalization of R&D systems may also become an obstacle to further industry upgrading. The general level of education in the labor force does not seem to be an obstacle to productivity improvements at shop-floor level. This explains why foreign investors are able to achieve rapid productivity improvements in CEECs, matching levels in their parent factories. However, a high share of the population with vocational training qualifications and a relatively small share with university education compared to EU levels is an obstacle to restructuring in the economy as a whole, particularly with regard to the adoption of IT technologies. The rapid integration of certain central European economies (Hungary, Czech Republic, Poland and Slovakia) into international production networks has not been followed by their technological integration as measured by improvements in the technological content of exports (Landesmann, 1997; 2000). In CEECs in which the share of employment in nominally high-tech industries is very high, the real value added and technology content is still very low. Figure 22.1 shows that the very high share of employment in high-tech manufacturing in Hungary and other Central European countries is accompanied by a very low share of business R&D in GDP.

All this suggests that development of the technology content of industry and the knowledge-based economy in general is the key to further industrial upgrading in the CEECs as well as to their technological integration into the world economy.

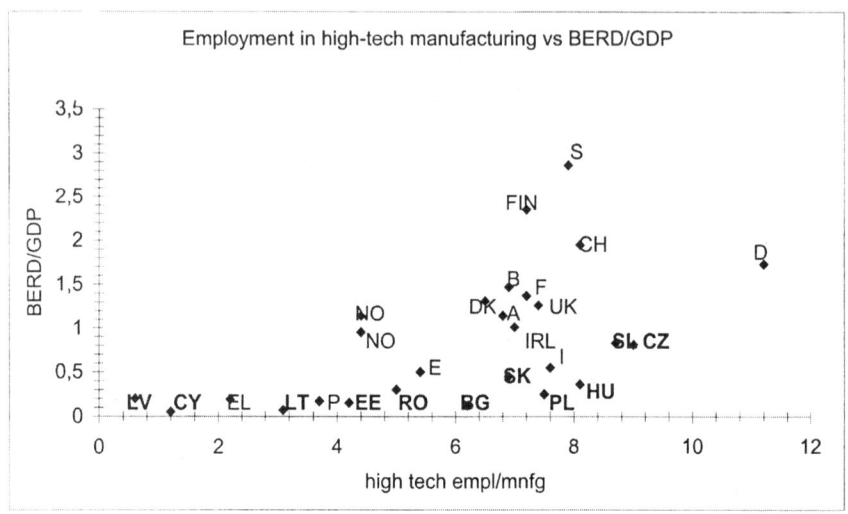

Source: Own compilation based on European Trendchart on Innovation—EC, 2002a

Fig. 22.1 Employment in high-tech manufacturing vs. BERD/GDP

3.2. THE KNOWLEDGE-BASED ECONOMY IN THE CEECs AND ITS EFFECTS ON S&T

There is increasing evidence in the OECD economies of shifts in the industrial and occupational structure towards a *knowledge-based* profile (OECD, 1996; 2000)[4]. Despite this, it is not certain whether the knowledge-based economy really is something new, in the sense that it entails structurally different processes of knowledge generation and distribution. In this chapter, it is assumed that, even without clear indicators of changes in processes of knowledge generation and distribution, it is still possible to talk of the knowledge-based economy based on increasing knowledge content of economic activities. The increased knowledge content comes from increasing importance of non-tangible investments which include R&D, investment in software, education and training, marketing and organisational change and from increased importance of stocks of human capital, organizational capital and intellectual capital. In this respect the CEECs are lagging behind the leading OECD economies. However, if trends and performance in the lead countries are critical to an understanding of the changing pattern and pace of technological progress in the follower nations (Chandler et al, 1997), we can expect that the shift towards a knowledge-based profile will eventually occur in the CEECs as well.

---

[4] Like the OECD (1996), we define knowledge-based economies as those that are based directly on the production, distribution and use of knowledge and information.

The activities that underpin the knowledge-based economy, are *knowledge creation,* or knowledge investments, and *knowledge diffusion,* or distribution (David and Foray, 1995). In an economy whose main production factor is knowledge, innovation is driven not only by intra-firm activities but equally by interaction between firms. In that respect, modern innovations are 'systemic' events in the sense that they involve interaction between different organizations and occur within communities of 'specialists'. Also, the blurring of the distinction between production and distribution makes innovation a much more 'collective' activity than it used to be in the past. From the economic point of view, effective knowledge distribution through formal and informal networks becomes essential for good economic performance (Lundvall, 1992; Nelson, 1993; David and Foray, 1995; Edquist, 1997).

The strengthening of the knowledge-based economy has important implications for S&T systems in CEECs. In our view, the two most important are, firstly, the shift towards diffusion-oriented activities within S&T systems and, secondly, the trend towards enterprise-based S&T systems.

(i) S&T systems have to go beyond R&D and the generation of new knowledge and embrace non-R&D activities, in particular utilization and diffusion activities. Historically, S&T systems in CEECs have been engaged solely in the generation of new knowledge. One of the key weaknesses of the Soviet-type economic system was its inability to diffuse innovations throughout the economy. With their mission-oriented S&T policies, CEECs were concerned primarily with major projects of national significance, often with an emphasis on national defense. Reorientation towards the knowledge-based economy requires the adoption of diffusion-oriented policies that seek to improve firms' ability to respond to new technologies. The emphasis is less on developing entirely new, cutting-edge technologies and more on promoting the widespread dissemination of technological capabilities throughout industry. Mostly, this involves the strengthening of institutional mechanisms for technology transfer, particularly with respect to education and vocational training systems, industrial standardization systems and cooperative research networks (Ergas, 1986, p. 4–5).

Because of its origins in the Cold War, Russian S&T policy shares most of the features of mission-oriented S&T policies. Its 'mission-based' nature comes from its strong defense orientation, exclusive focus on the R&D component and neglect of diffusion and its concentration on supply to the detriment of demand. The goal-oriented and predominantly R&D-focused policy of Russia is becoming incompatible with the need to develop the diffusion-oriented components of S&T policy. Although, the mission-oriented component of R&D is not so strong in other CEECs, their capacity for diffusion is still very weak.

The need to develop the diffusion component is becoming essential for the following reasons (Radosevic, 2003):

- Mission-type policies can be justified only in an environment in which demand for technology is strong and a rich network of market institutions and infrastructure exists to make use of the results generated in the R&D sector. Neither of these two conditions exists in CEECs. If S&T policy continues to retain its mission-oriented character this will only widen the gap between general economic conditions and the S&T system, which will further weaken the S&T system itself.
- The technology lag, particularly that which has developed in the last 10 years, cannot be reduced without a strong innovation policy focused on quality, training, diffusion, standards and organization. With their mission-oriented technology policies, CEECs, like Russia, might at best develop a few isolated sectors, while the rest of industry will lag behind. In other CEECs, even this is unlikely due to their smaller R&D systems and weak domestic demand. The new features of the emerging 'knowledge-based economy', which is focused around the Internet, business services and information technologies, require diffusion-oriented policies. If they persist with their mission-oriented policies, the CEECs will lag further behind in the application of new information and communication technologies.

The industrial and technological upgrading of CEECs must be based on significant, and in some sectors fundamental, changes to innovation processes and the knowledge base. The locus of technology efforts has moved from extra-mural R&D activities to firm-specific operations oriented around cost-efficiency and quality management activities. These are the areas in which the socialist enterprises were weakest in the past (Yudanov, 1997). As Freeman (1999) points out, the crucial weakness of the narrow NSI under socialism was the failure to develop R&D at enterprise level.

The post-socialist transformation of industry has not rectified this inherited problem. Examination of the restructuring process in six industrial sectors in CEECs shows that this restructuring did not involve domestic R&D (Bitzer and Hirschhausen, 1998). In-house R&D departments were drastically cut back while the industrial institutes were not integrated into large enterprises. The 1990s were years of stagnation and the erosion of R&D capacities. This erosion suggests that the potential for catching up is deteriorating.

Elsewhere (Dyker and Radosevic, 2000), we concluded that there was no strong synergy between transition policies in the narrow, macro-economic sense and the required shift towards the knowledge-based economy in CEECs. Privatization policies had the biggest effects on the knowledge-based economy, to the extent that they created the diversity of enterprise forms, sizes and strategies that is essential for knowledge diffusion and generation. Privatization models that inhibited the creation of dynamic small firms but also which aimed to break up large firms, as in East Germany for example, have slowed down innovation dynamics and knowledge diffusion. However, the issue is not the size of firms per se but the lack of organizational diversity that privatization may generate, as well as the absence of the interaction between large and small firms and specialized suppliers that is essential to innovation dynamics. This brings us to the question of the control of technical modernization, which we address in the next section.

3.3. CEECs AND MODERNIZATION: FOREIGN OR DOMESTIC CONTROL?

Our argument in this section is that the nature of the actors controlling technical modernization (i.e. whether they are foreign or domestic) can have significant effects on the knowledge-based economy and thus on S&T systems in CEECs. The balance between domestic and foreign-led technical modernization is essential for long-term growth and thus for S&T. Currently in the CEECs, two extreme modes of modernization can be observed: foreign-led modernization in most of the sectors in central Europe and domestically controlled modernization in most sectors in the European CIS (Russia, Belarus and Ukraine) with eastern Europe and the Balkans in an intermediate position. Both modes have their trade-offs, with important effects on S&T systems.

Domestically-led technical modernization is a process of technological upgrading and productivity improvements in which the key factors (assets, technology capability, distribution, supply, finance) are controlled by domestic actors. In foreign-led modernization, conversely, the key factors are under foreign control. The two terms denote analytical distinctions that may exist in pure, dichotomous forms or in hybrid forms in which control of key factors is distributed between domestic and foreign actors. Whether modernization is under domestic or foreign control depends on which key factors are driving technical modernization and who controls them.

Differences in the interaction between national industrial systems and international value chains between CEECs can explain to a great extent differences in the depth and breadth of industrial restructuring. The CEECs are becoming integrated into the world economy as markets as well as sites in international production networks. The forms of integration vary from country to country and from sector to sector, as well as in degrees of co-operation, competition, compliance, coalitions and the distribution of direct or structural control between domestic and foreign enterprises. Restructuring and technical modernization are to a great extent dependent on who is the main actor in the restructuring process and which elements of the process that actor controls (assets, labor process, supply, distribution, technology, finance). The various patterns of modernization are the result of these differences.

Countries like Belarus, Ukraine and Russia, where domestic actors tend to control technical modernization, have seen prolonged crises, low productivity improvements, excessive employment and very often failed attempts to trade access to its assets for access to foreign technology (Radosevic and Sadowski, 2003). In these countries, the industrial institutes have remained nominally linked to large enterprises but the relationships are devoid of real content. In most central European countries, where privatization gave full control to foreign owners, foreign-led modernization has raised the productivity of newly acquired enterprises, streamlined them and facilitated their integration into international production networks. The industrial institutes have been either dismantled or privatized and now operate as independent service companies (see contributions in this volume).

In sections 3.1. and 3.2., we noted that both technological integration as well as the building of a knowledge-based economy are the two key challenges for long-term growth, which by definition has to involve domestic S&T. The key issue is how to achieve a balance between integration through finance, trade, production and technology and the building of indigenous S&T capacities and the shift towards knowledge-based economic activities.

A country's modes of technology integration cannot be explained without taking account of its modes of technology accumulation and innovation system. In this respect, we can talk only about emerging systems of innovation in CEECs, primarily in areas with FDI. For the time being, national innovation systems are non-existent or highly fragmented and are confined to local competences in a few sectors (for example, pharmaceuticals).

The literature on technology in the global economy indicates that new trade and production patterns have not undermined the importance of national and local systems of innovation. Indeed, these matter even more in the context of the global economy (cf. Freeman, 1991; Porter, 1990). The capability of local and national systems to exploit the opportunities offered by trade and financial liberalization has also increased. Equally, globalization may weaken national and local innovation systems by technologically marginalizing whole sectors of national economies. Sectors in which foreign-led modernization is most advanced in the CEECs are closely integrated into international production networks via parent companies. However, production integration has led to the marginalization of local R&D institutions and the irrelevance and obsolescence of local knowledge. In countries and sectors where modernization is predominantly domestically led, companies are not integrated into international production networks and technological lags persist, although the S&T system nominally operates in its old form. In these countries and sectors technical modernization is unlikely without sectors being opened up to foreign owners.

After 10 years of transition, it seems that the pendulum in some CEECs has swung too much to the other side. From being completely inward-oriented, the openness of the CEECs has now revealed all the weakness of national production and technology networks. Non-existent or weak technology integration in the CEECs highlights the need for domestically generated technology and the construction of knowledge-based national economies. On the other hand, in other CEE economies and sectors international production and technology integration has barely begun.

CEECs have been reintegrated into the world economy at a time when trade patterns are strongly shaped by the complex integration strategies of MNCs and by changes in international trade and FDI regimes. CEECs, especially those in central Europe, have fully conformed to the new global economic regime and continue to adjust to the requirements of EU accession. The scale of the challenges they face in restructuring and the reduced scope for autonomous action in this respect have made them extremely dependent on foreign markets, capital and organizational capabilities. Even Russia cannot expect any fast restructuring of sectors like oil, gas, motor manufacturing or telecommunications without very considerable involvement by foreign capital and management. Central European economies have been very active in trying to lure foreign investors. Much of the FDI has been used to access local markets, export to the EU and take over assets in sectors such as telecoms and energy. The degree and pace of FDI penetration in some CEECs (Hungary, Czech Republic, Poland, Estonia) has been fast and deep and is now comparable to the East Asian economies. The direct effects of FDI have been very positive, as indicated by productivity data, especially when compared to domestically controlled firms. This has created concerns about the emergence of dual economies. Experience from other countries shows that furthering the positive effects of FDI (indirect effects) is not an automatic process. The degree to which FDI will benefit CEECs will depend to a great extent on their absorptive and innovative capabilities.

The fact that in many sectors in central European countries most of the critical functions are under the control of foreign enterprises is both a strength and weakness in the current stage of industrial transformation. As sectoral studies show, restructuring was fastest and productivity improvements were high in sectors where technical modernization was foreign-led. In the software sector, links with foreign software enterprises through different forms of international co-operative agreements (value-added resellers, customizers, system providers, etc.) are crucial in order to capture the domestic market. In PC assembly, good links with foreign components suppliers are essential.

In telecommunications equipment, domestic enterprises have become an integral part of MNC networks. In the automotive industry, domestic subcontracting networks depend quite heavily on foreign assemblers or first-tier suppliers. In all these cases, production efficiency has improved because of foreign takeovers or close cooperation with foreign firms. However, these productivity gains have been obtained at the cost of reduced strategic autonomy, especially in functions like finance, marketing and R&D that historically have never been strong in socialist enterprises. Further industrial upgrading is dependent on the mastering of business functions beyond production and assembly.

On the other hand, strategic autonomy and exclusively domestic control of technical modernization is not a viable option in most of the CEECs and sectors either. Case studies on successful domestic firms show that as a rule firms have grown through networking or linkages with foreign partners rather than through endogenous expansion (Radosevic and Sadowski, 2003[5]). The key long-term issue is how to achieve a balance between autonomy and the building of national innovation systems, on the one hand, and exposure to foreign markets, production linkages and technology, on the other. Those CEECs that manage to achieve complementarities and synergies between domestic and foreign control of modernization are poised for long-term growth.

4. THE RESTRUCTURING AND EUROPEANIZATION OF S&T SYSTEMS IN THE CEECs

In this section we focus on 'narrow NSI' or S&T systems in CEECs. Our argument, similar to that advanced in other chapters in this volume, is that the S&T systems in CEEEs have been transformed in such a way that they can now be described as post-socialist or post-Soviet R&D systems. In other words, they have adopted certain features of S&T systems in market economies but have also retained some of their unique post-socialist or post-Soviet features.

We use the terms "post-socialist" and "post-Soviet" to highlight important differences between S&T systems in the former USSR and other former socialist CEECs that have been carried forward into the post-1989 period. There were significant variations in the socialist countries in the extent to which R&D was extra-mural or intra-mural, in economic openness, in the importance of R&D in higher education and in the role of the Academies of Sciences. However, these differences, though very important, were a matter of degree rather than of substance. Similarly, the same forces have been driving the restructuring of S&T in the post-socialist period, but their strength has varied consid-

---

[5] For a series of working papers on foreign and domestic CEE firms from the perspective of industrial networks see: www.ssees.ac.uk/esrcwork.htm.

erably as they have been operating in contexts shaped by very different initial conditions. One of the most important factors affecting the restructuring of S&T systems in CEECs is the Europeanization of those systems, which is likely to become very marked in the EU candidate countries. We discuss the effects and trade-offs this will entail for S&T systems in CEECs in future.

4.1. THE TRANSFORMATION OF POST-SOCIALIST S&T SYSTEMS

The contributions to this book show the extent to which post-socialist S&T systems in individual CEECs have converged towards the S&T systems characteristic of market economies. At a general level, we have argued elsewhere (Radosevic, 1999b) that they still possess some socialist features (externalized R&D) as well as some peculiarly post-socialist features (ownership) that continue to set them apart from the market economy model of S&T. Firstly, enterprises have not yet become the main agents of innovation. Secondly, R&D is still externalized and there are significant management problems because the state's role as owner prevents it from exercising a cash flow control function in R&D. Thirdly, new gaps have emerged in R&D and innovation where applied and strategic R&D are absent from the spectrum of innovation activities. These differences are symptoms of as well as factors in the weak innovation capabilities of the post-socialist enterprises.

As Freeman (1999) notes, the key weakness of the socialist system was the failure to develop R&D at firm level. However, he also points out that this weakness is a common one in most developing countries. Contributions to this volume clearly show that this weakness has not been overcome in the market economy period. In fact, the absence of strong in-house R&D and the predominance of externalized R&D remains one of the key features of the post-socialist or post-Soviet R&D systems.

The elements of institutional divergence in S&T systems should be seen more as symptoms of a crisis in the transformation process than as organizational characteristics to technological development. At the core of these differences is the (in)ability of post-socialist enterprises to articulate a demand for technology. How long these divergent features will remain temporary remains an open question. Elsewhere (Radosevic, 2003) we have developed a framework, applied in that instance to Russia but of general relevance to all the post-socialist economies, which argues that the current R&D model is determined by the interaction of three forces: the attempts by policymakers to preserve S&T systems in their old forms, the survival strategies developed by enterprises and R&D organizations and the attempts to restructure R&D systems.

To date, the adjustment of S&T systems in the post-socialist (post-Soviet) period has been aimed at the 'preservation of S&T potential', and has been taking place in tandem with restructuring efforts and the development of survival strategies by researchers and R&D organizations. The 'preservation' aspect is an attempt by policymakers to preserve S&T in its old form in the face of inevitable functional, organizational and funding changes. Restructuring involves reform of the R&D system in an attempt to conform to the principles by which S&T operates in market economies. Survival strategies are the microstrategies adopted by individual institutes and researchers in an attempt to cope with shrinking public R&D budgets and the absence of effective demand for domestic R&D. These features characterize not only Russia but also other post-socialist (post-Soviet) states.

The interaction between preservation and restructuring policies, coupled with a variety of survival strategies at the micro level, have induced a degree of structural change that has produced a system that is quite different from the Soviet R&D system but does not resemble the R&D systems of OECD countries either. This post-Soviet R&D system has several specific features. Organizationally, the R&D system is still 'externalized' with:

- most R&D activities still taking place in the commercialized, but state-owned R&D sector;
- R&D institutes dependent on public funding remaining the dominant type of organization;
- reduced demand from industry leading to a polarization of the R&D spectrum, i.e. the share of applied research is shrinking in favor of basic research and development;
- the R&D system becoming internationalized, although but this is a sign of crisis rather than dynamism.

R&D in CEECs is still carried out in independent, commercialized, state-owned R&D institutes whose operations are based on R&D contracts from industry. The sector has not been integrated into industry and operates as a substitute for the limited in-house innovative activities of manufacturing firms.

The reduction in demand for R&D and technology has put those organizations that directly served enterprises (design bureaus, construction and project organizations) in considerable difficulties. A number of these downstream organizations have been closed, while the number of R&D institutes (upstream activities) has increased. The small number of R&D organizations in the higher education sector and in industrial enterprises remain an important feature of the post-Soviet R&D system. In some post-socialist central European countries, in particular in Hungary and Poland, and in the republics of the former SFR Yugoslavia, the universities' role in R&D had been always much more important and hence this feature is of limited relevance for these economies. However, even in these economies, in-house R&D has been and continues to be very limited.

It might be expected that the externalized R&D system would shift towards applied R&D in search of the pockets of R&D demand in industry with funding. In some CEECs this is the main trend, but in post-Soviet states, in particular in Russia, there is a clear trend towards the 'polarization of the R&D spectrum' (Radosevic, 1998). The share of basic research and development is increasing at the expense of applied research. This polarization reflects three factors. Firstly, the R&D system is becoming more geared to upstream activities, since government is the only secure source of funding for Academy of Sciences institutes. Secondly, industry and industrial R&D institutes are unable to fund applied R&D. The radical reduction in their planning and financial horizons has reduced their R&D activities to short-term development with potential for immediate commercialization. Thirdly, applied R&D, which in the past used to be financed by various ministries, government agencies and industry enterprises as well as from the defense budget, has now shrunk significantly.

This functional reorientation of the R&D system illustrates the main weaknesses in the post-Soviet R&D system. With the shrinking of applied research the system is polarized between basic science and short-term commercial developments. The issue of demand for innovation and linkages between different types of R&D activities is emerging as one of the key concerns in all post-socialist CEECs.

Faced with radically reduced demand for their services, R&D institutes are continually shifting towards foreign sources of funding. Russia and Hungary rank very high in this respect. However, this degree of internationalization has been forced on CEECs and is not the result of deliberate strategy. This enforced internationalization of R&D systems in CEECs is also reflected in the continuously rising number of external patents or applications by CEE residents for patents in other countries. The number of external patents is also rising fast in the OECD countries and reflects the globalization of the technology market. However, the number of resident patents in the OECD countries is stable, while in CEECs it has been dropping sharply. This suggests that the internationalization of R&D in the CEECs reflects not only the quality of its domestic inventions but also the financial crisis in the R&D system, which is forcing the sale of patents.

The internationalization of the CEECs' R&D systems should, in itself, be seen in a very favorable light. However, its restructuring potential is significantly undermined by the individualization of scientific cooperation, which is being conducted by researchers eager to bypass the institutions to which they are affiliated (Mayntz et al., 1998; Chapter 20 in this volume). The effects would have been much greater if internationalization had been used as a spur to the restructuring of research institutes instead of being seen as a means of preserving them purely in survival mode.

In our view, national differences and departures from the post-socialist (post-Soviet) S&T system arises out of the interaction between micro survival strategies, restructuring and attempts at preservation. These forces and the interplay between them have produced a relatively stable and slow-moving process of change that has led to the emergence of an R&D model that we characterize as the post-socialist (post-Soviet) R&D model. The contributions to this volume (Part II) and the synthesis chapters by Meske (Part III) show the extent to which these forces operate differently in different CEECs. For example, the balance between preservation and restructuring varies between Russia, Belarus and Ukraine. The evidence in this volume suggests that attempts at preservation have been strongest in Belarus and that restructuring has predominated in Russia, with Ukraine lying somewhere between the two. A comparative analysis by Meske in this volume (cf. Chapters 18 and 19) shows differences in terms of restructuring between central European (Hungary, Poland, Czech Republic and Slovenia) and eastern European countries (Bulgaria, Romania).

The next 10–30 years of this century will be marked by increasing divergences in the S&T systems among CEECs, whose main driving forces will come from differences in 'broad NSI' but also from impacts on S&T systems in candidate CEECs which we describe as the "Europeanization" of their S&T systems. Differences in 'broad NSI' will affect the main feature of the post-socialist S&T system. In those CEECs that will see recovery and high growth rates, a revival of domestic demand for R&D and a strengthening of in-house R&D are to be expected. This may lead to substantial institutional transformation in S&T systems, which will be organized around enterprises' innovation activities. In countries with sluggish growth rates or stop – go growth, further marginalization of domestic R&D for domestic innovative activities is to be expected. This may be the case even in EU candidate countries, where additional EU funding may actually deepen the gaps between international pockets of excellence in R&D and domestic innovation activities.

In the rest of this section, we analyze the effects that the Europeanization of S&T systems may have on the EU candidate countries among the CEECs.

4.2. THE EUROPEANIZATION OF S&T SYSTEMS IN CEECs

A common change in 'broad' and 'narrow' NSI in CEECs applying for EU membership accession is what we call Europeanization. This term is taken from political science and has a variety of different meanings. For our purposes, Europeanization is defined as the impact of EU obligations on state institutions and modes of governance as well as on wider state-society and state-economy relations (Featherstone and Kazamias, 2001). In S&T, Europeanization means that the dynamics of EC research, technology and development (RTD) policy is likely to become part of the organizational logic of national S&T and innovation policies [6].

CEE candidate countries have already become part of that dynamic as a result of pre-accession activities. In that respect, Europeanization can be seen as a major component of the forces driving the restructuring of their post-Soviet (post-socialist) R&D model. We assume that Europeanization will strengthen the restructuring component of the post-socialist (post-Soviet) S&T systems and that, as a result, S&T systems in the CEECs may diverge further from each other. However, it is difficult to predict whether Europeanization by itself can solve the key weakness of the post-socialist S&T system, which is the low level of demand for domestic R&D.

Europeanization will lead to an overhaul of the regulatory regime governing the private economy and public sector, i.e. a radical change in 'broad NSI'. This process is already underway in the form of transition programs and pre-accession activities. In that respect, the relationship between 'broad' and 'narrow' NSI may be expected to change in a way that will be conducive to the promotion of domestic R&D and innovation.

Much of the process of EU accession involves 'negative' integration or harmonization with the rules on deregulation, liberalization and competition (McGowan, 2002). Innovation activities are not the subject of accession negotiations. Positive integration or harmonization with specific institutions and organizations such as the Common Agricultural Policy, regional policy, structural funds or RTD policy is much more difficult to achieve, as in these cases EU rules prescribe a specific institutional model to which domestic arrangements have to be adjusted.

Although the main effect of pre-accession activities so far has been on the mutual opening of markets ('negative integration'), in the case of S&T the effect has been integration into Framework Programs and a new sense of vigor in S&T and innovation policy. This renewed vigor comes from the EU requirement for candidate countries to show that they are able to withstand the competitive pressures of the single market. Among other things, the candidate countries are asked to prepare National Development Programs in which innovation is one of the prominent chapters.

Analysis of the transfer of policy schemes and concepts from EU to candidate countries (CCs) shows that in a number of countries, recent initiatives in S&T policy accord with the EU communications, although they do not refer to them directly and officially. However, "overall it seems that the transfer of

---

[6] We use the term RTD here since it is common parlance in the EU.

innovation support schemes from EU Member States to Candidate Countries is rather scarce. Several initiatives are taken to support Candidate Countries to enhance their innovation policies. However, more needs to be done with respect to twinning or other bilateral transfer arrangements." (EC, 2002).

This external pressure for active policymaking comes at a time when the CEECs governments have realized that focusing exclusively on macroeconomic stability, privatization and attracting FDI may not be sufficient to ensure long-term growth. As macroeconomic policy becomes constrained by the need to achieve low inflation and the desire to join the European Monetary Union (EMU) soon after accession, the CEECs governments are left with only one option if they want to ensure growth and employment. They have to promote the transformation of supply-side conditions through public and EU-supported spending on human resources and S&T activities. The CEECs can be expected to focus increasingly on building up a stock of human, technology and physical capital to compensate for deregulation, increased competition in local markets and the lack of autonomy in other policy areas. As the focus on macroeconomic stability reaches its limits we may expect governments to turn to policy actions that will address the issue of growth. Domestic S&T and innovation will have a prominent place in such actions. However, as we pointed out in the introduction, there are no recipes or models for CEECs to emulate. The only example of catch-up within the EU (Ireland) may have too many idiosyncratic elements to be emulated, though some lessons from the Irish story are quite relevant for the CEE CCs (O'Connor, 2001).

Like the southern EU member states, the new CEEC EU members are likely, under the impetus of EU requirements, to develop new functions in several areas such as structural, vocational training, environmental protection and consumer protection policies and cross-border cooperation (Ioakimidis, 2001, p. 83 and 94). The studies carried out on innovation policy in the thirteen candidate countries concluded that none of the CCs could be considered to have a fully-fledged innovation policy (EC, 2001, 2003). EU accession is likely to push CEECs into developing innovation policy, including regional innovation policies, as one of the preconditions for the effective use of structural funds. Research and technology policy is likely to be expanded and modeled on EU arrangements and to be extended towards downstream activities such as knowledge diffusion, in particular through support for regional innovation policy. In R&D, EU support through Framework Programs will establish criteria of international excellence which will operate as reference points for the restructuring of domestic R&D groups and organizations. For example, EU support for centers of excellence, which is already being followed by domestic networking and selection, has this effect.

Europeanization can be expected to weaken the power of the central state in S&T policy and will enhance the power of regions in big CEECs like Poland and Romania. It will strengthen the innovation community, encourage new social associations and interest groups to participate in the process of developing RTD and structural policy to be supported by the EU. To judge from the Greek experience, it is likely that the policymaking process will become less bureaucratic and more transparent.

Whether all CEECs will exploit the opportunities created by Europeanization to modernize their S&T systems and integrate them into EU-wide S&T activities will depend on a variety of local factors. In some cases, Europeanization will elicit only passive responses or nominal conformance, with considerable derogation in practice. However, given the considerable opportunities that EU accession opens up for the CEECs, Europeanization can be expected to be the main instrument of modernization for CEECs. For this to be realized, Europeanization will have to involve not only top-down change but also bottom-up responses and strategies developed by firms, R&D organizations and regions. In S&T, Europeanization is already being perceived as modernization. S&T administrators from the CCs can now travel, exchange experiences and familiarize themselves with current developments in S&T and innovation policy in the EU. The domestic S&T policy community, like their counterparts in the southern EU countries, is likely to internalize the logic, norms, behavior and culture associated with integration (Featherstone and Kazamais, 2001, p. 17). However, whether we will see real or surrogate modernization of S&T systems through Europeanization will depend to a great extent on the structure and the role of national political elites as well as on the involvement of civil society in Europeanization.

We must be aware that Europeanization has limits and ambiguities. As the dynamic of EU RTD becomes part of the logic of national S&T policymaking, it is likely to impact strongly on the definition of policy priorities and may lead to the mechanical transfer of policy models that may not be the most relevant for the CEECs.

Experience of Europeanization in the southern EU countries shows that the strongest effects were on the definition of the relevant policy actions and mechanisms and of national priorities. In the case of the CEECs, this will be compounded by the great importance of funding streams from Framework Programs and, in future, structural funds. This is likely to lead to a sort of myopia, in which the importance of local problems and the search for local solutions is downgraded. The autonomy of CEECs in S&T policy may remain a theoretical possibility only, since in practice the EU may exert considerable influence over goals, cost allocation and the resource mobilization.

The automatic transfer of EU policy mechanisms may often be irrelevant to local S&T or not constitute the most effective policy actions. For example, the transfer of the science park model without regard for local demand makes such programs highly dependent on foreign funding and barely sustainable. The transfer of policy models to support domestic clusters in conditions where there are no strong domestic organizations that can operate as 'network organizers' or in whose interest it is to develop linkages usually has limited effects, if any at all. While Europeanization will enhance and legitimize the innovation community, this may at the same time become just one more layer of bureaucracy or civil society without domestic roots, which are than perceived as alien to the domestic S&T community. Although we are quite optimistic regarding the positive effects of Europeanization on S&T in CEECs, this by itself is no panacea but rather a great opportunity for CEE CCs to modernize their S&T systems and integrate them into the emerging EU-wide innovation system.

5. CONCLUSIONS AND POLICY IMPLICATIONS

We have analyzed the prospects for S&T and growth in CEECs from a broad perspective, in which S&T ('narrow' NSI) is seen as part of 'broad' NSI. 'Narrow' and 'broad' national systems of innovation are interrelated but 'narrow' NSI also has a certain degree of autonomy (Freeman, 1999). This is important for understanding why changes in 'narrow' NSI are not immediately reflected in 'broad' NSI and then in growth and recovery and *vice versa*. The relationship between these two systems in CEECs is very specific because of the considerable tension between a high 'catch-up' potential based on R&D capacities and human capital and still sluggish outcomes in terms of growth and restructuring.

'Narrow' NSI in CEECs is undergoing extensive functional, organizational and financial restructuring (see Meske et al., 1998 for evidence). However, despite these changes, the key weakness of the post-socialist (post-Soviet) systems remains the failure to reintegrate industrial institutes into enterprises. Industrial enterprises that are short of long-term finance and facing fierce competition in foreign markets are not able to generate demand for more upstream activities like R&D. In such a situation R&D is perceived as a liability rather than an asset (Meske, 2002).

During the 1990s, the main source of technology was via imports of capital goods and FDI. Endogenously generated R&D and technology played a marginal role in the industrial upgrading of the CEECs. The current patterns of industrial upgrading, which are most often led by foreign enterprises, will eventually reach their limits without domestically generated R&D and technology. The lack of domestic in-house R&D cannot be fully offset by extra-mural or foreign R&D. The weaknesses in 'narrow' NSI will become visible through inadequate in-house R&D, weak university - industry links and a lack of technological co-operation among enterprises. In order to grow, these economies will have to generate their own innovation dynamics in order to complement imported technologies. These innovation dynamics will have to be driven by local enterprises committed to R&D and innovation.

'Narrow' NSI cannot be ignored if CEECs are to continue to grow and restructure. It may be possible for a limited period, as was the case during the transformational crisis of the 1990s. However, it is unlikely that CEECs can continue their industrial upgrading without restructuring their 'narrow' NSI, which plays a very important role in the development of technological capability in any economy. Its role cannot be reduced to the direct provision of technical information to industry. Research systems have several functions that are important for industrial upgrading, of which the provision of new and useful information is only one. Other functions include the creation of new instrumentation and methodologies, the provision of skills developed by engaging in research, participation in research networks, the resolution of complex technological problems and the establishment of spin-offs (Martin and Salter, 1996).

What we find today in CEE are fragments of the old R&D systems which are trying to adjust by adopting a variety of survival strategies, together with new pockets of innovation activities. We describe this system as the post-socialist or post-Soviet R&D system. Industrial institutes have been left to their own devices and are slowly reinventing themselves as service firms or industrial enterprises. Academies of Sciences institutes, attracted by government funding as the only stable source, are shifting towards basic research. Universities are trying to build a new position based on the stability that comes from teaching and to reorient towards research. Where they exist, in-house R&D departments are oriented towards their own needs and are trying to build up links with foreign sources of innovation. Domestic subsidiaries of foreign MNCs are entirely oriented towards the parent company in all the most important functions, including R&D, finance and marketing. Intra-organizational restructuring, that is the splitting of institutes into smaller organizations or the creation of spin-offs attached to institutes, has prevailed over inter-organizational restructuring involving several organizations from different sectors such as manufacturing industry, university, academy or industrial institutes.

In terms of the institutional superstructure, all CEECs have a developed S&T system with a large number of R&D institutes. In terms of organizational structure, however, their S&T systems are far from fully developed. However, the extensive institutional infrastructure in CEE still has to contend with very low demand for its activities due to weakness at the enterprise level. Weakness in the reconstitution of enterprises as the main network organizers of innovation processes is hindering the restructuring and development of 'narrow'

NSI. The building of the future NSI will depend on how this process progresses in the various countries. The increasing divergence in terms of growth and restructuring between 'western' (central Europe and Baltic) and 'eastern' CEECs (Bulgaria, Romania and European CIS) suggests that the reconstitution of enterprises as the main actors in the innovation process may lead to a faster emergence of NSI in central Europe. The reason for that is partly historical, as these countries, especially Hungary, the Czech Republic and Slovakia, have inherited a larger share of enterprises with in-house R&D activities from the socialist period. Moreover, 'in-house' R&D in some large enterprises survived the period of drastic cuts in R&D activities at the start of the transition process. Another reason is that these economies have experienced significantly higher levels of FDI and the process of Europeanization is far more advanced in their R&D systems and in their economies in general, as a result of the pre-accession process.

The establishment of a conducive environment by putting in place the necessary elements of 'broad' NSI (privatization, finance, legal protection, communication infrastructure etc.) strongly influences enterprises' innovation activities. In the transition period, they were actually more decisive in this respect than 'narrow' NSI. On the other hand, the new NSI is also likely to be shaped by the way enterprises embody innovation activities. However, this process is not entirely micro or macro-driven. As Nelson (1997) argues, it is a mistake to ask whether it is national factors or strong firms that create comparative advantages, since in those cases where the national institutional environment or legal structures, or specific policies, seem to have made a big difference, one also sees firms effectively taking advantage of the potential. While firms take advantage of favorable national factors they themselves also upgrade national factors. This explains why it is difficult to foresee which countries among the CEECs will catch up and which will fall behind.

It is not yet clear what national systems of innovation are emerging in the CEECs. These systems are far from being fully formed and it would be more appropriate to search first for signs of the emergence of sectoral innovation systems. Sectoral innovation systems are groupings of enterprises and their related networks of public and private institutions that are involved in the development, diffusion and utilization of innovation. These systems will strongly shape the character of NSIs in CEE. Based on the current patterns of production networks in CEECs it seems that these systems will be very heterogeneous. In some countries, such as Hungary, NSI may be based more on foreign enterprises. In Russia, they may be formed around large domestic industrial groups (Freinkman, 1995; Perotti and Gelfer, 1999; Popova, 1998). In countries like Estonia, they may be formed around small enterprises. In other countries, the NSI could be dualistic in character, with subsectors of small and large firms being unrelated to each other or with weak links between domestic and foreign firms. In some cases, they may be based on a few strong regions which are the drivers of growth. In these cases, the NSI could be strongly shaped by a few regional systems of innovation. Alternatively, NSIs could be formed around one or two sectors in which the innovation process is developed on a collective basis, while in the rest of the economy the innovation links are very weak. For the time being, the innovation dynamic is strongest among foreign enterprises. Our conclusion is that this is the greatest strength but also, potentially, the greatest long-term weakness of the CEECs that have attracted large volumes of FDI. The way CEECs integrate into international production and innovation networks will strongly shape their NSI.

During the 1990s, the integration process evolved between the two extremes of strongly foreign and strongly domestically-led technical modernization. Long–term growth can be achieved only when there is balance and complementarity between these two modes of modernization. This balance has been and will continue to be influenced by the way the state influences the interaction between domestic firms and 'narrow' NSI and MNCs, especially in sectors where regulations are important (pharmaceuticals, telecoms, energy). At present, it is foreign enterprises that are exerting the strongest influence on the shaping of production networks in almost all CEECs. Moreover, innovation activities are emerging through various forms of alliances with foreign firms. However, this process of interaction between domestic and foreign capital is mediated by the state. This introduces an important political or control dimension to the process of technical modernization, which will have implications for the nature of the emerging NSI in CEECs (Kuznetsova and Kuznetsov, 1999; Hayri and McDermott, 1998).

Transition policies have been far from sufficient for building 'narrow' NSIs, which in all countries are hybrid systems and require public-private cooperation. So far, the dominant response in most of the CEECs has been radically to reduce public funding but without any clear idea of what the new public R&D system should look like. The lack of active restructuring and the inability to formulate a coherent long-term policy in R&D could have been justified in the early years of transition, when a sharp decline in funding made orderly restructuring impossible. However, a wait-and-see policy on 'narrow' NSI, especially in relation to industrial R&D, has become counterproductive because of the costs incurred.

In all CEECs, the responses to restructuring are still weak although they vary greatly from country to country (see contributions in this volume). In CEE CCs, Europeanization is the major component of the restructuring process. It has already reinvigorated S&T and innovation policy in these countries and is likely to have significant positive effects on the restructuring of their S&T systems. In that respect, Europeanization may have a greater effect on the building of NSIs in CEECs than state policy.

After 10 years of pursuing the transition policy agenda, CEECs are now searching for alternative policy solutions that will also address the problem of their technological competitiveness. In that respect, Europeanization comes as a time when the reinvigoration of policy under EU influence can be effectively coupled with the endogenous search for new policy solutions that try to address technology and innovation.

Given the current role of the state in these countries, it is very unlikely that we will see the implementation of highly selective structural (industrial and technological) policies (along the lines of East Asian economies thirty years ago) aimed at a strengthening of inter-firm and inter-sectoral technological linkages. However, we still do not have a clear picture of the 'market-friendly' policies that might emerge and would be capable of implementation by individual states (Stiglitz and Ellerman, 2000). The key problem of this 'third way' in policy is how to better integrate structural and macroeconomic policies, to induce economic growth and to initiate structural change (Tunzelmann, 2002; Dailami and ul Haque, 1998).

Most of the central European countries are small economies with little bargaining power that want to join the EU but have already exposed their market to heavy EU competition. Consequently, the scope for pursuing technological objectives through foreign trade policies is greatly reduced. In general, policy options range from sector-specific or vertical policies (industrial policies) to horizontal policies (technology policy). In accordance with the EU policy philosophy, CEE CCs are likely to pursue exclusively horizontal policies.

Whether CEECs should pursue industrial policies that have immediate effects but are also much more demanding in terms of administrative requirements and finance is an issue only for non-EU candidate countries. In view of the negative experiences of CEE with state-run policies, there is a natural reluctance to promote policies with an imminent danger of government failure, whereby policy is easily turned into lobbying. An alternative would be to pursue technology (horizontal) policies that do not address specific sectors but target deficient capabilities, like R&D, engineering, production quality, or deficiencies in technical infrastructure (IT, testing and measurement facilities, etc.). The drawback of these policies is that they work slowly and with unclear sectoral effects. When allocating limited public funding, governments are seeking tangible and much quicker results. Again, CEE CCs could be in a very advantageous position in that respect, since many of the horizontal policies linked to Framework Program and structural funds will be pursued without regard to domestic political cycles.

The historical experience of CEE shows the considerable limitations of exclusive supply-type R&D measures. On the other hand, structural difficulties on the demand side are such that key bottlenecks cannot be resolved through S&T policy alone. A new phase of transformation in the CEECs, in which basic economic reforms, economic stability, and privatization are relatively settled, calls for much more innovative solutions in industrial and innovation policy, particularly in terms of low-cost policy measures.

As we have argued elsewhere (see Radosevic, 1994, 1997), the problem is probably not the type of policy *per se* but the ability of government to implement it in co-operation with industry. The (in)appropriateness of a specific policy is possible to assess only within the specific industry and country context and should include an assessment of the role of the state and of business-government interactions. Again, we would expect that CEE CCs will enjoy a considerable advantage in that respect as Europeanization will stimulate the modernization of the administrative apparatus and bring diverse professional and other social groups into the policy-making process.

REFERENCES

Amsden, A.H., and T. Hikino (1994) "Project execution capability, organisational know-how and conglomerate corporate growth in late industrialization." *Industrial and Corporate Change* 3: 111–148.

Bitzer, J., and C. von Hirschhausen (Eds.) (1998) *Conceptual Framework, Summary and Industry Studies*. Final Report C on the TSER Project 'Industrial Restructuring'. Berlin: Deutsches Institut für Wirtschaftsforschung (DIW).

Brzeski, A., and E. Colombatto (1999) "Can Eastern Europe Catch Up?" *Post-Communist Economies* 11, no. 1: 5–25.

Chandler, A.D. Jr., F. Amatori, and T. Hikino (Eds.) (1997) *Big Business and the Wealth of Nations*. Cambridge: Cambridge University Press.

Chavance, B., and E. Magnin (1998) "National Trajectories of Post-Socialist Transformation: Is there a convergence towards western capitalism?" *Prague Economic Papers* no. 3: 227–237.

Dailami, M., and N. ul Haque (1998) *What Macroeconomic Policies Are 'Sound'?* World Bank Working Paper no. 1995/October 1998.

David, P.A., and D. Foray (1995) *Accessing and expanding the science and technology knowledge base*. STI (Science, Technology and Industry) Review. Paris: OECD.

Djankov, S., and B. Hoekman (1999) *Foreign Investment and Productivity Growth in Czech Enterprises*. World Bank Working Paper no. 2115/May 1999.

Dosi, G., C. Freeman, and S. Fabiani (1994) "The Process of Economic Development: Introducing some stylized facts and theories of technologies, firms and institutions." *Industrial and Corporate Change* 3, no. 1: 1–45.

Dyker, D.A. (2000) "The Structural Origins of the Russian Economic Crisis." *Post-Communist Economies* 12, no. 1: 5–24.

Dyker, D.A., and S. Radosevic (1999) "What can Quantitative Analysis of Trends in Science and Technology Tell us about Patterns of Transformation and Growth in the Post-Socialist Countries?" *Innovation and Structural Change in Post-Socialist Countries: A Quantitative Approach.* Ed. D.A. Dyker and S. Radosevic. Dordrecht: Kluwer Academic Publishers.

Dyker, D.A., and S. Radosevic (2000) "Building the knowledge-based economy in countries in transition – from concepts to policies." *Journal of Interdisciplinary Economics* 12, no. 1: 41–70.

EBRD (1999) *Transition report 1999 – ten years of transition.* London: European Bank for Reconstruction and Development (EBRD).

EBRD (2000) *Transition report 2000 – employment, skills and transition.* London: EBRD.

EBRD (2001) *Transition report 2001.* London: EBRD.

EC (2001) "Innovation Policy in Six Candidate Countries: the Challenges – Cyprus, Czech Republic, Estonia, Hungry, Poland and Slovenia. Final Report September 2001." Study funded by the European Commission Directorate Generale (EC DG) Enterprise. Internet: www.cordis.lu/innovation-smes/src/studies3.htm#studies_candidate_countries.

EC (2002) "Transfer of Innovation Policy Schemes to Candidate countries." EC DG Enterprise. Internet: trendchart.cordis.lu/Reports/Documents/Transfer_Policy_Schemes_Candidate_Countries_March_2002.pdf.

EC (2002a) *European Trendchart on Innovation – 2002 European Innovation Scoreboard.* Internet: trendchart.cordis.lu/Reports/Documents/report2.pdf.

EC (2003) "Innovation Policy Issues in Seven Candidate Countries: Romania, Bulgaria, Slovakia, Latvia, Lithuania, Malta and Turkey." Study funded by the EC DG Enterprise, forthcoming.

Edquist, C. (Ed.) (1997) *Systems of Innovation: Technologies, Institutions and Organisations.* London: Pinter Publishers.

Eichengreen, B., and R. Kohl (1998) "The External Sector and Development in 'Eastern' Europe." *Journal of International Relations and Development* I (1–2), no. 20–45, Special Issue on FDI and Industrial Modernization in Central Europe.

Ergas, H. (1986) *Does Technology Policy Matter?* CEPS paper no. 29. Brussels: Centre for European Policy Studies (CEPS).

Ernst, D. (1998) "Catching-up, Crisis and Truncated Industrial Upgrading: Evolutionary Aspects of Technological Learning in East Asia's Electronics Industry." Paper presented at the United Nations University (UNU) INTECH Lisboa Conference, September.

Freeman, C. (1991) "Networks of Innovators – A Synthesis of Research Issues." *Research Policy* 20, no. 5: 499–514.

Freeman, C. (1999) "'Catching-up' and innovation systems: Implications for Eastern Europe." Unpublished book chapter, mimeo.

Featherstone, K., and G. Kazamias (Eds.) (2001) *Europeanization and the southern periphery.* London: Frank Cass.

Freinkman, L. (1995) "Financial-industrial Groups in Russia: Emergence of Large Diversified Private Companies." *Communist Economies & Economic Transformation* 7, no. 1: 51–66.

Frydman, R., C. Gray, M. Hessel, and A. Rapaczynski (2000) "The limits of discipline. Ownership and hard budget constraints in the transition economies." *Economics of Transition* 8, no. 3: 577–601.

Grabher, G. (1997) "Adaptation at the Cost of Adaptability? Restructuring the Eastern German Regional Economy." *Restructuring Networks in Post-Socialism: Legacies, Linkages, and Localities.* Ed. G. Grabher and D. Stark. Oxford: Oxford University Press, 107–136.

Grabher, G., and D. Stark (Eds.) (1997) *Restructuring Networks in Post-Socialism: Legacies, Linkages and Localities.* Oxford: Oxford University Press.

Granovetter, M. (1995) "Coase Revisited: Business Groups in the Modern Economy." *Industrial and Corporate Change* 4, no. 1: 93–130.

Guerrieri, P. (1999) "Technology and structural change in trade of the CEE countries." *Innovation and Structural Change in Post-Socialism: A Quantitative Approach.* Ed. D. Dyker and S. Radosevic. Dordrecht: Kluwer Academic Publishers.

Havrylyshyn, O., I. Izvorski, and R. van Rooden (1998) *Recovery and Growth in Transition Economies 1990–1997 – A Stylized Regression Analysis.* Working Paper of the International Monetary Fund (IMF) no. 98/141. Internet: www.imf.org/external/pubs/ft/wp/wp98141.pdf.

Hanson, P. (1997) "What sort of capitalism is developing in Russia?" *Communist Economies and Economic Transformation* 9, no. 1: 27–42.

Hayri, A., and G.A. McDermott (1998) "The network properties of corporate governance and industrial restructuring: a post-socialist lesson." *Industrial and Corporate Change* 7, no. 1: 153–194.

Hotopp, U., S. Radosevic, and K. Bishop (2002) *Trade and industrial upgrading in countries of central and eastern Europe: patterns of scale and scope based learning.* Working Paper No. 23. London: School of Slavonic and East European Studies (SSEES) at the University College London (UCL). Internet: www.ssees.ac.uk/publications/working_papers/wp23.pdf.

Hunya, G. (1997) "Large privatisation, restructuring and foreign direct investment". *Lessons from the economic transition: central and eastern Europe in the 1990s.* Ed. Z. Salvatore. Dordrecht: Kluwer Academic Publishers, 275–300.

Hunya, G. (Ed.) (2000) *Integration Through Foreign Direct Investment: Making Central European Industries Competitive.* Cheltenham: Edward Elgar.

Ioakimidis, P.C. (2001) "The Europeanization of Greece – an overall assessment." *Europeanization and the southern periphery.* Eds. K. Featherstone and G. Kazamias. London: Frank Cass, 12–29.

Khanna, T., and K. Palepu (1997) "Why focused strategies may be wrong for emerging markets." *Harvard Business Review* July/August: 41–51.

Khanna, T., and K. Palepu (1999) "The right way to restructure conglomerates in emerging markets." *Harvard Business Review* July/August: 125–134.

Kubielas, S. (2003) "Polish macroeconomic and S&T policies: interlinkages for growth and decline." *Journal of International Relations and Development* (forthcoming).

Kuznetsova, O., and A. Kuznetsov (1999) "The State as a Shareholder: Responsibilities and Objectives." *Euro-Asia Studies, Vol. 51, No.3*: 433–446.

Landesmann, M. (1997) *Emerging patterns of European Industrial Specialization: Im-*

plications for Labour Market Dynamics in Eastern and Western Europe. WIIW Research Report no. 230. Vienna: The Vienna Institute for Comparative Economic Studies (WIIW).

Landesmann, M. (2000) "Structural Change in the Transition Economies 1989–1999." *Economic Survey of Europe* no. 2/3, chapter 4: 95–123. Internet: www.unece.org/ead/survey.htm. (also: WIIW Research Reports no. 269)

Lankes, H.P., and A.J. Venables (1996) "Foreign direct investment in economic transition: the changing pattern of investments." The *Economics of Transition* 4, no. 2: 331–347.

Lundvall, B.-A. (Ed.) (1992) *National Systems of Innovation – Towards a Theory of Innovation and Interactive Learning*. London: Pinter Publishers.

Martin, B., and A. Salter (1996) *The relationship between publicly funded basic research and economic performance*. A SPRU Review, Report prepared for HM Treasury, April. Brighton: University of Sussex.

Mayntz, R., U. Schimank, and P. Weingart (1998) *East European Academies in Transition*. Dordrecht: Kluwer Academic Publishers.

McGowan, F. (2002) *State Strategy and Regional Integration: the EU and Enlargement*. Project Working Paper. Internet: www.ssees.ac.uk/publications/working_papers/pwp22.pdf.

Meske, W. (2002) *Science in Formerly Socialist Countries – Asset or Liability within New Societal Conditions?* WZB-Paper P 02–401. Berlin: Wissenschaftszentrum Berlin für Sozialforschung (WZB).

Meske, W., J. Mosoni-Fried, H. Etzkowitz, and G.A. Nesvetailov (Eds.) (1998) *Transforming Science and Technology Systems – The Endless Transition? NATO Science Series 4: Science and Technology Policy – Vol. 23*. Amsterdam: IOS Press.

Nelson, R. (Ed.) (1993) *National Systems of Innovation: A Comparative Study*. New York: Oxford University Press.

Nelson, R. (1997) *The Evolution of Competitive or Comparative Advantage: A Preliminary Report on a Study*. Working Paper 96–21. Laxenburg: International Institute for Applied Systems Analysis (IIASA).

O'Connor, P.T. (2001) *Foreign Direct Investment and Indigenous Industry in Ireland – Review of Evidence*. Project Working Paper No. 5/January. London: UCL, SSEES. Internet: www.ssees.ac.uk/ireland.pdf.

OECD (1996) *Employment and Growth in the Knowledge-based Economy*. Paris: OECD.

OECD (2000) *Knowledge Based Economy*. Paris: OECD.

Perotti, E.C., and S. Gelfer (1999) *Red Barons or Robber Barons? Governance and Financing in Russian FIG*. Centre for Economic Policy Research (CEPR) Discussion Paper no. 2204. Internet: www.cepr.org/pubs/dps/DP2204.asp.

Popova, T. (1998) "Financial–Industrial Groups (FIGs) and Their Roles in the Russian Economy." *Review of Economies in Transition* no. 7: 5–28.

Porter, M.E. (1990) *The Competitive Advantage of Nations*. New York: Free Press.

Radosevic, S. (1994): "Strategic technology policy for Eastern Europe." *Economic Systems* 18, no. 2: 87–116.

Radosevic, S. (1997) "Strategic policies for growth in post-socialism: theory and evidence based on the case of Baltic states." *Economic Systems* 21, no. 2: 165–196.

Radosevic, S. (1998) "National Systems of Innovation in Economies in Transition: Between Restructuring and Erosion." *Industrial and Corporate Change* no. 1: 77–108.

Radosevic, S. (1999) *International Technology Transfer and Catch-Up in Economic Development.* Cheltenham: Edward Elgar Publishers.

Radosevic, S. (1999a) *Science, technology and growth: issues for Central and Eastern Europe.* Summary of the project 'Restructuring and Reintegration of Science and Technology Systems in Economies in Transition', funded by DGXII of the EC under the Targeted Socio-Economic Research (TSER) Programme, 1996–98. Brighton: SPRU. Internet: www.sussex.ac.uk/spru.

Radosevic, S. (1999b), "Divergence or Convergence in R&D and Innovation Between "East" and "West". *Innovation and Transformation. Eds.* Brzezinski, H. and M. Fritsch, Edward Elgar, Cheltenham.

Radosevic, S. (2003) "Patterns of preservation, restructuring and survival: science and technology policy in Russia in the post-Soviet era." *Research Policy*, forthcoming.

Radosevic, S. (2003a) "Growth of enterprises through alliances in central Europe: the issues in controlling access to technology, market and finance".*International networks and industry restructuring in central Europe, Russia and Ukraine.* Eds. S. Radosevic and B. Sadowski, Book manuscript.

Radosevic, S., and L. Auriol (1998) "Patterns of Restructuring in Research, Development and Innovation Activities in Central and Eastern European Countries: Analysis Based on S&T Indicators." *Research Policy* no. 28: 351–376.

Radosevic, S., and B. Sadowski (Eds.) (2003) *International networks and industry restructuring in central Europe, Russia and Ukraine.* Book manuscript.

Rodrik, D. (2000) *Can Integration into the World Economy Substitute for a Development Strategy?* Paper presented at the World Bank ABCDE-Europe Conference in Paris, June 26–28, 2000.

Stephen, J. (2000) *The productivity gap between East and West Europe: What Role for Sectoral Structures During Integration?* Working Paper. Halle: Institute for Economic Research (IWH).

Stiglitz, J., and D. Ellerman (2000) *New bridges across the chasm: macro- and microstrategies for Russia.* Paper presented at the ECAAR panel on the Russian economy held in January. Boston: World Bank.

Szalavetz, A. (1996) *Restructuring and Export Performance. The Role of Foreign Partners in the Adaptation Process of Hungarian Companies.* Working Paper No. 64. Budapest: Institute for World Economy.

Tunzelmann, N. von (2002) *Final Report* of the EU funded project MACROTEC 'Integration of macroeconomic and S&T policies'. Brighton: SPRU, University of Sussex.

UN ECE (2000) *Economic Survey of Europe.* Geneva: United Nations Economic Commission for Europe (UN ECE).

Yudanov, A.Y. (1997) "USSR: Large Enterprise in the USSR – The Functional Disorder." *Big Business and the Wealth of Nations.* Ed. A.D. Jr. Chandler, et al. Cambridge: Cambridge University Press, 397–432.

Zukowski, R. (1998) "From transformational crisis to transformational recovery: the case of Poland's industries." *Economic Systems* 22, no. 4: 367–397.